AN INTRODUCTION TO BIOLOGICAL MEMBRANES
From Bilayers to Rafts

AN INTRODUCTION TO BIOLOGICAL MEMBRANES
From Bilayers to Rafts

WILLIAM STILLWELL
Professor Emeritus of Biology
Indiana University
Purdue University
Indianapolis

AMSTERDAM • BOSTON • HEIDELBERG • LONDON
NEW YORK • OXFORD • PARIS • SAN DIEGO
SAN FRANCISCO • SINGAPORE • SYDNEY • TOKYO
Academic Press is an imprint of Elsevier

ELSEVIER

Academic Press is an imprint of Elsevier
32 Jamestown Road, London NW1 7BY, UK
225 Wyman Street, Waltham, MA 02451, USA
525 B Street, Suite 1800, San Diego, CA 92101-4495, USA

British Library Cataloguing-in-Publication Data
A catalogue record for this book is available from the British Library

Library of Congress Cataloging-in-Publication Data
A catalog record for this book is available from the Library of Congress

ISBN: 978-0-444-52153-8

For information on all Academic Press publications
visit our website at www.store.elsevier.com

Typeset by TNQ Books and Journals

Printed and bound in Europe

Transferred to Digital Printing in 2014

Working together
to grow libraries in
developing countries

www.elsevier.com • www.bookaid.org

Dedication

This book is dedicated to my wife Penelope who has tolerated me all these years, my son Max, my daughter Jessica, and Cosmo, the brains of the operation.

Contents

Preface   ix

1. Introduction to Biological Membranes

A. What is a Biological Membrane?   1
B. General Membrane Functions   3
C. Eukaryote Cell Structure   4
D. Size of Domains   8
E. Basic Composition of Membranes   10
Summary   11
References   11

2. Membrane History

A. Oil On Water: Interface Studies   13
B. The Lipid Bilayer Membrane   20
Summary   27
References   28

3. Water and the Hydrophobic Effect

A. Water – Strength in Numbers   29
B. Structure of Water   31
C. Properties of Water – No Ordinary Joe   32
D. Surface Tension   34
E. The Hydrophobic Effect   37
Summary   40
References   41

4. Membrane Lipids: Fatty Acids

A. What are Lipids?   43
B. Why are there so Many Different Lipids?   44
C. Lipid Classification Systems   45
D. Fatty Acids   46
Summary   55
References   55

5. Membrane Polar Lipids

A. Bonds Connecting to Acyl Chains   57
B. Phospholipids   60
C. Sphingolipids   71
D. Sterols   74
E. Membrane Lipid Distribution   76
F. Plant Lipids   76
G. Membrane Lipids Found in Low Abundance   78
Summary   81
References   82

6. Membrane Proteins

A. Introduction   85
B. The Amino Acids   86
C. How Many Membrane Protein Types
   are there?   91
Summary   103
References   104

7. Membrane Sugars

A. Introduction   107
B. Glycolipids   111
C. Glycoproteins   112
D. GPI-Anchored Proteins   114
Summary   115
References   115

8. From Lipid Bilayers to Lipid Rafts

A. Development of Membrane Models   117
B. The Danielli-Davson Pauci-Molecular
   Model   118
C. The Robertson Unit Membrane Model   119
D. Benson and Green's Lipoprotein Subunit
   Models   120
E. The Singer-Nicolson Fluid Mosaic Model   122

F. Simons' Lipid Raft Model   125
Summary   127
References   128

9. Basic Membrane Properties of the Fluid Mosaic Model

A. Size and Time Domains   132
B. Membrane Thickness   134
C. Membrane Asymmetry   137
D. Lateral Diffusion   144
E. Lipid Trans-Membrane Diffusion (Flip-Flop)   151
F. Lipid Melting Behavior   157
G. Membrane 'Fluidity'   170
Summary   172
References   172

10. Lipid Membrane Properties

A. Complex Lipid Interactions   176
B. Non-Lamellar Phases   182
C. Lipid Phase Diagrams   190
D. Lipid–Protein Interactions   194
E. Lipid Interdigitation   206
Summary   210
References   211

11. Long-Range Membrane Properties

A. Membrane Bilayer Lipid Packing   215
B. Membrane Protein Distribution   222
C. Imaging of Membrane Domains   223
D. Homeoviscous Adaptation   232
Summary   235
References   235

12. Membrane Isolation Methods

A. Introduction   239
B. Breaking Open the Cell: Homogenization   241
C. Membrane Fractionation: Centrifugation   248

D. Membrane Fractionation: Non-Centrifugation Methods   253
E. Membrane Markers   260
Summary   261
References   262

13. Membrane Reconstitution

A. Detergents   266
B. Membrane Protein Isolation   271
C. Membrane Lipid Isolation   275
D. Model Lipid Bilayer Membranes — Liposomes   282
E. Membrane Reconstitution   292
F. Lipid Rafts   295
Summary   300
References   300

14. Membrane Transport

A. Introduction   305
B. Simple Passive Diffusion   309
C. Facilitated Diffusion   311
D. Active Transport   317
E. Ionophores   324
F. Gap Junctions   328
G. Other Ways to Cross the Membrane   329
Summary   334
References   335

15. Membranes and Human Health

A. Liposomes as Drug-Delivery Agents   339
B. Effect of Dietary Lipids on Membrane Structure/Function   348
Summary   352
References   353

Chronology of Membrane Studies   357
Index   361

Preface

Virtually all biological processes are in some fashion involved with membranes. However, despite their unquestioned importance, many aspects of complex membrane structure and function remain unresolved. Often not only are the answers not available, but even formulating the appropriate questions is difficult. A single membrane is composed of hundreds of proteins and thousands of lipids, all in constant flux. Methodologies to study membranes must span time and size domains that range over a million fold! A good case can be made that understanding membranes will be the next great frontier in the life sciences.

From the inception, it became obvious to me that trying to write a book on membranes is a near hopeless endeavor. No-one can possibly be an expert in all aspects of membrane science. Membrane studies range from structural and theoretical biophysics at one extreme to nutritional, cell biology, and membrane metabolic studies at the other. Bridging the two extremes are countless highly technical methodologies that are difficult to explain without getting hopelessly bogged down in technical minutia. Choosing what should be included and what excluded from this book reflects my personal 40 years of research experience on the biophysics and biochemistry of model membrane systems.

Through the years there have been a large number of books written about membranes. Most of these have been multi-authored, highly technical compendiums that are essentially "preaching to the choir". They are indecipherable to anyone who is not already an expert in the narrow topic covered in the book. While these multi-authored books have their place in membrane studies, they are essentially a non-cohesive collection of loosely related review articles that are written by experts from very different backgrounds and writing styles. As a result they are comprehensive, but very hard to read. In sharp contrast, excellent general discussions covering membranes can now be found in very well written chapters in almost all current biochemistry, cell and molecular biology, and physiology textbooks. These brief chapters are of necessity too abbreviated to adequately cover a field as expansive as membranes. The multi-authored compendiums are too detailed while the chapters from general textbooks lack sufficient detail.

Between the extremes of edited compendiums and general textbook chapters are a smaller number of general membrane books that have the advantage of being written by a single (or at most a few) authors. *An Introduction to Biological Membranes: From Bilayers to Rafts* is such a book. This book is an attempt to write a broad textbook covering many aspects of membrane structure/function that bridges membrane biophysics and cell biology. It was my basic contention that it is not necessary to understand the intricacies of NMR, single particle tracking, X-ray crystallography and so on, to appreciate the contributions that these techniques have made to membrane studies. The book primarily targets advanced undergraduates and beginning graduate students. However, even membranologists with a good knowledge of contributions by their contemporaries often have very limited knowledge of fundamental contributions by early membrane pioneers, a recurring theme of this book. The appendix to this book is a time line of 100 of the most important discoveries in membrane science dating from ~540 BC to present.

In the process of cleaning out my office before fading into retirement in 2010, I found that I had accumulated about 60 membrane books, and this is only a fraction of what is available. For 30 years I taught a beginning graduate level course entitled *Biological Membranes*. For most of those years I could not find a suitable textbook for the course. In 1988 Robert B. Gennis published his book *Biomembranes: Molecular Structure and Function*. To this day I consider Gennis' book to be the best membrane structure book ever written and for about 10 years I adopted this book for my course. Unfortunately, after 25 years, a second edition of this book has yet to be published. Something is needed to fill this gap. From my perspective, the new book should be a single-authored, broad book that covers many aspects of membrane structure and function without getting lost in unnecessary details — in other words, 'membrane biophysics light'. It is clear that each topic covered in *An Introduction to Biological Membranes: From Bilayers to Rafts* could be greatly expanded, but this will have to wait for another day.

Through the years other single-authored general membrane books have appeared, each of which has its advantages and disadvantages. A few examples are listed below:

Gennis, R.B. 1988. Biomembranes: Molecular Structure and Function. Springer-Verlag. New York, NY. 533 pp.

Luckey, M. 2008. Membrane Structural Biology. With Biochemical and Biophysical Foundations. Cambridge University Press, New York, NY, 332 pp.

Mouritsen, O.G. 2005. Life as a Matter of Fat. The Emerging Science of Lipidomics. Springer, Berlin, Heidelberg, 276 pp.

Jones, M.N. and Chapman, D. 1995. Micelles, Monolayers and Biomembranes. John Wiley and Sons, New York. 252 pp.

Yeagle, P.L. 1993. The Membranes of Cells, 2nd Edition. Academic Press, New York. 349 pp.

Jain, M.K. 1988. Introduction to Biological Membranes, 2nd Edition. John Wiley & Sons, New York. 423 pp.

Finean, J.B., Coleman, R. and Michell, R.H. 1984. Membranes and Their Cellular Functions, 3rd Ed. Blackwell Scientific Publications, Oxford, New York. 227 pp.

Malhotra, S.K. 1983. The Plasma Membrane. John Wiley and Sons, New York. 209 pp.

Houslay, M.D. and Stanley, K.K. 1982. Dynamics of Biological Membranes: Influence on Synthesis, Structure and Function. John Wiley & Sons, New York. 330 pp.

William Stillwell
Sand Key, Florida
January, 2013

CHAPTER

1

Introduction to Biological Membranes

OUTLINE

A. What is a Biological Membrane?   1
B. General Membrane Functions   3
C. Eukaryote Cell Structure   4
Endomembrane System   4
Plasma Membrane   4
Nuclear Envelope (Membrane)   6
Endoplasmic Reticulum (ER)   7
Golgi Apparatus   7
Lysosome   7
Peroxisome   7

Mitochondria   7
D. Size of Domains   8
Can we see a membrane?   9
Can we see a cell?   9
What can you see with a light
microscope?   9
E. Basic Composition of Membranes   10
Summary   11
References   11

A. WHAT IS A BIOLOGICAL MEMBRANE?

The American Heritage Dictionary defines a membrane as 'a thin pliable layer of plant or animal tissue covering or separating structures or organs.' The impression this description leaves is one of the plastic wrap covering a hamburger. By this definition, membranes are static, tough, impenetrable, and visible. Yet, nothing could be farther from the truth. The entire concept of *dynamic* behavior is missing from this definition, yet dynamics is what makes membranes both essential for life and so difficult to study.

If we could somehow instantaneously freeze a membrane and learn the composition and location of each of the countless numbers of molecules comprising the membrane, and then instantly return the membrane back to its original unfrozen state for a microsecond before re-freezing, we would find that the membrane had substantially changed while unfrozen. Although the molecular composition would remain the same over this short time, the molecular locations and interrelationships would be altered. Therefore membranes must have both static and dynamic components. While static describes what is there, dynamics describes how the components interact to generate biological function.

Every cell in the human body is a tightly packed package of countless membranes. The human body is composed of ~63 trillion cells (6.3×10^{13} cells), each of which is very small. For example, a typical liver cell would have to be 5X larger to be seen as a speck by someone with excellent vision (it is microscopic). Each liver cell has countless numbers of internal membranes. If you could somehow open one single liver cell and remove all of the internal membranes and sew them together into a quilt, the quilt would cover ~840 acres, the size of New York's Central Park! And that is from one single cell. Therefore, there are enough membranes in a human body (6.3×10^{13} cells) to cover the earth millions of times over!

All life on Earth is far more similar than it is different. Living organisms share a number of essential biochemical properties, collectively termed the 'thread of life'. Included in these essential properties is ownership of a surrounding plasma membrane that separates the cell's interior from its external environment. It is likely that all living things inhabiting planet Earth today arose from a single common ancestor more than 3.5 billion years ago. The first cell probably contained minimally a primitive catalyst (a pre-protein), a primitive information storage system (a pre-nucleic acid), a source of carbon (perhaps a primitive carbohydrate) and this mixture had to be surrounded by a primitive plasma membrane that was likely made of polar lipids. Membranes were therefore an essential component of every cell that is alive today or has ever been alive.

With 3.5 billion years of biological evolution, the complexity of membranes in cells has greatly expanded from that of a simple surrounding plasma membrane to where they now occupy a large portion of a eukaryote's interior space. An electron microscopic picture of a 'typical' eukaryotic (liver) cell is shown in Figure 1.1 [1]. It is evident from the complexity of this micrograph that identifying, isolating, and studying membranes will be a difficult task.

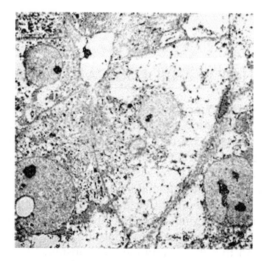

FIGURE 1.1   Transmission electron micrograph of a liver cell, a 'typical' cell [1].

B. GENERAL MEMBRANE FUNCTIONS

It is now generally agreed that biological membranes are probably somehow involved in all cellular activities. The most obvious function of any membrane is separating two aqueous compartments. For the plasma membrane this involves separation of the cell contents from the very different extra-cellular environment. Membranes are therefore responsible for containment, ultimately delineating the cell. Separation, however, cannot be absolute, as the cell must be able to take up essential nutrients, gases, and solutes from the exterior, while simultaneously removing toxic waste products from the interior. A biological membrane therefore must be selectively permeable, possessing the ability to distinguish many chemically different solutes and knowing in which direction to redistribute them. Biological membranes must therefore house a variety of specific, vectorial transport systems (discussed in Chapter 14).

A characteristic of all living cells is the establishment and maintenance of trans-membrane gradients of all solutes. Of particular interest are large ion gradients typically associated with the plasma membrane. Table 1.1 is a comparison of the mean concentration of selected ions inside and outside a typical mammalian cell, and the magnitude of each gradient. To maintain gradients of this size, efficient energy-dependent transport systems must be employed (discussed in Chapter 14). Directional trans-membrane structure is required to generate these ion gradients.

In addition to trans-membrane structure, it is now believed that biological membranes are composed of countless numbers of very small, transient, lateral lipid microdomains. Each of these domains is proposed to have a different lipid and resident protein composition. Thus the activity of any membrane must reflect the sum of the activities of its many specific domains. One type of lipid microdomain, termed a 'lipid raft', has received a lot of recent attention as it is reputed to be involved in a variety of important cell signaling events. If the lipid raft story (discussed in Chapter 8) holds up, this new paradigm for membrane structure/function may serve as a model for other types of as yet undiscovered non-raft domains. Each of these domains might then support a different collection of related biochemical activities. Therefore, membranes have both trans-membrane and lateral structures that are just beginning to be understood.

All membranes possess an extreme water gradient across their very thin (~5 nm) structure. In the membrane aqueous bathing solution water concentration is ~55.5 M water in water (1,000 g of water per liter divided by 18, the molecular weight of water), while the membrane

TABLE 1.1   Trans-membrane Ion Gradients of a 'Typical' Mammalian Cell.

Ion	Inside	Outside	Gradient
Na^+	10 mM	140 mM	14-fold
K^+	140 mM	4 mM	35-fold
Ca^{2+}	1.0 μM	1.0 mM	1,000-fold
Cl^-	100 mM	4.0 mM	25-fold

interior is quite dry (< 1 mM water). The aqueous interface provides a charged or polar physical surface to help arrange related functional enzymes, known as pathways, in one plane for increased efficiency. In contrast, the dry interior provides an environment for dehydration reactions. Both the aqueous interface and dry interior are responsible for maintaining the proper conformation of membrane proteins (discussed in Chapter 6).

In addition to transport, biological membranes are also the site of many other biochemical or physiological processes including: inter-cellular communication, cell—cell recognition and adhesion, cell identity and antigenicity, regulation (resident home of many receptors), intracellular signaling, and some energy transduction events.

C. EUKARYOTE CELL STRUCTURE

While the plasma (cell) membrane defines cell boundaries, internal membranes define a variety of cell organelles. In eukaryotes, the internal membranes also separate very different internal aqueous chambers resulting in compartmentation into membrane-bordered packets called organelles. Each organelle supports different sets of biological functions. Figure 1.2 is a cartoon depiction of a 'typical' animal (eukaryote) cell [2]. All of the important membrane-bound organelles are depicted. Throughout this book, specific examples demonstrating aspects of membrane structure or function will be selected from these organelle types. Below is a very brief description of the major cellular membranes. More detailed descriptions can be readily found in many cell biology [3—6], and biochemistry [7,8] textbooks.

Endomembrane System

With the exception of mitochondria, peroxisomes and, in plants, chloroplasts, the intracellular membranes are suspended together in the cytoplasm where they form an interconnected complex. Although membrane types are unique and can be separated from one another, all are related structurally, chemically, functionally, and developmentally. This strong inter-relationship is referred to as the 'endomembrane system'. The endomembrane system divides the cell into many compartments, primarily organelles, which are distinct yet interconnected. The connected membranes include the nuclear envelope, the rough and smooth endoplasmic reticulum, the Golgi apparatus, lysosomes, vacuoles, many types of small vesicles, and the plasma membrane (Figure 1.3) [9]. Transport back and forth between these various membranes and compartments is continuous and under strict regulation [10]. The endomembrane system is the foundation of cell trafficking.

Plasma Membrane

All cells are surrounded by a plasma membrane (PM) that separates the cell contents from the rest of the world [11]. The PM is the most dynamic and busiest of all cellular membranes and more is known about the PM than any other membrane. The majority of examples described in this book were obtained from PM studies. The PM defines the cell's

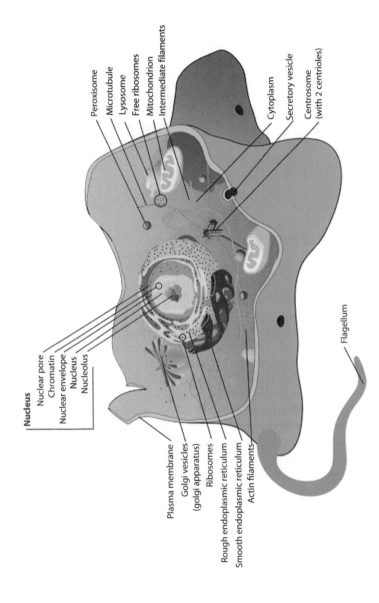

FIGURE 1.2 Cartoon depiction of the major components of an animal cell [2].

FIGURE 1.3   The endomembrane system of a 'typical' animal cell. The various membranes and compartments are connected to one another structurally, chemically, functionally, and developmentally [9].

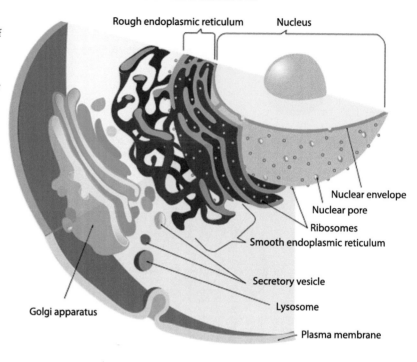

boundary and its interaction with the external environment. It is responsible for transporting nutrients into the cell while allowing waste products to leave. Thus, it prevents unwanted materials from entering the cell while keeping needed materials from escaping. It maintains the pH of the cytosol, and preserves the proper cytosolic osmotic pressure. Proteins on the PM surface assist the cell in recognizing and interacting with neighboring cells. Other proteins on the plasma membrane allow attachment to the cytoskeleton and extracellular matrix, functions that maintain cell shape and fix the location of membrane proteins. The PM contains several characteristic functions and structures that can be used to identify PM fractions (Chapter 12). Tight junctions seal contacts between cells while desmosomes are adhesion sites between adjacent cells. Gap junctions (Chapter 14) contain hexagonal arrays of pores that allow communication between adjacent cells. Caveolae and coated pits (Chapter 14) are similar shaped PM invaginations that are involved in cell signaling and solute uptake, respectively. Plasma membranes are indeed complex entities.

Nuclear Envelope (Membrane)

The nuclear envelope is a double membrane surrounding a perinuclear space [12]. This space is probably contiguous with the lumen of the endoplasmic reticulum. The envelope has large nuclear pores (about 600 Å) that allow passage of large RNA-protein complexes out of the nucleus into the cytoplasm and movement of regulatory proteins from the cytoplasm into the nucleus.

Endoplasmic Reticulum (ER)

The endoplasmic reticulum (ER) is a complex network of cisternae or tube-like structures occupying a considerable percentage of the cell's internal volume [13]. The portion of the ER with attached ribosomes is known as the rough ER. It is the site for biosynthesis of non-cytoplasmic proteins that are either secreted, internalized into the lysosome, or become PM proteins. The portion of ER devoid of ribosomes is known as the smooth ER. Functions of the smooth ER include sterol biosynthesis, drug detoxification, calcium regulation, and fatty acid desaturation. A specialized ER, the sarcoplasmic reticula has but one function — regulation of intra-cellular calcium levels. The endoplasmic reticula were first seen by Keith R. Porter, Albert Claude, and Ernest F. Fullam in 1945 [14].

Golgi Apparatus

The Golgi apparatus [15,16] is a series of stacked, disk-shaped tubules. The Golgi is named after the man who discovered it in 1898, Carmillo Golgi. For this, in 1906 Golgi was awarded one of the first Nobel Prizes. It is in the Golgi that post-translational modification of glycoproteins synthesized originally in the ER and destined for secretion takes place. Other possible final destinations of proteins from the Golgi include incorporation into the PM or to the lysosome. Characteristic Golgi-resident enzymes involve sugar modification of proteins and include glucosidases and glycosyl transferases.

Lysosome

The lysosome contains some 40 hydrolytic enzymes whose function is to degrade macromolecules into component parts for re-use in the cell [17]. As a result lysosomes are often referred to as the 'cell's garbage disposal'. The organelle was discovered in the early 1950s and, in 1955, it was named lysosome, by Belgian cytologist Christian de Duve, for its ability to lyse membranes.

Peroxisome

Peroxisomes contain a battery of oxidative enzymes that are involved in breakdown of small molecules [18]. Important peroxisomal enzymes include D-amino acid oxidase and catalase, the enzyme responsible for the degradation of dangerous peroxides. Peroxisomes are also involved in drug detoxification and the biosynthesis of essential ether-phospholipids known as plasmalogens (discussed in Chapter 5). Peroxisomes were also discovered by Christian de Duve in 1967 [19]. For discovering lysosomes and peroxisomes, de Duve was awarded the 1974 Nobel Prize for Medicine (Figure 1.4).

Mitochondria

Mitochondria are the 'powerhouse' of the cell, producing most cellular ATP. The processes of electron transport and oxidative phosphorylation are housed in the highly folded mitochondrial inner (Cristae) membrane [20]. The mitochondrial aqueous interior chamber, called

FIGURE 1.4   Christian de Duve (1917–). *Courtesy of Christian de Duve Institute of Cellular Pathology [24].*

the matrix, houses most of the enzymes involved in the Krebs Cycle (terminal steps in sugar oxidation) and β-oxidation (fatty acid oxidation) [21]. The inner mitochondrial membrane is surrounded by a second membrane, the outer mitochondrial membrane, which is very different and far less dynamic than the inner membrane. Mitochondria have been defined as 'semi-autonomous, self-replicating organelles', meaning they grow and replicate independently of the cell in which they are housed. This is a vestige left over from their origin as freely living prokaryotes that took up refuge inside larger prokaryotes about 1.5 billion years ago. This concept, known as Endosymbiont Theory [22,23], was originally ridiculed, but is now generally accepted. Mitochondria contain their own, albeit small, genome and code for a handful of mitochondrial membrane-protein components.

D. SIZE OF DOMAINS

Since membrane studies span a wide range of size and time domains, a variety of often esoteric instrumentation must be employed. The studies addressed in this book will range in size from Angstroms (Å) to microns (μm): Å (10^{-10} m), nm (10^{-9} m) and μm (10^{-6} m).

We will first address the question of size by asking whether someone with excellent vision can actually see a membrane. A person with excellent vision can resolve two spots about 0.1 mm apart.

$$0.1 \text{ mm} = 10^{-4} \text{m} = 10^2 \text{ μm} = 10^5 \text{ nm} = 10^6 \text{ Å}$$

Can we see a membrane?

Since a membrane bilayer is less than 50 Å thick, it would have to be about 20,000 times bigger than it is, just to be seen as a speck by someone with excellent vision. Is it then possible that someone can see a much larger cell?

Can we see a cell?

Resolving power of the eye	100 μm	1,000,000 Å
Erythrocyte	7 μm	70,000 Å
Bacteria	2 μm	20,000 Å
Mitochondria	2 μm	20,000 Å
Liver cell	20 μm	200,000 Å

Even the relatively large liver cell would have to be ~5X bigger to be seen as a speck by the human eye.

What can you see with a light microscope?

The resolving power of a good light microscope is wavelength-dependent. Shorter wavelength (blue) light has a better resolving power than does longer wavelength (red) light. Although the entire visible light spectrum is quite narrow, from 380–750 nm, it is possible to push resolution of the light microscope down to about 300 nm or 3,000 Å. Since a biological membrane bilayer is ~50 Å thick, the membrane would have to be 60X bigger than it is to be *directly* detectable, even by a light microscope.

Resolution of a membrane was not possible until discovery of the electron microscope (EM) in 1931 by Ernst Ruska (Figure 1.5) who later won the 1986 Nobel Prize in Physics for this achievement. The EM, however, did not become commonly employed until the 1950s. Most EM use on biological samples has a resolution of ~10 Å and so can readily distinguish a biological membrane. Even higher resolution is possible in certain cases.

FIGURE 1.5   Ernst Ruska (1901–1988). *Courtesy of Siemens AG, http://www.siemens.com/history/en/personalities/scientists_and_engineers.htm*

It is evident that living cells, and therefore the membrane that surrounds them, are very small, making their study all the more difficult. It is now believed that if living cells are found elsewhere in the universe, they too will be small. The size of a living cell is limited by a universal constant, the rate of solute diffusion in water.

$$\text{Diffusion rate} = K(T/m)^{\frac{1}{2}}$$

Where K is a constant that is related to the size of the solute, T is the temperature in °K and m is the molecular weight of the solute. Even diffusion of water (a tiny solute) in water is quite low:

Diffusion Distance for Water in Water

Time	Distance
0.1 sec	10 μm
1 sec	100 μm
100 sec	1 mm

Diffusion works well over very short distances but is far too slow to be effective over long distances. This limits the size of cells. Diffusion is further discussed in Chapter 9. Coupled with the small sizes associated with membranes are the rapid times involved in membrane dynamics. Size domains spanning the μm to Å range and time domains ranging from μs to ns, necessitates use of combinations of biophysical instrumentation discussed in Chapter 14.

E. BASIC COMPOSITION OF MEMBRANES

A striking feature of biological membranes is that while at first glance all biological membranes may appear to be quite similar in size, structure and basic composition, they support very different functions. The conundrum of how membranes can be similar, yet different, forms the central theme of this book. The structural similarity of all membranes is to a large extent due to the polar lipids that comprise the lipid bilayer (Chapters 4 and 5) while biochemical diversity is due to the proteins that reside in the bilayer (Chapter 6). As a general rule, the more biochemical functions a particular membrane supports, the higher its protein content will be. Table 1.2 demonstrates this concept. The myelin sheath

TABLE 1.2   General Composition, Expressed as Percent by Weight, of Three Very Different Mammalian Membranes.

Membrane	Lipid	Protein
Myelin Sheath	80%	20%
Plasma Membrane	50%	50%
Mitochondrial Inner Membrane	25%	75%

has perhaps the least biochemical functions of any mammalian membrane (it provides insulation for nerves), while the mitochondrial inner membrane has many functions including a major connected pathway (electron transport and associated oxidative phosphorylation). The plasma membrane falls in between the two extremes. The many types of membrane lipids and proteins are discussed in Chapters 5 and 6, respectively.

SUMMARY

Understanding membrane structure and function is one of the major unsolved problems in life science. Membranes are intimately involved in almost all biological processes including; establishing and maintaining trans-membrane gradients; compartmentalizing biochemical reactions into distinct functional domains; controlling transport into and out of cells; inter- and intra-cellular communication; cell-cell recognition; and energy transduction events. What makes biological membranes so difficult to study is their small size (<10 nm in width) and compositional complexity, being composed of hundreds of proteins, thousands of lipids and numerous surface carbohydrates, all in constant flux. A microscopic liver cell is so packed with internal membranes that the total membrane surface area of a single cell is ~840 acres, the size of New York's Central Park. Remarkably, all of the cell membranes are related to one another structurally, chemically, functionally and developmentally. A basic conundrum is that while at first glance all membranes appear to be quite similar in size, structure and basic composition, they support very different functions.

Chapter 2 will discuss the chronology of membrane studies from their murky beginnings centuries ago to the classic Gorter and Grendel experiment (1925) that proposed the membrane lipid bilayer.

References

[1]   Reyes H. Acute fatty liver of pregnancy: A cryptic disease threatening. Clin Liver Dis 1999;3(1):69–81.
[2]   Mariana Ruiz Villarreal: http://en.wikipedia.org/wiki/File:Animal_cell_structure_en.svg; public domain.
[3]   Lodish H, Berk A, Matsudaira P, Kaiser CA, Kreiger M, Ploegh H. et al. Molecular Cell Biology. 6th ed. New York, NY: WH Freeman; 2007.
[4]   Pollard TD, Earnshaw WC, Lippincott-Schwartz J. Cell Biology. 2nd ed. Elsevier; 2007.
[5]   Becker WM, Kleinsmith LJ, Hardin J, Bertoni GP. The World of the Cell. 7th ed. San Francisco, CA: Benjamin Cummings; 2009.
[6]   Alberts B, Johnson A, Lewis J, Raff M, Roberts K, Walter P. Molecular Biology of the Cell. 4th ed. Garland Science; 2007.
[7]   Nelson DL, Cox MM. Lehninger Principles of Biochemistry. 5th ed. New York, NY: W Freeman; 2008.
[8]   Berg JM, Tymoczko JL, Stryer L. Biochemistry. 6th ed. New York, NY: W Freeman; 2007.
[9]   Mariana Ruiz Villarreal: http://en.wikipedia.org/wiki/Endomembrane_system; public domain.
[10] Lippincott-Schwartz J, Phair RD. Lipids and cholesterol as regulators of traffic in the endomembrane system. Ann Rev Biophys 2010;39:559–78.
[11] Cooper GM. The Cell: A Molecular Approach. 2nd ed. Sinauer Associates; 2000.
[12] Chi YH, Chen ZJ, Jeang KT. The nuclear envelopathies and human diseases. J Biomed Sci 2009;16:96.
[13] Becker WM, Kleinsmith LJ, Hardin J, Bertoni GP. The World of the Cell. San Francisco, CA: Benjamin Cummings; 2009. p. 333–9.
[14] Porter KR, Claude A, Fullam EF. A study of tissue culture cells by electron microscopy. J Exp Med 1945;81:233–46.

[15] Pavelk M, Mironov AA. The Golgi Apparatus: State of the art 110 years after Camillo Golgis discovery. Berlin: Springer; 2008.

[16] Glick BS. Organization of the Golgi apparatus. Curr Opin Cell Biol 2000;12:450–6.

[17] Luzio JP, Pryor PR, Bright NA. Lysosomes: fusion and function. Nat Rev Mol Cell Bio 2007;8:622–32.

[18] Wanders RJ, Waterham HR. Biochemistry of mammalian peroxisomes revisited. Annu Rev Biochem 2006;75:295–332.

[19] de Duve C. The peroxisome: a new cytoplasmic organelle. Proc R Soc Lond B Bio Sci 1969;173:71–83.

[20] Mannella CA. Structure and dynamics of the mitochondrial inner membrane cristae. Biochim Biophys Acta 2006;1763:542–8.

[21] McBride HM, Neuspiel M, Wasiak S. Mitochondria: more than just a powerhouse. Curr Biol 2006;16:R551.

[22] Sagan L. On the origin of mitosing cells. J Theor Bio 1967;14:255–74.

[23] Margulis L. Symbiosis in cell evolution. New York, NY: W Freeman; 1981. p. 452.

[24] Christian de Duve Institute of Cellular Pathology: http://www.icp.ucl.ac.be/about/deduve.htm

Membrane History

OUTLINE

A. Oil On Water: Interface Studies   13
   Pliny the Elder: (55 A.D.)   13
   Benjamin Franklin: (1772)   15
   Lord Rayleigh: (1890)   16
   Agnes Pockels: (1891)   16
   Irving Langmuir: (1917)   18

B. The Lipid Bilayer Membrane   20
   William Hewson: (1773)   20

C. H. Schultz: (1836)   20
Karl von Nageli: (1855)   21
Wilhelm Pfeffer: (1877)   23
Charles Ernest Overton: (1899)   23
Evert Gorter: (1925)   25

Summary   27

References   28

The current concept of membrane structure is based on the Fluid Mosaic Model outlined by Singer and Nicolson in 1972 (see Chapter 8) [1]. Not surprisingly, this model did not just spring to life in a fully developed form, but instead was conceived and developed slowly over centuries [2,3]. In this chapter, two seemingly unrelated, but parallel, historical paths are discussed, the study of oil on water and the study of the cell outer barrier (the plasma membrane). These two approaches did not merge until the classic experiment of Gorter and Grendel in 1925 [4].

A. OIL ON WATER: INTERFACE STUDIES

Although not appreciated until the 20th century, the study of membrane physical properties had its origin in prehistoric observations by ancient mariners of oils floating on the surface of water. Some of science's most revered figures contributed to this field.

Pliny the Elder: (55 A.D.)

The concept of the oily nature of a membrane had its origins in the pre-historic past. The first written description concerned a practice commonly used in ancient times of stilling water surfaces by the use of cooking oils. The practice was used by ancient seafarers and was described in the first century A.D. by the naturalist Pliny the Elder (Figure 2.1) in his 37 volume encyclopedic series *Naturalis Historia*. In Book 2, Chapter 106, Pliny describes the ancient common knowledge about the effect of oils on rough seas, stating 'all sea water is made smooth by oil and so divers sprinkle oil on their face because it calms the rough element and carries light down with them'. Pliny died in the Mount Vesuvius eruption of 79 A.D. when he decided to sail his boat into Pompeii to get a better look. Later it was reported that oyster fishermen poured oil onto the surface of rough seas to calm them so they could better see shell fish sitting on the bottom. In a similar fashion, ancient fishermen applied oil to better see schools of herring at long distances. When whaling ships returned to port with their catch, the entire harbor was found to be covered with oil, and the water was no longer choppy. Although the calming effect of oil on the sea was common knowledge in ancient times, it remained little more than an observation until the pioneering experiments of Benjamin Franklin.

Pliny the Elder, actual name Gaius Plinius Secundus, was a prolific writer. His 37 volume encyclopedia *Naturalis Historia* is a collection of just about everything he found interesting in his surroundings including important descriptions of manufacturing processes for papyrus, purple dyes, and for gold mining. Because Pliny was probably the most famous person ever to die in a volcanic eruption, current volcanologists use the term 'Plinian' when referring to a sudden volcanic eruption and 'ultra-Plinian' for describing an exceptionally large and violent eruption. The 1883 eruption of Krakatoa was ultra-Plinian. Although no real image

FIGURE 2.1   Pliny the Elder, 23–79 A.D. *Courtesy Prints and Photographs Division, Library of Congress, LC-USZ62-66932*

or description of Pliny the Elder survives today, this highly imaginative 19th century portrait is often passed off as being authentic.

Benjamin Franklin: (1772)

Benjamin Franklin was an early American statesman, an internationally renowned scientist, and a world-class tinkerer (Figure 2.2). Scientifically he is best known as being one of the discoverers of electricity based on his courageous kite experiment from 1750. However, Franklin had an exceptional curiosity and dabbled in many other ventures including creating the first public library and fire department, inventing a new musical instrument, the glass harmonica, that Mozart wrote a piece for, accurately mapping the Gulf Stream, suggesting Daylight Saving Time and becoming 'the Father of Metrology'. One of Franklin's lesser known contributions involved 'the stilling of waves by oil'. Franklin's oil slick work is described in an interesting 1989 book, *Ben Franklin Stilled the Waves*, by Charles Tanford [5]. Tanford is better known for his development of the 'Hydrophobic Effect Theory' that describes the physics behind membrane stabilization (Chapter 3). Franklin's wave-stilling experiment was first done on a mission to London representing the American colony, from Pennsylvania. In a 1772 visit to the Lake District in northern England, he poured a single teaspoon of olive oil (primarily triolein) in the Derwent Water lake. The oil rapidly spread, smoothing the water surface over an incredible ½ acre! He recognized that the oil layer must be very thin, since it displayed 'prismatic colors,' as described by Isaac Newton in his book *Opticks*. Franklin also noted that if a similar drop of oil was placed on a polished

FIGURE 2.2   1777 Portrait of Benjamin Franklin (1706–1790) by Jean-Baptiste Greuze.

marble table, it did not spread, indicating the importance of both oil and water. Franklin's observations were not published in a formal scientific paper written by him, but instead appeared as extractions of personal letters written to friends in 1774. These letters appeared the same year in *Philosophical Transactions of the Royal Society of London*. Unfortunately Franklin missed a golden opportunity when he failed to estimate how thin the oil slick must have been. He knew the volume of the olive oil as well as the lake area it covered. This simple calculation would have estimated the oil layer at about 10 Å, a good estimate of the molecular length of triolein. This would have been the first correct measurement of molecular size, a parameter that would have to wait another 120 years until Lord Rayleigh repeated Franklin's experiment.

Benjamin Franklin's work on oils on water was extracted from a letter to a friend, Dr. William Brownrigg, dated November 7, 1773. The extract was published in *Philosophical Transactions of the Royal Society of London*, in 1774 [6]. A quote from this letter follows:

> 'At length being at Clapham where there is, on the common, a large pond, which I observed to be one day very rough with the wind. I fetched out a cruet of oil, and dropt a little of it on the water. I saw it spread itself with surprising swiftness upon the surface;.... the oil, though not more than a teaspoonful, produced an instant calm over a space several yards square, which spread amazingly, and extending itself gradually till it reached the lee side, making all that quarter of the pond, perhaps half an acre, as smooth as a looking glass.'

Lord Rayleigh: (1890)

Lord Rayleigh (real name John William Strutt) was a well educated, very high profile gentleman scientist at the turn of the 19th century (Figure 2.3). In 1904 he became one of the first Nobel Prize winners for his discovery of the noble gas argon and was the first to explain the age-old question of why the sky is blue [7]. During his long and prolific career, he published more than 500 papers, primarily on optics and sound, and all were personally hand written! Although, as his title implies, he was quite wealthy, he did not use a secretary. In 1890 Rayleigh repeated the Franklin experiment, but took it a step farther. He precisely measured both the volume of olive oil added to his experimental water bath and the surface area over which it spread. From these measurements he calculated the molecular size of triolein, the major component of olive oil, to be 16 Å.

Agnes Pockels: (1891)

Perhaps the most compelling of the major players in the colorful history of membranes was Agnes Pockels (Figure 2.4). Miss Pockels, as she was always referred to, overcame almost insurmountable odds to become one of the earliest women to make a lasting impact on science. She lived in Braunschweig, in north central Germany and, as was typical of girls in the late 19th century, had no formal education. Her job was to cook, sew, and take care of the house. However, she had a keen interest in science and dabbled in her kitchen with pots and pans, sewing threads, and buttons to study the phenomena of oil on water. Her now famous kitchen experiments began in 1880 when Miss Pockels was only 18. In these primitive conditions she worked out the basic methods of making and measuring lipid monolayers. Her basic methods are still in practice today. The Langmuir Troughs used today

FIGURE 2.3   John William Strutt Lord Rayleigh (1842–1919), from *Popular Science Monthly,* Volume 25, 1884.

closely resemble those she developed in the late 19th century and the sewing buttons she used to measure surface tension later became the du Nuoy ring (Chapter 3). In 1891 she gathered her results together and sent an unsolicited letter to Lord Rayleigh, one of the most famous scientists of his time. Despite her total lack of credentials, Lord Reyleigh realized the novelty of her work, and with his help, she got her paper published in *Nature* within 2 months (1891) [8]. It is hard to imagine that someone with no position or formal education could dabble in her kitchen and have a first scientific paper immediately published in the prestigious journal *Nature*! Miss Pockels' methodology was so superior to Rayleigh's that he subsequently adopted her procedure. During her 'career' she published 14 articles over 35 years. She also repeated the Franklin experiment in 1892 and obtained a molecular size for triolein of 13 Å. Her science career faded with World War I and the post war problems in Germany. Her life finally ended much as it began, as a homemaker in 1935.

The first description of Ms Pockels' experiments was in the form of a letter from her to Lord Rayleigh dated January 10, 1891. This letter was later included in the journal *Nature*, published on March 12, 1891. The following quote describes what would decades later be known as a 'Langmuit Trough', although it was first invented by Agnes Pockels.

'A rectangular tin trough, 70 cm long, 5 cm wide, 2 cm high, is filled with water to the brim, and a strip of tin about 1 ½ cm wide laid across it perpendicular to its length, so that the underside of the strip is in contact with the surface of the water, and divides it into two halves. By shifting this partition to the right or left, the surface on either side can be lengthened or shortened in any proportion, and the amount of the displacement may be read off on a scale held along the front of the trough.'

FIGURE 2.4   Agnes Pockles (1862–1935). *Courtesy of the Archive of Braunschweig Technical University.*

Soon after her death in 1935 Nobel Laureate Irving Langmuir, 'The Dean of Surface Chemists', stated that her methodology 'laid the foundation for nearly all modern work with films on water'.

Irving Langmuir: (1917)

At a time when nearly all world-renowned scientists were European, Irving Langmuir (Figure 2.5) was born in America and did his work while employed by General Electric (GE) in Schenectady, New York. While at GE he became famous for developing the nitrogen filled incandescent light bulb. Because of the large income the new bulb made for GE, he was allowed free reign to experiment on just about whatever he pleased, including oil/water interface physics.

In 1917, Langmuir also repeated the Franklin experiment and reported the molecular size of triolein to be 13 Å [9]. Langmuir is credited with refining air/water interface studies into a precise science. For this contribution he was awarded the 1932 Nobel Prize in Chemistry [10]. Langmuir was one of the first to understand that molecules were not just simple spheres. He envisioned the asymmetric molecular shape of surfactants (molecules that accumulate at the air/water interface) having a dual personality — with one region being hydrophobic and another being hydrophilic. In his 1917 *Journal of the American Chemical Society* paper he wrote 'Oleic acid on water forms a film one molecule

FIGURE 2.5   Irving Langmuir (1881–1957).

deep, in which the hydrocarbon chains stand vertically on the water surface with the COOH groups in contact with the water.' He was also the first to predict that hydrocarbon chains are very flexible and that a fatty acid with a double bond (oleic acid) occupies a larger cross sectional area and is more compressible than is a similar straight chain saturated fatty acid (stearic acid). For his many contributions Langmuir was also awarded the Franklin Medal in 1934, although ironically he did not mention Benjamin Franklin in either his acceptance speech for the Medal, a later publication of the speech, or his seminal 1917 paper. The surface science journal *Langmuir* is named after him. Remarkably, Langmuir's iconic 1917 paper was the only one he ever published on lipid monolayers.

Langmuir had a very long and, for the most part, distinguished career at General Electric. However, in 1947 he dabbled in salting clouds to affect rain. In collaboration with the U.S. military, they dropped 80 kg of dry ice into a hurricane that was safely off shore in the Atlantic Ocean. Unfortunately the hurricane abruptly changed direction and came ashore doing extensive damage in Georgia. To frustrate potential lawsuits, the U.S. military classified the data, burying evidence of the failed project for decades.

B. THE LIPID BILAYER MEMBRANE

Our current understanding of membrane structure, the Fluid Mosaic Model, is based on a lipid bilayer. A lipid bilayer is essentially two oil/water interface monolayers, (described above), placed back to back. The first biological membrane to be appreciated, studied, and characterized is the plasma membrane. We now know that every cell currently living on planet Earth and probably every cell that has ever existed on our planet is surrounded by a plasma membrane. The plasma membrane is the barrier that separates the cell interior from the rest of the external environment. The study of membranes began indirectly in 1665 with Robert Hooke's discovery of the cork cell and subsequently followed a long and tortuous path. Outlined below are only a few early but important steps in this process.

William Hewson: (1773)

The earliest experiments on the plasma membrane were osmotic studies on red blood cells. The first important investigator was the under-appreciated Englishman William Hewson, (Figure 2.6) a good friend and colleague of Benjamin Franklin. In fact, in 1770 Franklin sponsored Hewson for Fellow of the Royal Society. Hewson's work involved microscopic studies of erythrocyte shape. He noted that upon addition of water, erythrocytes changed from flat (discord) to spherical (globular). With too much water erythrocyte simply dissolved, a process now known as hemolysis. Hewson was the first to show osmotic swelling and shrinking of erythrocytes and from this deduced the existence of a cell (plasma) membrane as a structure surrounding a liquid protoplasm [11]. Unfortunately, however, this important work on membranes was largely ignored. Hewson was highly respected and made major contributions to several fields. He has sometimes been referred to as the 'father of hematology'. He is given credit for isolating fibrin and in defining the lymphatic system for which he was given the Copley medal in 1769.

In 1998 Hewson received renewed interest, but for an entirely different reason. While in London (1757—1762 and 1764—1775), Benjamin Franklin stayed at a house at 36 Craven Street, which is now the home of the Benjamin Franklin House Museum. In 1772 Hewson also ran an anatomy school at this same location. In 1998, workmen restoring the Franklin Museum dug up the remains of six children and four adults hidden below the home. Most of the bones showed signs of having been 'dissected, sawn or cut'. One skull had been drilled with several holes. The bones were from the same years Franklin lived there. The question arose, could this have been a very old crime scene? And what did Benjamin Franklin know about the happenings?

C. H. Schultz: (1836)

Although membranes are too thin to be seen directly by the human eye, or even by light microscopy (Chapter 1), their presence can be detected as an invisible barrier by plasmolysis or by staining the outer membrane surface with various dyes. In 1836 C.H. Schultz used iodine to visualize the erythrocyte plasma membrane. He was then able to estimate the erythrocyte membrane thickness to be about 220 Å, only a little larger than the currently accepted measurement of ~50—100 Å. The importance of this very early date in membrane studies

FIGURE 2.6   William Hewson (1739–1774). *Courtesy of the National Library of Medicine.*

must be put in historical context. Schultz's work came only 8 years after Wohler's classic 1828 experiment synthesizing the first organic molecule (urea) by heating an inorganic molecule (ammonium cyanate). This was the beginning of the end for the theory of 'vitalism', a concept that believed organic molecules were different from inorganic molecules by possessing a 'vital spirit' imparted by God. Three years later, in 1839, Theodor Schwann established the Cell Theory. Incidentally, Schwann's promising career, that included the first use of the term 'membrane' ended by age 30, as he was ridiculed for having the audacity to suggest that alcohol fermentation was the result of living organisms. Schultz's use of iodine foreshadowed later experiments from the mid to late 19th century employing vital dyes in the study of membranes.

Karl von Nageli: (1855)

The botanist Karl von Nageli is usually given credit for laying the foundation for cell membrane osmotic studies on plant cells (Figure 2.7). Indeed, it is much easier to accurately measure plasmolysis as a function of external osmotic strength with plant, as opposed to animal cells due to their thick and easily observable cell walls. Simple microscopic observations of plasmolysis on thin onion peels (only a few cells thick) are often used in high school biology classes. In a solution of sufficient salt or sugar, the cytoplasm is observed to pull away from the cell wall (Figure 2.8). In animal cells the plasma membrane barrier can be clearly distinguished due to differences in light refraction across the membrane. Light travels more slowly in the dense cytoplasm than it does in the much less dense bathing solution.

FIGURE 2.7   Karl von Nageli (1817–1891).

Using thin leaves of the common aquarium plant Elodea, the location of the plasma membrane can be more readily observed than by light refraction. The outer plasma membrane barrier defines a track around which tiny green chloroplasts are carried by cytoplasmic streaming. This can be very easily detected by simple light microscopy.

In his day von Nageli was a highly respected botanist. Among his discoveries were chromosomes and the function of many plant parts including antheridia and spermatozoids of

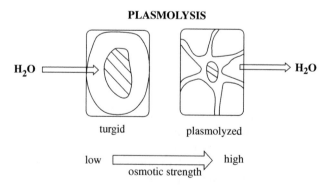

FIGURE 2.8   Plasmolysis in a plant cell. At a critical osmotic strength of the media, water leaves the cell and the plasma membrane pulls away from the cell wall. At this point the cell is said to be 'plasmolyzed'.

FIGURE 2.9   Wilhelm Pfeffer (1845–1920), from *Popular Science Monthly*, Volume 50, 1896.

ferns. Unfortunately he was stubborn and refused to accept new ideas proposed by others and he believed in spontaneous generation! From 1866–1873 von Nageli exchanged letters with the Austrian monk Gregor Mendel, whom we now know as the 'father of modern genetics'. Von Nageli did not appreciate or even understand Mendel's work and discouraged him from continuing. It was von Nageli who rejected the original paper by Mendel that, when rediscovered 40 years later, revolutionized genetics.

Wilhelm Pfeffer: (1877)

Membrane Theory, the first general theory of cell physiology, is often attributed to Wilhelm Pfeffer (Figure 2.9). From his studies on the osmotic behavior of then invisible plasma membranes, he concluded the cell barrier must be thin and semi-permeable. He proposed that the plasma membrane was similar to an artificial copper ferrocyanide membrane. Some 20 years later Overton demonstrated that the permeability barrier (plasma membrane) was in fact lipid in nature.

Charles Ernest Overton: (1899)

Charles Ernest Overton (Figure 2.10) is credited with being the first true 'membranologist'. Although Overton was born in England, he worked in Germany and Sweden and published

FIGURE 2.10   Charles Ernest Overton (1855–1933). *Courtesy of Oregon State University Libraries Special Collections.*

exclusively in German. Overton's major contribution to the understanding of membranes involved permeability studies [12,13]. In the late 19th century it was believed that membranes were semi-permeable, meaning at that time they were only permeable to water. In sharp contrast, Overton found that other molecules also crossed membranes, with neutral molecules crossing faster than charged molecules. Overton also showed that very different types of cells exhibited similar permeabilities, implying a common set of properties. He also found a parallel between a solute's membrane permeability and its solubility in olive oil. He concluded that membranes must include a lipid-like barrier. He also guessed, quite correctly, that membranes might contain cholesterol and lecithin (phosphatidylcholine), two molecules that were known at the time. The permeability Overton measured for most tested solutes was 'passive', meaning diffusion occurred down the solute's concentration gradient by dissolving into and crossing the membrane (Chapter 14). He also recognized the need for 'uphill' (now called active) transport against a gradient for some solutes that did not follow his lipid-solubility rule (Chapter 14). Overton's work, published in 1899, was so far ahead of its time that it was not truly appreciated for decades. As an extension of his permeability studies, Overton later found an important relationship between an anesthetic's efficacy, its membrane permeability, and its solubility in olive oil. This correlation is now referred to as the Meyer-Overton Theory [14].

Evert Gorter: (1925)

During his entire and long scientific career (born in 1881, he worked until he died in 1954), Evert Gorter (Figure 2.11) was a highly respected Dutch pediatrician [15]. Gorter must have been a very busy man as he balanced two jobs, a paying one (pediatrician) and a non-paying one (scientist). In addition, he suffered from arthritis and was wheelchair-bound. In a 1925 paper in the *Journal of Experimental Medicine*, Gorter and his research assistant F. Grendel published a very short report consisting of only 5 pages with no figures and one brief data table (shown in Figure 2.12). This paper [4], is now considered to be the most significant work ever done on membranes. However, as is often the case, this monumental insight was unappreciated when it was published and was only rediscovered many years later. Gorter and Grendel were the first to offer experimental proof that membranes are 'lipid bilayers'. They extracted erythrocyte membrane lipids with benzene and floated them on a Langmuir Trough. Upon rapid evaporation of the benzene, a lipid monolayer was formed. They then used the earlier 1917 paper by Langmuir [9] as a guide to compress the membrane lipid monolayer until it resembled a normal compressed lipid monolayer. Unfortunately this was a guess. They knew how many extracted erythrocytes produced a measured monolayer area and they calculated the surface area of a typical bi-concave-shaped erythrocyte. Knowing each of these parameters, they discovered the area of a monolayer from extracted erythrocyte lipids was twice the surface area of the erythrocytes, hence the membrane was a lipid bilayer. To prove

FIGURE 2.11   Evert Gorter (1881–1954). *Courtesy of Digitaal Wetenschapshistorisch Centrum (DWC).*

Animal.	Amount of blood used for the analysis.	No. of chromocytes per c.mm.	Surface of one chromocyte	Total surface of the chromocytes (a).	Surface occupied by all the linoids of the chromocytes (b).	Factor a:b.
	gm.		sq.µ	sq.m.	sq.m.	
Dog A	40	8,000,000	98	31.3	62	2
	10	6,890,000	90	6.2	12.2	2
Sheep 1	10	9,900,000	29.8	2.95	6.2	2.1
	9	9,900,000	29.8	2.65	5.8	2.2
Rabbit A	10	5,900,000	92.5	5.46	9.9	1.8
	10	5,900,000	92.5	5.46	8.8	1.6
	0.5	5,900,000	92.5	0.27	0.54	2
Rabbit B	1	6,600,000	74.4	0.49	0.96	2
	10	6,600,000	74.4	4.9	9.8	2
	10	6,600,000	74.4	4.9	9.8	2
Guinea Pig A	1	5,850,000	89.8	0.52	1.02	2
	1	5,850,000	89.8	0.52	0.97	1.9
Goat 1	1	16,500,000	20.1	0.33	0.66	2
	1	16,500,000	20.1	0.33	0.69	2.1
	10	19,300,000	17.8	3.34	6.1	1.8
	10	19,300,000	17.8	3.34	6.8	2
	1	19,300,000	17.8	0.33	0.63	1.9
Man.	1	4,740,000	99.4	0.47	0.92	2
	1	4,740,000	99.4	0.47	0.89	1.9

FIGURE 2.12 Data table from the Gorter and Grendel 1925 paper indicating that there was enough lipid in the plasma membrane of erythrocytes to surround the cell exactly twice. This is the first experimental support for the existence of a lipid bilayer. *The table is taken from [4].*

the universal importance of their conclusion, they obtained erythrocytes from different animals including man, rabbit, dog, guinea pig, sheep, and goat. All produced the same results. The paper ends, 'It is clear that all our results fit in well with the supposition that the chromocytes (erythrocytes) are covered by a layer of fatty substances that is two molecules thick'.

Although their conclusion that the lipid bilayer is the fundamental building block of membranes is now universally recognized to be at the heart of membrane structure, it is evident how lucky Gorter and Grendel were. The erythrocyte is perhaps the only eukaryotic cell that would give a ratio of monolayer area to cell surface area of 2. All other eukaryotic cells have extensive internal membranes that would significantly affect these measurements. The prediction that there is exactly enough lipid to surround a cell twice means the lipid bilayer had to constitute 100% of the membrane surface. There is no room for integral membrane proteins! Benzene is a poor solvent for extracting polar lipids. Gorter probably extracted only ~70% of the membrane lipids. These authors did not compress their monolayer to biological pressures and they also miscalculated the surface area of an erythrocyte bi-concave disk! All of these errors conveniently canceled each other out resulting in a nice, interpretable ratio of 2. The only data table reported in their seminal paper is shown in Figure 2.12. Gorter went on to publish more than 50 additional papers, but all on proteins. Grendel ceased publishing in 1929.

1925 turned out to be a crucial year for the beginning of lipid bilayer studies. While Gorter and Grendel used the general, undefined term 'lipoid' to describe the nature of the bilayer component, J.B. Leathes and H.S. Raper in their 1925 book *The Fats* [16], suggested phospholipids might be essential structural elements of cell membranes. Also in 1925, the American, Hugo Fricke, used electrical impedance measurements to determine the thickness of the erythrocyte membrane [17]. His value of 33 Å was essentially correct but he did not recognize it as a bilayer. This is surprising since his value is approximately twice the lipid monolayer values of ~13 to 16 Å reported by Pockels, Rayleigh, and Langmuir. In 1926 James B. Sumner reported isolation of the first enzyme, urease, an astonishing feat in its day [18]. For this work, Sumner was a co-recipient of the 1946 Nobel Prize in Chemistry. Sumner accomplished his work despite having lost his left arm (he was left handed) in a hunting accident when he was 17 years old. His wife, Cid Ricketts Sumner, wrote a book that was adapted into the popular 1957 movie *Tammy and the Bachelor*.

It is ironic that urea was the central compound in two of the major breakthroughs in biochemistry, Wohler's first synthesis of an organic molecule (urea) from an inorganic molecule (ammonium cyanate) in 1828 and Sumner's 1926 purification of urease. So, by the late 1920s, all pieces were in place to establish a realistic, working model for a biological membrane. This membrane model first appeared in a 1935 article by Davson and Danielli (Chapter 8).

SUMMARY

Membrane history has a very colorful past that followed two distinct paths. One path investigated lipid monolayers (oil on water), while the other monitored the plasma membrane of living cells. The first oil on water experiment is attributed to Ben Franklin in 1772. The Franklin experiment was greatly refined and quantified in the 1880s by Agnes Pockels, a most remarkable woman. Despite being an uneducated homemaker, at the age of 18 Pockels used old pots and pans, sewing threads and buttons to measure lipid monolayer properties. Pockel's methodologies were further refined in 1917 by Irwin Langmuir, "The Dean of Surface Chemists".

The earliest plasma membrane experiments demonstrated a semi-permeable barrier using osmotic measurements with erythrocytes (William Hewson, 1772). In the late 1890's Charles Ernest Overton, the first true "membranologist", found a parallel between a solute's membrane permeability and its solubility in olive oil. From this he concluded that membranes are a lipid-like barrier. In 1925 Gorter and Grendel brought together both approaches with a seminal paper proposing a lipid bilayer as the cornerstone of membrane structure.

The most important molecule in membrane structure/function, and even life itself, is water. Chapter 3 will discuss the multitude of physical properties that account for water's essential role in membranes.

References

[1]   Singer SJ, Nicolson GL. The fluid mosaic model of the structure of cell membranes. Science 1972;175:720–31.
[2]   Eichman P. From the Lipid Bilayer to the Fluid Mosaic: A Brief History of Model Membranes. SHiPS Resource Center for Sociology, History and Philosophy in Science Teaching.
[3]   Ling GN. Life at the Cell and Below-Cell Level. The Hidden History of a Fundamental Revolution in Biology. Pacific Press; 2001.
[4]   Gorter E, Grendel F. On bimolecular layers of lipids on the chromocytes of the blood. J Exptl Med 1925;41:439–43.
[5]   Tanford C. Ben Franklin Stilled the Waves. Duke University Press; 1989.
[6]   Franklin B. Philosophical Transactions of the Royal Society of London, 1774. Vol 64, p. 445–7.
[7]   From Nobel Lectures, Physics 1901–1921. Amsterdam: Elsevier; 1967.
[8]   Pockels A. Surface tension. Nature 1891;43:437–9.
[9]   Langmuir I. The constitution and fundamental properties of solids and liquids. II,. J Amer Chem Soc 1917;39:1848–906.
[10] From Nobel Lectures, Chemistry 1922–1941. Amsterdam: Elsevier; 1966.
[11] Kleinzeller A. William Hewsons studies of red blood corpuscles and the evolving concept of a cell membrane. Amer J Physiol 1996;271:C1–8.
[12] Kleinzeller A. Ernest Overtons contribution to the cell membrane concept: A centennial appreciation. Am J Physiol Cell Physiol 1998;274:C13–23.
[13] Kleinzeller A. Membrane permeability − 100 years since Ernest Overton. Curr Top Mem 1999;48:1–22.
[14] Janoff AS, Pringle MJ, Miller KW. Correlation of general anesthetic potency with solubility in membranes. Biochim Biophys Acta 1981;649:125–8.
[15] Dooren LJ, Wiedemann HR. The pioneers of pediatric medicine: Evert Gorter (1881–1954). Eur J Pediatr 1986;145:329.
[16] Leathes JB, Raper HS. Monographs on Biochemistry, The Fats. 2nd ed. London: Longmans, Green; 1925.
[17] Fricke H. The electrical capacity of suspensions with special reference to blood. J Gen Physiol 1925;9:137–52.
[18] Sumner JB. The isolation and crystallization of the enzyme urease. J Biol Chem 1926;69:435–41.

CHAPTER

3

Water and the Hydrophobic Effect

OUTLINE

A. Water — Strength in Numbers   29

B. Structure of Water   31

C. Properties of Water — No Ordinary Joe   32

D. Surface Tension   34

E. The Hydrophobic Effect   37

Summary   40

References   41

A. WATER — STRENGTH IN NUMBERS

What is so special about water? It is colorless, tasteless, odorless, and extremely abundant. At first glance it would appear that water is as bland and ordinary as any compound could possibly be. Yet water is the most studied material on Earth and has been for a very long time. It has been known from early historical records that water is the most important biochemical in life. Hidden by its deceptively unexciting cloak, water houses a plethora of subtle, but extraordinary, properties. The basic properties of water can be found in any general Biology or Biochemistry book and are summarized nicely in Wikipedia [1].

Water is the most abundant molecule on earth, comprising 70–75% of the Earth's surface as liquid and solid. The total volume of the Earth's water is approximately 1,360,000,000 km^3 with 97.2% in oceans, 1.8% as ice, 0.9% in ground water, 0.02% in fresh water, and 0.001% as water vapor. The abundance of liquid water leaves one with the presumption that the fluid state is common on earth. In fact fluids are quite rare. The second most abundant fluid on this planet is petroleum, but even petroleum has its origins in water-based life.

An Introduction to Biological Membranes
http://dx.doi.org/10.1016/B978-0-444-52153-8.00003-9

From early Greek times through to modern space exploration it has been believed that water is essential for life as we know it. The Greek natural philosopher Thales of Miletus (ca. 640–548 B.C.) is reputedly the first to have emphasized the fundamental importance of water, assigning it as the first principle of the cosmos, being 'the origin of all things'. Hippo of Samos (450 B.C.) perceived life as water. NASA now bases its search for life in the universe on the assumption that without liquid water, life is not possible. Therefore the search for extra-terrestrial life is predicated on the search for water.

Although not yet proven, it is likely that water is abundant in all corners of the universe [2,3]. Since H and O are among the most abundant elements in the universe, it is reasonable to assume that water will be found everywhere as well. In our solar system, evidence of water has been found on Earth and its moon, Mercury, Mars, Neptune, Pluto, Jupiter's moons Triton and Europa, and Saturn's moon Enceladus [4]. Of particular interest to the question of extra-terrestrial life are Mars, Europa, and Enceladus. Speculation about the existence of water and even life on Mars has existed since the first reports of 'canals' on the red planet by Angelo Secchi in 1858 and Giovanni Schiaparelli in 1877. Percival Lowell took the questionable descriptions of long straight channels and the unquestioned existence of Martian seasons to another level in the 1890s when he adopted the view they were 'canals' built by intelligent beings. Intelligent life on Mars soon became ingrained in the public conscience after publication in 1897 of H.G. Wells' *The War of the Worlds*. More recently, NASA's Martian exploration has shown the planet to be dry, with no pooled surface water. However, the Mars Global Surveyor spacecraft found evidence for mostly sub-surface water and that water clearly flowed on Mars in the distant past, as evidenced by erosion channels. Also, strong evidence for past water on the Martian surface was provided by the robotic rovers, Spirit and Opportunity. More recently, (July, 2008), water has been physically found in Martian soil by the Phoenix Lander.

Perhaps of even more importance than Mars is Jupiter's moon Europa and Saturn's moon Enceladus. Europa is unlike other planetary moons. Its surface is smooth and un-cratered, indicating the existence of a giant subterranean ocean made of water that is covered by a frozen ice crust. The frozen ice surface (−260°F) cracks as a result of Jupiter's enormous gravitational pull. The Jovian moon's surface is therefore covered with stress fracture lines that are colored, indicating the presence of slushy water that is extruded from the moon's interior, refreezing on the surface. The extrudate contains organic solutes, accounting for its color. The underlying water ocean has been estimated to be perhaps as deep as 100 Km. Warmed by Jupiter's gravitational field, Europa's interior is full of liquid water and may in locations be of sufficient temperature to support life. Even if life did not have enough time to evolve on Europa, the moon may be a museum of a chemical evolution 'laboratory' frozen in time.

Enceladus is a small moon revolving just outside Saturn's rings. Like Europa, Enceladus is home to a vast underground ocean. The ocean appears to be 'fizzy like a soft drink and could be friendly to microbial life' [5]. Figure 3.1 shows a Cassini spacecraft image of vaporous, icy jets emerging from fissures on Enceladus' surface [5]. If water is the most important biochemical for the development of life, Europa and Enceladus are the places to explore.

FIGURE 3.1   A Cassini spacecraft image of vaporous, icy jets emerging from fissures on Enceladus.

B. STRUCTURE OF WATER

The chemical composition of water, (2 hydrogens, 1 oxygen), was discovered by Henry Cavendish around 1781. Water is a polar molecule having both positive and negative poles, and exists in a tetrahedral arrangement, extending out in all dimensions (Figure 3.2). The partially positively charged hydrogens (each H is +0.41) are at an angle of 104.5 degrees from each other with respect to the partially negatively charged oxygen (−0.82). The charge distribution produces a dipole moment of 1.85 D (see Figure 3.2). This polar structure is responsible for water's exceptionally strong self-attraction (cohesion) and numerous unusual properties that are responsible for its central role in life. Water is both small in size (MW 18) and highly mobile. The concentration of water in water (1,000 gl^{-1}/18 $gmol^{-1}$) is 55.5 M, making water both life's solvent, as well as being an essential reagent in life processes.

An important feature of water is its ability to readily form hydrogen bonds (H-bonds) with itself and other polar entities. H-bonding in water is the major driving force for membrane stability as first suggested by Latimer and Rodebush in 1920. As discussed in Chapter 2, the first quarter of the 20th century was a critical time period that produced many seminal membrane studies. Included in these was a first understanding of the importance of water structure.

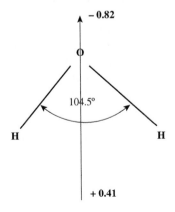

FIGURE 3.2   The structure of water.

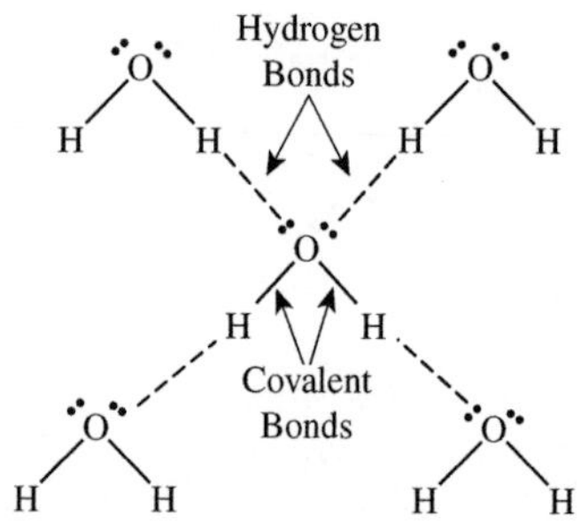

FIGURE 3.3   Hydrogen bonding in water.

Hydrogen bonding occurs when an atom of hydrogen is shared between two electronegative atoms instead of residing on just one (Figure 3.3). In the case of water, the hydrogen is shared unequally between two oxygens of adjacent waters. Water is unique in that it can form 4 H-bonds, while other H-bonding molecules like HF, ammonia, and methanol can only form 1 H-bond. In fact liquid water contains by far the densest H-bonding of any solvent. H-bonding accounts for the anomalous physical properties of water that are particularly important over a narrow temperature range (0°C–100°C), where water exists as a liquid and is therefore suitable for biological processes. For example, a similar compound, H_2S, has much weaker H-bonding than water and is a gas at biological temperatures.

At any one instant in time a hydrogen is more closely associated with one water's oxygen (covalent bond length 0.101 nm) than the other water's oxygen to which it is H-bonded (bond length 0.175 nm). As a result, a covalent bond (~492 KJ/mol) is stronger than an H-bond (~23.3 KJ/mol). However, even the weaker H-bond is almost 5x stronger than the average thermal collision fluctuation energy at 25°C and both are far stronger than weak van der Waals interactions. The hydrogen residence is rapidly alternated between the two adjacent water oxygens with an equilibrium time scale of less than a femtosecond (10^{-15} sec).

In the solid ice state every water molecule is H-bonded to 4 other water molecules producing an enormous, highly organized extended water cluster. Upon melting, an initial assumption would be that the organized water clusters would be completely destroyed, hence resulting in a fluid. However, as shown in Table 3.1, this is not the case. The solid is transformed into a liquid as a result of breaking a surprisingly small number of H-bonds. Even near physiological temperature (40°C) water exists primarily in extended clusters where most waters are still H-bonded to 4 other water molecules. It is not until water is in the gaseous state (100°C) that the extended clusters disappear.

C. PROPERTIES OF WATER – NO ORDINARY JOE

The structure of water and its facility to form H-bonds dictates its extraordinary properties. For example, water is small, stable, abundant, and is the only pure substance on earth that exists in all three states of matter: solid; liquid; and gas. Below are listed a few of water's special properties that are related to its central role in life [6–8].

TABLE 3.1   Size of Water Clusters as a Function of Temperature.

Temp °C	Number of water molecules in a cluster
0 (ice)	All (100)
0 (liquid)	91
40 (liquid)	38
100 (gas)	0

1. Most solids expand when they melt but water expands when it freezes. In fact water is the only substance on Earth where the maximum density of its mass does not occur when it becomes solidified. The density of liquid water (1.000 g/cm^3) is greater than solid ice (0.917 g/cm^3) causing ice to float in water and resulting in an ~9% increase in volume upon freezing. If ice were denser than liquid water, bodies of water would freeze from the bottom up, vastly reducing the amount of liquid water on Earth. Also, it is expansion during freeze/thaw cycles that turns inhospitable rock into soil.

2. Water has a boiling point almost 200°C higher than that expected on the basis of boiling points of similar compounds. While the boiling point of H_2O is 100°C, that of H_2S is only -60°C. Therefore H_2S is a gas at biologically important temperatures and could not serve in place of water as 'life's solvent'.

3. Water's high latent heat of evaporation (2.270 kJ/kg) gives resistance to dehydration and provides considerable evaporation cooling.

4. Water has a melting point (0°C) at least 100°C higher than expected on the basis of melting points of similar compounds and also has a high heat of fusion (melting, 334 kJ/kg).

5. Water is an excellent solvent and is often referred to as a 'miracle liquid' or the 'universal solvent'. Water is such a good solvent it almost always has some solute dissolved in it (about 5 ppb impurities in distilled water). Almost every substance known has been found dissolved in water to some extent. Water efficiently dissolves salts like NaCl that have very limited solubility in most other liquids. Seawater for example is ~3.5% salt. This property allows for facile transfer of nutrients necessary for life. Liquid water is an excellent solvent due to its polarity, high dielectric constant, and small size. In contrast, solid ice is a very poor solvent.

6. Although water has an unusually high viscosity, a measure of resistance to flow (0.001 Pa·s at 20°C), it is still 10^4 to 10^5 less viscous than the interior of a membrane.

7. Water conducts heat more easily than any liquid except mercury.

8. Water has an unusually high heat capacity. It takes more heat to raise the temperature of 1 g of water 1°C than any other liquid. Its specific heat is 1 cal/g°C or 4.186 J/g°C. As a result water can store enormous amounts of thermal energy, minimizing the volume of body coolant required to remove metabolic heat. Due to its high heat capacity, high thermal conductivity, and abundance in cells, water is the major thermal regulator in life. Liquid water is truly special as it has over twice the specific heat capacity of ice or steam. Thermal properties of water are due to the ease of intra-molecular hydrogen-bonding.

9. Water's pKa is 15.74 making the pH of unmodified water ~7.0, neither acidic nor basic. However, absolutely pure water is very hard to produce. CO_2 is rapidly absorbed producing carbonic acid and dropping the pH of freshly distilled water to ~5.7.
10. Water is highly non-compressible. Even in a deep ocean at 4,000m, under a pressure of 4×10^7 PA, there is only a 1.8% change in water's volume.
11. Pure water is an excellent electrical insulator and in fact does not conduct electricity at all! But water is such a good solvent it almost always has some solute dissolved in it, meaning that in a living cell where the salt concentration is high, water can readily conduct electricity.

D. SURFACE TENSION

Another important property of water stemming from its structure is surface tension [9]. Surface tension is of tremendous importance in the historical study of membranes (Chapter 2) and is essential in understanding membrane structure and stability. Water has the largest surface tension (72.8 dynes/cm at 25°C) of any common liquid except mercury. For example, ethanol, a much weaker H-bonding molecule than water, has a surface tension of only 22 dynes/cm. Surface tension is responsible for holding raindrops together, for making ocean waves possible and for capillary action that brings water up a tree through the xylem. Water's surface tension is even high enough to support the weight of some insects like the water strider. An extreme example is the Basilisk lizard that can actually 'walk on water' for 10 to 20 minutes and so is more commonly referred to as the 'Jesus lizard' (Figure 3.4).

Surface tension is a measure of the strength of the water surface film that reflects the strong intra-molecular H-bond-driven cohesive forces of water. The force of cohesion (sticking together) between water molecules is the same in all directions (Figure 3.3). In bulk solution every water molecule can be H-bonded to 4 other water molecules, but waters that are unfortunate enough to find themselves at the air—water or oil—water interface can only form 2 H-bonds (see Figure 3.5). The increase in number of H-bonds in the bulk solution compared

FIGURE 3.4   The Basilisk or 'Jesus lizard' can 'walk on water' due to water's high surface tension. *Reproduced with permission [13]*

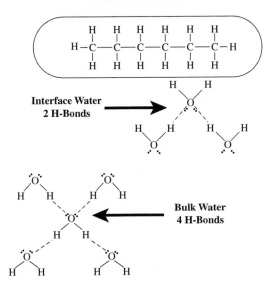

FIGURE 3.5   Dispersion of hexane into water is highly unfavorable due to the difference in H-bonds available to bulk-waters (4) compared to hexane-interface waters (2).

to the interface makes the bulk water structure more thermodynamically favorable. As a general rule, the more H-bonds that can be formed, the more stable the structure will be. Therefore water molecules at the air—water interface feel a net force of attraction that pulls them away from the interface and back into the bulk of the liquid. As a result the entire liquid tries to take on a shape that has the smallest possible surface contact area with air or oil — a sphere. Thus, ignoring gravitational effects, the shape of a raindrop would be a sphere. The stronger the H-bonds between molecules, the higher will be the surface tension. The intermolecular force of adhesion between water and a hydrocarbon that comprise the interior of a membrane is much lower than the intra-molecular force of cohesion between the water molecules themselves. Therefore water tries to minimize its surface contact area by separating from the hydrocarbon. This is known as the Hydrophobic Effect [10,11] and is the major stabilizing force for membrane structure (discussed below). Hydrophobic means fear of (phobia) water (hydro).

Surface tension was one of the first quantitative measurements that was possible on a 'model membrane' and so was heavily employed by the early monolayer experimental physicists including Agnes Pockels and Irwin Langmuir (Chapter 2). Earliest measurements were done employing a Du Nouy Ring, named after the French physicist who developed the technique in the late 1890s. Actually, Agnes Pockels predated Du Nouy by using common shirt buttons! By this method, a platinum ring (now replaced by a platinum-iridium alloy) is attached to a sensitive tensiometer incorporating a precision micro-balance. The ring is immersed in the aqueous solution just beneath the air—water interface and the force required to pull the ring slowly through the interface is measured (Figure 3.6). The force slowly increases until a maximum value is reached. At this point, the ring breaks free from the surface, leaving the solution, and the pulling force drops to zero. Surface tension is directly related to the maximal force required to free the ring from the solution. A similar method in use today employs a Wilhelmy

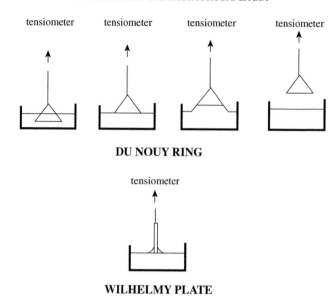

FIGURE 3.6   Du Nouy Ring and Wilhelmy Plate methods to measure surface tension.

Plate in place of the Du Nouy Ring (Figure 3.6). With this method, the plate (usually made of platinum) is slowly lowered to the aqueous interface whereupon the surface 'jumps' onto the plate and the weight of the 'jumped' solution is measured. It is remarkable that the simple century-old technique to measure surface tension, developed before the concept of a lipid bilayer had even been imagined, has changed very little through the years.

Anything that disrupts the surface packing decreases surface tension. One class of molecules that is very effective at reducing the surface tension of water is known as surfactants (surface active reagents). Surfactants are amphipathic molecules, having both a polar and a non-polar end (see Figure 3.7). Surfactants accumulate at the air–water interface where their non-polar tails stick into the air, avoiding unfavorable interaction with water, while the polar end of the molecule sits comfortably in the water. The surfactants replace some of the surface interface waters, decreasing the net surface tension. For example, in an early study, the high surface tension of water (72.8 dynes/cm) was reduced to ~7–15 dynes/cm by adding membrane phospholipids to the interface. The implication is that a membrane surface would have a low surface tension.

Surfactants are often detergents (see Chapter 13) that employ reduction in surface tension in order to function. In one simple demonstration, metal needles are carefully floated on the air–water interface. Although metal is denser than water, the needles can float due to the surface tension of the water. Upon the addition of a small drop of detergent to the aqueous sub-phase, the needles sink to the bottom. The heavier needles sink before the lighter needles as the surface tension drops. Detergent surfactants play a critical role in disrupting or even dissolving membranes. They are essential for extracting proteins from membranes in a non-denatured form for subsequent reconstitution studies. Detergents are a vital tool in membrane studies and are discussed in detail in Chapter 13.

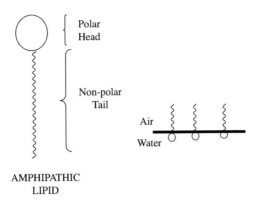

FIGURE 3.7   Structure of an amphipathic lipid and its orientation at the air–water interface.

E. THE HYDROPHOBIC EFFECT

It is quite remarkable that membranes are so stable without any covalent bonds holding them together! The subtle forces driving membrane bilayer stability must indeed be strong, if not obvious. These forces are referred to as the Hydrophobic Effect and have been elegantly discussed by Charles Tanford (Figure 3.8) in his two books, *Hydrophobic Effect: Formation of*

FIGURE 3.8   Charles Tanford (1921–2009) [14].

Micelles and Biological Membranes in 1973 and *The Hydrophobic Effect* in 1980 [10,11]. The Hydrophobic Effect is attributed to intra-molecular H-bonding between water. Charles Tanford is also the author of *Ben Franklin Stilled the Waves*, an interesting account of the early history of membrane studies [12].

H-bonds are the result of sharing a hydrogen between two electronegative atoms (e.g. F, Br, O, N, S etc.). H-bonds cannot form between water and a hydrocarbon as C and H have nearly equal affinity for electrons. If the hydrocarbon hexane is placed in a box of water in the absence of gravity, hexane will form a perfect sphere indicating minimal contact surface area with water. Entropy, a measurement of the degree of randomness, would predict that without any other forces, hexane by itself should disperse, decreasing its order. However, observation clearly shows that in water hexane prefers to exist in what appears to be an unfavorable, highly organized spherical structure. Remove water from the box and hexane would disperse. Therefore the cohesive forces that keep water together (H-bond forces) are much stronger than the hexane entropic forces that would disperse the sphere. Very weak van der Waals forces that would keep the hexane molecules together are also insignificant when compared to the cohesive forces of water. Hexane molecules stay together in water because it is essentially more favorable for the water molecules to hydrogen bond to each other than to engage in very weak interactions with non-polar hexane molecules. The bulk waters are surrounded by, and H-bonded to, 4 other waters, while the waters at the hexane interface are surrounded by, and H-bonded to, only 2 other waters (Figure 3.5). The reduced number of H-bonds available to waters at the hexane interface makes them energetically less favorable. As a result, it takes work to drive a hydrocarbon (fatty acyl) chain into water, creating a new, unfavorable interfacial surface.

The Hydrophobic Effect is the major driving force for all biological structure, including that of membranes. The hydrocarbon tails of membrane phospholipids segregate away from water, making the membrane interior dense with hydrocarbon chains, but devoid of water. Energetics of membrane formation is similar to the previously discussed case of a hexane sphere separating from water.

Figure 3.9 depicts the 5 major forces contributing to membrane bilayer stability in water: the Hydrophobic Effect; head group — water interactions; head group — head group interactions; entropy of the caged tails; and van der Waals forces. The hydrophobic effect is by far the most important in membrane stabilization. In contrast to the hexane sphere discussed above, membranes are composed of amphipathic lipids that also have a polar, often charged head group that can favorably interact with water, further stabilizing the membrane. Additional stabilizing forces can come from ionic interactions between the head groups of adjacent phospholipids. Another force driving membrane formation is related to the motion (entropy) of the bilayer acyl chains. The hydrocarbon chains of phospholipids exposed to water would be totally surrounded by a stiff water cage high surface tension where their motion would be highly restricted (low entropy, unfavorable). Upon shedding the water cage, the acyl chains are segregated into the hydrophobic bilayer interior, resulting in a large increase in favorable acyl chain motion (increase in entropy).

Although the Hydrophobic Effect is by far the major force stabilizing membranes, the membrane interior is also held together very weakly by van der Waals forces. Van der Waals forces are the result of a collection of very close molecular and surface interactions discovered

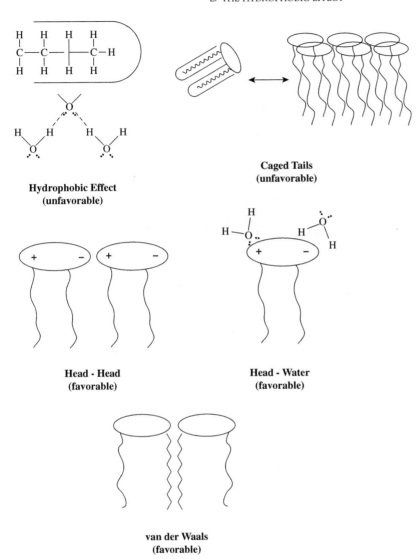

FIGURE 3.9 Five forces that stabilize the membrane lipid bilayer in water.

Hydrophobic Effect
(unfavorable)

Caged Tails
(unfavorable)

Head - Head
(favorable)

Head - Water
(favorable)

van der Waals
(favorable)

by the Dutch scientist Johannes Diderik van der Waals (Figure 3.10) in the late 19th century. For his achievements van der Waals received the 1910 Nobel Prize in Physics. In regard to membranes, the van der Waals force is more correctly termed the van der Waals-London force. This very weak force is neither covalent nor ionic. Instead it is the result of induced dipoles that form instantaneously between two very close molecular surfaces. This force increases with the length of the non-polar part of the adjacent surfaces and so is particularly important to the densely packed acyl chains that comprise the membrane hydrophobic interior. Therefore, there are at least 5 forces stabilizing the membrane bilayer, the most important being the Hydrophobic Effect.

FIGURE 3.10   J.D. van der Waal (1837–1923) [15].

TABLE 3.2   Bond Energies Expressed as Dissociation Energies (kcal) for
Several Types of Bonds.

Bond Type	Bond Energies Dissociation Energy (kcal)
Covalent	400
Hydrogen bonds	12–16
Permanent dipole–dipole	0.5–2
Van der Waal-London (induced dipole–dipole)	<1

The relative strengths of several types of bonds are listed in Table 3.2. From this table it is obvious that covalent bonds are more than an order of magnitude stronger than hydrogen bonds and hydrogen bonds are more than an order of magnitude stronger than van der Waals-London forces.

SUMMARY

Water is the most important biochemical in life and is now considered to be a fingerprint of life. Despite its apparent simplicity, water houses a plethora of subtle, but extraordinary properties. Liquid water is abundant, small in size, and highly mobile. From a membrane perspective, it is water's propensity to form hydrogen bonds that is responsible for its strong self-attraction (cohesion) and many remarkable properties. Even at physiological temperature, liquid water exists as large extended clusters held together by hydrogen bonds. Water is life's solvent, an essential reagent in life, the major thermal regulator in life, and the driving force for all biological structure. The aversion of hydrocarbons for water, known as the Hydrophobic Effect, is the major stabilizing factor for membranes.

Membranes are composed primarily of lipids and proteins, often with carbohydrates adorning their (outer) surface. Chapter 4 will discuss simple membrane lipids, with an emphasis on fatty acids.

References

[1] Wikipedia, the Free Encylopedia. Properties of Water: http://en.wikipedia.org/wiki/Water
[2] Hanslmeier A. Water in the universe. In: Series: Astrophysics and Space Science Library. 1st ed. Vol. 368. Springer; 2011.
[3] Kotwicki V. Water in the Universe. 1991–2005, www.u24u.com
[4] Encrenaz T. Water in the solar system. Ann Rev Astronomy Astrophysics 2008;46:57–87.
[5] NASA/JPL/SSI. Cassini Flyby Shows Enceladus Venting. Mosaic by Emily Lakdawalla 2011.
[6] Nelson DL, Cox MM. Water. In Lehninger Principles of Biochemistry. 5th ed. New York, NY: W Freeman; 2008. p. 43–66.
[7] Matthews R. Wacky water. New Scientist 1997;154:40.
[8] Wiggins PM. Role of water in some biological processes. Microbiol Rev 1990;54:432–49.
[9] Surface tension by the ring method (Du Nouy method). Laboratory Experiments, Physics. Gottingen, Germany: PHYWE SYSTEME GMBH; 2007.
[10] Tanford C. Hydrophobic Effect: Formation of Micelles and Biological Membranes. John Wiley; 1973.
[11] Tanford C. The Hydrophobic Effect. John Wiley; 1980.
[12] Tanford C. Ben Franklin Stilled the Waves. Duke University Press; 1989.
[13] Gibbons Whitfield J, Luhring TM. Reptiles in Encyclopedia of Inland Waters, Editor-in-Chief:Gene E. Likens. Oxford: Elsevier Inc.; 2009. 497–505. Photograph by Cris Hagen.
[14] Gratzer, Walter. Retrospective:Charles Tanford (1921–2009). ASBMB Today; Dec 2009. 15–17.
[15] Dutch National Archives. The Hague, Fotocollectie Algemeen Nederlands Persbureau (ANEFO); 1945–1989. Bestanddeelnummer 119-0447.

Membrane Lipids: Fatty Acids

OUTLINE

A. What are Lipids?   43

B. Why are there so Many Different
Lipids?   44
  Lipid Functions   44

C. Lipid Classification Systems   45

D. Fatty Acids   46
  Chain Length   47
  Saturated Fatty Acids   47
  Double Bonds   48
  Fatty Acid Nomenclature   49

Monoenoic Fatty Acids   50
Polyenoic Fatty Acids   50
Omega Designation   51
Most Abundant Membrane Fatty Acids   53
Trans Fatty Acids   53
Branched Chain (Isoprenoid) Fatty Acids   54
Unusual Fatty Acids   55

Summary   55

References   55

A. WHAT ARE LIPIDS?

The term 'lipid' is derived from the Greek word 'lipos' for fat. Lipids are therefore by definition insoluble in water but soluble in non-polar organic solvents. Water insolubility is conferred by lipid molecular structure, having large regions of their surface composed of hydrocarbons with very few polar groups. A subset of lipids, those of interest in membrane studies, are structurally schizophrenic, containing segments that are polar and so prefer dissolution into water, and more extensive regions that are totally non-polar and which would avoid water at all costs. These lipids are referred to as 'amphipathic', containing both hydrophobic and hydrophilic moieties (Figure 3.7), and are the focus of this book.

An Introduction to Biological Membranes
http://dx.doi.org/10.1016/B978-0-444-52153-8.00004-0

B. WHY ARE THERE SO MANY DIFFERENT LIPIDS?

Lipids have proven to be very difficult to study due to their overwhelming diversity [1]. Although the human body may contain about 1,000 major lipids, countless minor lipids also exist. As a result, lipid studies lag far behind protein and nucleic acid studies and it is often considered that lipids are the poor relations among wealthier biological macromolecules. At first glance there may appear to be an almost infinite number of lipid species, and new ones are being discovered each day. It has been estimated that a typical animal cell contains about 1,000 different major lipid species, which is a large number. But this is a pittance when minor lipids are added into the calculation. For example, it has been reported that human tears contain >30,000 lipid molecular species [2,3]! And this is undoubtedly just the tip of the lipid iceberg. We probably know about only a fraction of the total lipids in the biosphere. Comprehending such large numbers of very different and often strange molecules makes scientists uncomfortable. The question has often been raised, why are there so many different lipids [1], and can such an array of structures be organized into an understandable and useful classification system?

Lipid Functions

It is clear that after ~4 billion years of chemical and biological evolution, nature does not make mistakes. There must be a reason, and hence a discoverable function, for each and every lipid species. These functions can be roughly divided into three general categories: storage lipids, structural lipids and active lipids. Storage lipids are generally non-polar and so are normally found sequestered away from water in lipid droplets and are not commonly found in membranes. They are storage forms of fatty acids (triacyl glycerols, see Chapter 5) that are densely packed biochemical energy sources. Triacyl glycerols contain 9 kcal/g of energy compared to only 4 kcal/g for sugars. Structural lipids form the amphipathic matrix of membranes and are the focus of this book. Active lipids come in many shapes and sizes and possess potent biochemical activities. Often they are found in membranes at much lower levels than structural lipids. Membrane lipids make up 5% to 10% of the dry mass of most cells and storage lipids account for more than 80% of the dry mass of fat storing adipocytes. Lipids exhibit a wide diversity of biological functions including:

1. Energy source (fats and oils).
2. Membrane bilayer component defining the cell permeability (lipid bilayer) barrier.
3. Providing the matrix for assembly and function of many catalytic processes.
4. Chaperones for protein folding.
5. Light absorber (e.g. chlorophyll).
6. Energy transduction (e.g. component attached to rhodopsin).
7. Electron carrier (e.g. coenzyme Q).
8. Hormone (e.g. testosterone, progesterone).
9. Vitamins (e.g. vitamin A, E, D).
10. Anti-oxidant (e.g. vitamin E).
11. Signaling molecules (e.g. ceramide).

C. LIPID CLASSIFICATION SYSTEMS

With so many different lipids, it is not surprising that several different classification systems have been proposed. However, any classification system is artificial, and so has both advantages and disadvantages. It does not matter what the system is, some lipids will not fit cleanly into a single category, and so may be placed into several categories, or may even fall entirely between categories. One good example of this conundrum is classifying the membrane lipid sphingomyelin (SM, Chapter 5). SM contains sphingosine and so can be classified as a sphingolipid, but it also contains phosphate, making it a phospholipid. So, where does SM belong? Each year countless numbers of previously unknown lipids are isolated and described and so any classification system must be flexible and expandable to accommodate the newcomers. It has become evident that a comprehensive nomenclature system is desperately needed before the task becomes too immense for the scientific community to tackle.

One important theme that runs through all of the proposed membrane lipid classification systems is the fundamental importance of fatty acids. The simplest division of lipids begins with the process of hydrolysis (discussed in Chapter 5). If the hydrolyzed lipid releases a free fatty acid it is deemed 'complex'. If it does not, it is 'simple'. Ironically, while esterified fatty acids, found in complex lipids, are an essential component of membrane structure, un-esterified free fatty acids are at most a minor component of membranes and in fact are harmful to membrane stability (Chapter 5). The two basic lipid types, simple and complex, are then further subdivided. Two classification systems are shown in Tables 4.1 and 4.2. The earlier system published by Hauser and Poupart in 1991 [4] is presented in Table 4.1.

TABLE 4.1   The 1991 Classification System of Hauser and Poupart [4].

1. Nonhydrolyzable (Nonsaponifiable) Lipids
     Hydrocarbons
          Simple alkanes
          Terpenes (isoprenoid compounds)
     Substituted hydrocarbons
          Long-chain alcohols
          Long-chain fatty acids
          Detergents
          Steroids
          Vitamins

2. Simple esters
     Acylglycerols
     Cholesterol esters
     Waxes

3. Complex lipids
     Glycerophospholipids
     Sphingolipids

4. Glycolipids
     Glycoglycerolipids
     Glycosphingolipids
          Cerebrosides
          Gangliosides
     Lipopolysaccharides

TABLE 4.2   The 2005 'Comprehensive Classification System' of Fahy et al. [5]. This System Divides all Lipids into 8 Major Categories and Further Subdivides the Categories, Assigning a 12-Digit Identifier to Each Lipid.

Major categories
FA. Fatty Acids
GL. Glycerolipids
GP. Glycerophospholipids
SP. Sphingolipids
ST. Sterol Lipids
PR. Prenol Lipids
SL. Saccharolipids
PK. Polyketides

The newer and more complex system (Table 4.2, from Fahy, E. et al., 2005 [5]) is based around 'lipidomics' a loosely defined term encompassing several diverse fields of lipid study. Lipidomics is a large-scale study of non-water-soluble molecules (lipids). The term was first introduced in 2001 as an analog to the well-established fields of proteomics (based on proteins) and genomics (based on nucleic acids). The goal of lipidomics is to relate lipid composition to lipid physical properties and biological activities [6]. While the lipidomics-based classification system is comprehensive and compatible with informatics, from a membrane structure perspective it is basically similar to the simpler system of Hauser and Poupart and is unnecessarily complex for the membrane studies outlined in this book. Only a small fraction of the lipids classified by Fahy et al. have a significant role in membranes. Here we focus only on the lipids that *significantly* affect membrane structure and function. This book will not generally consider non-membrane lipids or many other important aspects of lipids such as metabolism and nutrition.

The discussion of membrane lipids will begin with a lengthy description of fatty acids and then will be followed in Chapter 5 by the most important membrane structural lipids, complex fatty acid-containing glycerolipids (GL), glycerophospholipids (GP), sphingolipids (SP), and sterols (ST).

D. FATTY ACIDS

Fatty acids are the major building block of complex membrane lipids and so their properties are fundamental to understanding membrane structure and function. A fatty acid is a mono-carboxylic acid often with a long un-branched aliphatic (hydrocarbon) tail which may be either saturated or unsaturated [7].

Chain Length

Fatty acids commonly have a chain of 4 to 36 carbons (usually un-branched and even numbered), which may be saturated or unsaturated. Chains of carbon length 4 to 6 are referred to as short chain; 8 to 10 as medium chain; and 12 to 24 (or longer), as long chain. Short chain fatty acids are biochemically more closely related to sugars than to fats and so are not major components of membranes. Most membrane fatty acyl chains are 14, 16, 18, 20, 22, or 24 carbons in length. While present, odd carbon chain fatty acids are far less abundant. Preponderance of even carbon number fatty acids is the result of their biosynthesis from 2-carbon acetyl units. Acyl is a general term meaning an esterified fatty acid of unspecified chain length.

Saturated Fatty Acids

Acyl chains may contain from 0 to 6 double bonds [9]. Fatty acids with no double bonds belong to the class known as saturated. Saturated fatty acids, unlike unsaturated fatty acids, cannot be modified by hydrogenation or halogenation. They are also highly resistant to oxidation since they have no susceptible double bonds. Saturated fatty acids are named for the saturated hydrocarbon with the same number of carbons. The series of even-carbon, saturated fatty acids from C-2 to C-24 are shown in Table 4.3. Listed are the fatty acids' melting points (T_ms) that, with the exception of acetic acid, increase with chain length, and their water solubility, that decreases with chain length. Also note

TABLE 4.3 The Series of Even Carbon Saturated Fatty Acids from C-2 to C-24 with Melting Point and Water Solubility.

Systematic Name	Trivial Name	Structure	Melting Point (°C)	Water Solubility (g/l, 20°C)
Acetic	Acetic	2:0	16.7	infinite
Butanoic	Butyric	4:0	−7.9	infinite
Hexanoic	Caproic	6:0	−3.4	9.7
Octanoic	Caprylic	8:0	16.7	0.7
Decanoic	Capric	10:0	31.6	0.15
Dodecanoic	Lauric	12:0	44.2	0.055
Tetradecanoic	Myristic	14:0	53.9	0.02
Hexadecanoic	Palmitic	16:0	63.1	0.007
Octadecanoic	Stearic	18:0	69.6	0.003
Eicosanoic	Arachidic	20:0	75.3	insoluble
Docosanoic	Behenic	22:0	79.9	insoluble
Tetracosanoic	Lignoceric	24:0	84.2	insoluble

that at physiological temperature (37°C) fatty acids of 10-C or less are fluid (their T_m is below 37°C) while those of 12-C or more are solid (their T_m is above 37°C). The very short chain fatty acids (C-2 and C-4) are completely miscible with water, while the very long chain fatty acids (C-20, C-22 and C-24) are essentially insoluble in water. These properties will be important in later discussions of membrane 'fluidity' (order), thickness, interdigitation, and permeability (Chapter 9). The three most common saturated fatty acids found in membranes are myristic (14-C), palmitic (16-C), and stearic (18-C). These three fatty acids have been known for a long time as they were discovered before 1850! Stearic acid, for example, was first identified in 1823 by the early French biochemist Chevreul, a full century before Sumner in 1925 isolated the first enzyme, urease, from jackbean meal.

All long chain fatty acids, those important in membrane structure, have very limited water solubility, but are soluble in organic solvents. The solubility of palmitic acid (16:0) in various organic solvents is presented in Table 4.4. The best solvent, chloroform, has historically been the solvent of choice for efficiently dissolving lipids. In fact, mixtures of chloroform/methanol/water have been the primary solvents used for membrane lipid extractions since 1957 [8]. The large differences in lipid solubility for different organic solvents are an important property in separation of the various fatty acids for compositional analysis (Chapter 13).

Double Bonds

Membrane acyl chain double bonds are primarily *cis* (also designated Z) and are almost never conjugated. Normally, the double bonds are separated by intervening methylene groups that result in exceptional chain flexibility characteristic of fatty acids with multiple double bonds (Chapter 9). The intervening methylene group is also a major target for lipid peroxidation. Many common membrane fatty acids are therefore readily oxidized and must be effectively protected by membrane anti-oxidants (e.g. vitamin E).

TABLE 4.4   Solubility of Palmitic Acid in Organic Solvents (g/l, 20°C).

Chloroform	151
Benzene	73
Cyclohexane	65
Acetone	53.8
Ethanol (95%)	49.3
Acetic acid	21.4
Methanol	37
Acetonitrile	4
(water)	0.007

Conjugated double bonds	$-C-C=C-C=C-C-$
Methylene-interrupted double bonds	$-C-C=C-C-C=C-C-$

If a single double bond is present, the fatty acid belongs to the monoenoic or mono-unsaturated class (MUFA), while fatty acids with more than one double bond are referred to as polyenoic or polyunsaturated fatty acids (PUFA).

Fatty Acid Nomenclature

In order to define the structure of fatty acids, it is necessary to describe the number of carbons in the chain, as well as the number, type (*cis* or *trans*), and position of each double bond. The nomenclature employed in this book will first list the number of carbons in the chain, followed by a colon, and then the number of double bonds. Next the position (superscript Δ followed by the carbon position of the double bond from the carboxyl terminus) is listed. It is assumed the double bond is *cis*, unless a *t* indicating *trans* is added. For example, elaidic acid, an 18-carbon *trans* monoeonic acid with the double bond between carbons 9 and 10 is written $18:1^{\Delta 9t}$, while the homologous 18-C *cis* fatty acid, oleic acid, is designated $18:1^{\Delta 9}$. The homologous sequence of 18-C fatty acids is shown in Table 4.5 with the alternate omega designation (discussed below) included.

Melting points (T_ms) are an important general property of fatty acids that will have significant implications for later discussions of membrane 'fluidity' (order), thickness, and permeability (Chapter 9). Using the 18-C sequence as an example, upon adding a first double bond (stearic $\rightarrow$ oleic) the T_m decreases by a very large $-53.4°C$ (Table 4.6). Addition of a second double bond (oleic $\rightarrow$ linoleic) substantially further decreases T_m, by $-21.2°C$. The addition of a third double bond (linoleic $\rightarrow$ α-linolenic) decreases T_m, but only by an additional $-6°C$. Compare this to the saturated fatty acid series listed in Table 4.3. Saturated fatty acid T_ms increase with every additional 2-carbons in chain length, but only by a small amount, similar to that noted for the addition of a third double bond in the 18-C sequence. T_m increases by $+9.2°C$ in going from 14-C myristic to 16-C palmitic, and by $+6.5°C$ in going from palmitic to 18-C stearic. These T_m changes indicate that the effect of double bonds, particularly the first two double bonds, is much greater than changing carbon chain length.

TABLE 4.5 T_ms of the 18-Carbon Series of Fatty Acids.

Name	Structure	T_m	Omega
Stearic	18:0	69.6	0
Oleic	$18:1^{\Delta 9}$	16.2	9
Linoleic	$18:2^{\Delta 9,12}$	-5	6
α-Linolenic	$18:3^{\Delta 9,12,15}$	-11	3
γ-Linolenic	$18:3^{\Delta 6,9,12}$		6
Elaidic	$18:1^{\Delta 9t}$	43.7	9

TABLE 4.6   The Most Important Monoenoic Fatty Acids Found in Membranes [9].

Systematic Name	Trivial Name	Structure	Melting Point (°C)
cis-9-hexadecenoic	palmitoleic	16:1(n-7)	0.5
cis-6-octadecanoic	petroselinic	18:1(n-12)	30
cis-9-octadecanoic	oleic	18:1(n-9)	16.2
trans-9-octadecanoic	elaidic	18:1(n-9)	43.7
cis-11-octadecanoic	vaccenic	18:1(n-7)	39
cis-11-eicosenoic	gadoleic	20:1(n-9)	25
cis-13-docosenoic	erucic	22:1(n-9)	33.4
cis-15-tetracosenoic	nervonic	24:1(n-9)	39

Monoenoic Fatty Acids

The monoenoic fatty acid (MUFA), oleic acid, was also discovered by Chevreul in his pioneering studies on pork fat in 1823 [10]. Now over 100 monoene fatty acids are known. The most important monoenoic fatty acids found in membranes are listed in Table 4.6 with their corresponding melting point (T_m). The omega designation is referred to as n and is discussed below. It is evident in comparing the T_ms that saturated fatty acids of the same chain length have much higher T_ms than do fatty acids with one double bond. For example, stearic acid (18:0) has a melting point of 69.6°C while oleic acid ($18:1^{\triangle 9}$) is 16.2°C.

Polyenoic Fatty Acids

All polyenoic fatty acids (PUFAs) have very low melting points (Table 4.7) and are highly susceptible to oxidative degradation (peroxidation) [11]. UV radiation, high temperature, oxygen, metals, and alkaline conditions are sufficient to alter these molecules by enhancing not only oxidation but also double bond migration and isomerization, resulting in a multitude of complex positional geometric isomers. Positional isomerization can occur in PUFAs

TABLE 4.7   Most Important Membrane Polyenoic Fatty Acids [11].

Systematic Name	Trivial Name	Structure	Melting Point (°C)
9,12-octadecadienoic	linoleic	18:2(n-6)	−5
6,9,12-octadecatrienoic	γ-linolenic	18:3(n-6)	
9,12,15-octadecatrienoic	α-linolenic	18:3(n-3)	−11
5,8,11,14-eicosatetraenoic	arachidonic	20:4(n-6)	−50
5,8,11,14,17-eicosapentaenoic	EPA	20:5(n-3)	−54
7,10,13,16,19-docosapentaenoic	DPA	22:5(n-3)	
4,7,10,13,16,19-docosahexaenoic	DHA	22:6(n-3)	−44

yielding structures with double bonds adjacent to one another (conjugated double bonds). Conjugated linoleic acid (CLA) is a mixture of such isomers (more than 28 have been found to date) and has been used as a measure of free radical activity (see Figure 4.1). CLA also has anti-bacterial properties and can even inhibit leukemic cell growth.

A wide variety of human health benefits have been reputed for the two long chain omega-3 PUFAs, eicosapentaenoic acid (EPA) and docosahexaenoic acid (DHA). In fact it is almost impossible to find a single human affliction that has not been tested with EPA or DHA. Of particular importance are various cancers, heart disease, and immunological and neurological disorders [12,13]. Below are the ethyl esters of EPA ($20:5^{\Delta 5,8,11,14,17}$) and DHA ($22:6^{\Delta 4,7,10,13,16,19}$), both of which are employed in enteral (tube-feeding) applications (see Figure 4.2).

Omega Designation

The omega nomenclature for fatty acids was proposed by R.T. Holman in 1964 [14]. This classification divides fatty acids into metabolic families reflecting how they were synthesized biochemically. The system is based on location of the last double bond in the chain, specifically how many carbons it is from the terminal or omega methyl group. The terminal carbon is designated omega, n or ω. Figure 4.3 presents the 18-C sequence of fatty

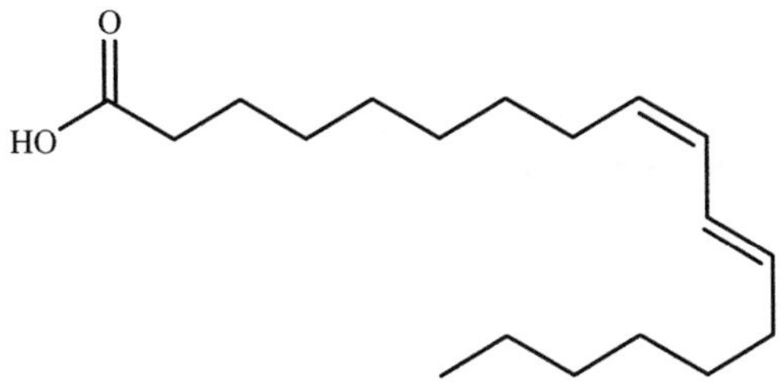

FIGURE 4.1 A conjugated linoleic acid (CLA, $18:2^{\Delta 9,11t}$).

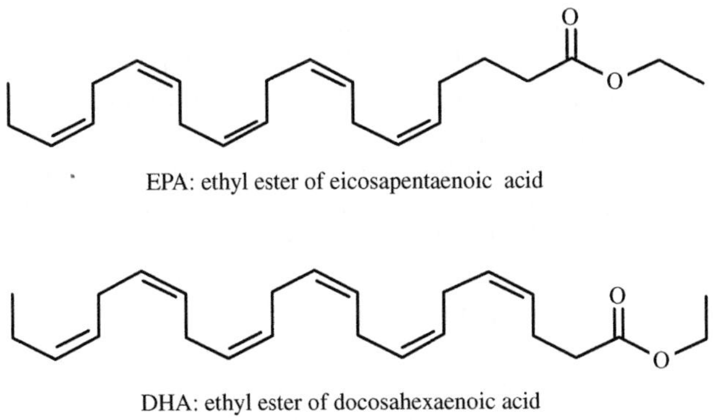

EPA: ethyl ester of eicosapentaenoic acid

DHA: ethyl ester of docosahexaenoic acid

FIGURE 4.2 EPA: Ethyl ester of eicosapentaenoic acid and DHA: Ethyl ester of docosahexaenoic acid.

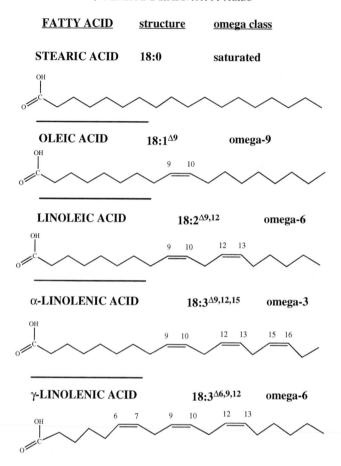

FIGURE 4.3   Structure of the 18-carbon sequence of important membrane fatty acids.

acids with their omega designations. For example, α-linolenic acid ($18:3^{\Delta 9,12,15}$) has the last of its 3 double bonds 3 carbons from the terminal (omega) end and so is an omega-3 fatty acid. Although docosahexaenoic acid ($22:6^{\Delta 4,7,10,13,16,19}$) has 4 more carbons in its chain than α-linolenic acid, it too is an omega-3 fatty acid. The omega designation is therefore independent of fatty acid chain length. It is now known that the 18-C omega-3 fatty acid α-linolenic acid is far more similar to the 22-C omega-3 fatty acid docosahexaenoic acid in supporting human health than it is to the saturated 18-C stearic acid (18:0) or the monounsaturated 18-C oleic acid ($18:1^{\Delta 9}$). By this system the major classes of fatty acids are saturated, omega-9, omega-6 and omega-3, although other minor classes of fatty acids also exist.

A strong case has been made for the number and location of double bonds being more important for human health than chain length. The class known as omega-3s has been of particular importance. Omega-3 fatty acids play crucial roles in cardiac, brain, and neurological function, as well as having roles in normal growth and development [12,13]. Since man cannot make double bonds (desaturate) beyond the Δ9 position, essential omega-3 fatty acids including α-linolenic, EPA, and DHA must be primarily obtained from the diet. The long chain omega-3 fatty acids EPA and DHA are often obtained from fish, such as salmon, tuna, and halibut, and other marine life including algae and krill [15]. The shorter chain omega-3, α-linolenic acid, is a common component of plants that in man is only inefficiently converted to the longer chain omega-3s by chain desaturation before the $\triangle 9$ position followed by chain elongation. Although rarely found in plants, arachidonic acid (AA, $20:4^{\triangle 5,8,11,14}$) is the most abundant of the omega-6 series in animals. In man AA is a major component of membranes and is the principal precursor of eicosanoids including the prostaglandins, isoprostanes, and isofurans. It is believed that the omega-6/omega-3 ratio is prognostic of human health. Ideally this ratio should be between 1:1 to 4:1, but in the American diet this ratio is currently between 10:1 to 30:1, resulting in high cardiovascular disease.

Most Abundant Membrane Fatty Acids

The fatty acids most commonly found in mammalian membranes are: palmitic (16:0), stearic (18:0), palmitoleic (16:1), oleic (18:1), linoleic (18:2), arachidonic (20:4), and docosahexaenoic (22:6). Table 4.8 presents a 'typical' mammalian membrane fatty acid composition profile [16]. The presented fatty acid compositions are of phosphatidylcholine (PC) isolated from rat liver membranes. Fatty acids are expressed as weight percent.

Trans Fatty Acids

Although far less common, some *trans* (also designated E) double bonds are routinely found in human membranes [17]. They enter the diet through dairy products, meat, and partially hydrogenated plant oils and are suspected to disrupt normal membrane functions. Although *trans* fatty acids are missing in plant lipids, they are present in animal fats, particularly those of ruminants. The most common *trans* fatty acid found in man is elaidic acid ($18:1^{\triangle 9t}$). However, a bewildering array of *trans* fatty acids is produced as byproducts of

TABLE 4.8   Fatty Acid Composition of Phosphatidylcholine (PC) Isolated from Rat Liver Membranes. The Data was Taken from [16].

Membrane	Fatty acid								
	14:0	16:0	16:1	18:0	18:1	18:2	18:3	20:4	22:6
Rat Liver	0.5	29.7	1.0	16.8	10.4	16.8		18.3	3.4
Mito (outer)	0.4	27.0	4.1	21.0	13.5	13.5		15.7	3.5
Mito (inner)	0.3	27.1	3.6	18.0	16.2	15.8		18.5	3.8
Plasma Membrane	0.9	36.9		31.2	6.4	12.9	tr	11.1	
SER	0.4	28.6	3.1	26.5	10.6	14.9		14.0	0.7
RER	0.5	22.7	3.6	22.0	11.1	16.1		19.7	2.9
Golgi	0.9	34.7		22.5	8.7	18.1	tr	14.5	

partial hydrogenation of plant lipids, a commonly employed procedure in manufacturing processed foods. Catalytic hydrogenation results in double bond reduction, migration up and down the chain and partial conversion of normal *cis* to deleterious *trans* double bonds. Increased incidence of heart disease [18] associated with lipid hydrogenation has resulted in banning *trans* fats from many parts of the world. The question remains why *cis* fatty acids are essential for human health while *trans* fatty acids are harmful. A clue may be found in the T_ms reported in Table 4.6. *Trans* fatty acids have a much higher melting point than do the homologous *cis* fatty acids. For example, the *trans* monoenoic fatty acid elaidic acid T_m is 43.7°C while the homologous *cis* oleic acid T_m is 16.2°C. In comparison, the saturated 18-C stearic acid has a T_m of 69.6 °C (Table 4.3). Therefore, at least by T_m analysis, although both elaidic and oleic acid have 18 carbons and a $\triangle 9$ double bond, elaidic acid is more similar to saturated stearic acid than it is to oleic acid. From the T_ms it can be concluded that *cis* double bonds have a larger effect on lipid packing than do *trans* double bonds that exhibit considerable saturated fatty acid properties (Chapter 9).

Branched Chain (Isoprenoid) Fatty Acids

The primary function of double bonds is to disrupt tight lipid packing in membranes, thus increasing the packing free volume ('breathing space') required for normal membrane protein function (Chapter 9) [19]. However, nature has found other ways to duplicate the function of double bonds. In lower organisms bulky methyl branches often replace double bonds. Higher organisms also contain repeating methyl branches in long hydrophobic structures called isoprenoids. These long hydrocarbon chains appear to be polymers of a 5-carbon branched compound called isoprene (see Figure 4.4). However, biological isoprenoids are not actually made from isoprene. Isoprene is a structural component of natural rubber and ~95% of industrial isoprene is used to produce *cis*-1,4-polyisoprene, a synthetic rubber used in the production of diving suits.

In plants the hydrophobic side chain of chlorophyll is phytol, a 20-carbon (4 isoprene units) isoprenoid. Many other important biochemicals including the family of retinoids (vitamin A) and ubiquinone (coenzyme Q) are anchored to membranes through long chain isoprenoids (Chapter 5).

In 1984 Schmidt, Schneider, and Glomset [20] reported that many proteins are anchored to membranes via long chain isoprenoids. This family of proteins has since been termed 'prenylated proteins' and has been shown to facilitate cellular protein−protein interactions and membrane-associated protein trafficking. The largest group of prenylated proteins is the plasma membrane GTP-binding proteins that transduce extracellular signals into

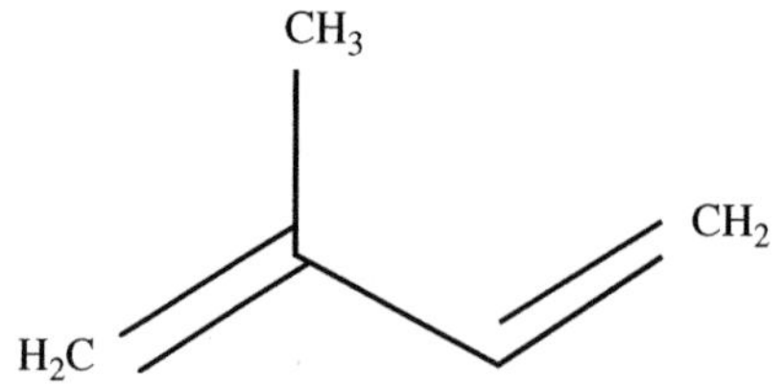

FIGURE 4.4   Isoprene.

intracellular changes via downstream effectors. Prenylated proteins are produced by post-translational modifications where two types of isoprenoids are covalently attached to cysteine residues at or near the terminal carboxyl group (Chapter 6). The two isoprenoids are farnesyl (15 carbons, 3 isoprene units) or geranylgeranyl (20 carbons, 4 isoprene units).

Unusual Fatty Acids

The number of frequently occurring fatty acids is <10 in plants and ~20 in animals. However, there is a tireless search for other rare and unusual fatty acids in various hidden corners of the biosphere. Unlike mammals, plant and bacterial fatty acids often have odd numbered carbon chains. Bacteria commonly have methyl branches or even other functional groups including cyclopropyl, cyclobutyl, cyclopentyl, and cyclohexyl rings or epoxy or hydroxyl groups in their fatty acid chains. At one extreme is the most unsaturated fatty acid found in nature, 28:8 isolated from a dinoflagellate. There have now been thousands of strange fatty acids identified that are currently considered to be a curiosity in search of a function, and so will not be discussed in this book.

SUMMARY

Lipids are insoluble in water and have large regions of their surface composed of hydrocarbons with very few polar groups. Their aversion to water is responsible for the hydrophobic effect that drives membrane stability. Membrane lipids are 'amphipathic', with a polar end anchoring the lipid to the aqueous interface and a long hydrophobic segment that forms the oily membrane interior. In human cells there are over 1,000 major lipid species and countless minor lipids. There have been several attempts to classify lipids, the most recent being a lipidomics-based system by Fahy et al. from 2005. The most important lipid component of membranes is fatty acids. A fatty acid is a mono-carboxylic acid with a long (normally 14−24 carbons), un-branched, hydrophobic tail which may be either saturated or unsaturated (usually 0−6 *cis* non-conjugated double bonds). Fatty acids are the basic building blocks of complex membrane lipids and so are largely responsible for membrane structure, fluidity, and function.

Although fatty acids comprise the bulk of the membrane hydrophobic interior, they rarely exist in free, non-esterified form. Chapter 5 discusses complex, polar membrane lipids where fatty acids are normally found esterified to phospholipids, sphingolipids or some membrane proteins. A major non-esterified class of membrane lipid, sterols (particularly cholesterol), is also discussed.

References

[1] Dowhan W. Molecular basis for membrane phospholipid diversity: Why are there so many lipids? Ann Rev Biochem 1997;66:199−232.
[2] Nicolaides N, Santos EC. The di- and triesters of the lipids of steer and human meibomian glands. Lipids 1985;20:454−67.
[3] Borchman D, Foulks GN, Yappert MC, Tang D, Ho DV. Spectroscopic evaluation of human tear lipids. Chem Phys Lipids 2007;147:87−102.

[4] Hauser H, Poupart G. Lipid structure. In: Yeagle P, editor. The Structure of Biological Membranes. Boca Raton, FL: CRC Press; 1991. p. 3—71 [Chapter 1].

[5] Fahy E, Subramaniam S, Brown HA, Glass CK, Merrill Jr AH, Murphy RC, et al. A comprehensive classification system for lipids. J Lipid Res 2005;46:839—62.

[6] Wenk MR. The emerging field of lipidomics. Nat Rev Drug Discov 2005;4:594—610.

[7] Cyberlipid Center. Fatty Acids. online, non-profit scientific organization, www.cyberlipid.org/fa/acid0001.htm.

[8] Folch J, Lees M, Stanley GHS. A simple method for the isolation and purification of total lipids from animal tissues. J Biol Chem 1957;226:497—509.

[9] The AOCS Lipid Library. Fatty Acids: Straight-Chain Saturated. Structures, Occurrence and Biosynthesis, Lipidlibrary.aocs.org; 2011.

[10] The AOCS Lipid Library. Fatty Acids: Straight-Chain Monoenoic. Structures, Occurrence and Biochemistry, Lipidlibrary.aocs.org; 2011.

[11] The AOCS Lipid Library. Fatty Acids: Methylene interrupted Double Bonds. Structures, Occurrence and Biochemistry, Lipidlibrary.aocs.org; 2011.

[12] University of Maryland Medical Center. Contempory Medicine. Omega-3 fatty acids, www.umm.edu.

[13] Holman RT. The slow discovery of the importance of omega 3 essential fatty acids in human health. J Nutr 1998;128:427S—33S.

[14] Holman RT. Nutritional and metabolic interrelationships between fatty acids. Fed Proc 1964;23:1062—7.

[15] Holub B. Dietary sources of omega-3 fatty acids. DHA/EPA Omega-3 Institute, www.dhaomega3.org; 2011.

[16] White DA. The phospholipids composition of mammalian tissues. In: Ansell GB, Hawthorne JN, Dawson RMC, editors. Form and Function of Phospholipids. 2nd ed. New York: Elsevier; 1973. p. 441—82.

[17] IUFoST Scientific Information Bulletin. Trans fatty acids; 2010.

[18] Mozaffarian D, Katan MB, Ascherio A, Meir J, Stampfer MJ, Walter C, et al. Trans fatty acids and cardiovascular disease. N Engl J Med 2006;354:1601—13.

[19] The AOCS Lipid Library. Fatty Acids: Branched Chain. Structures, Occurrence and Biosynthesis, Lipidlibrary.aocs.org; 2011.

[20] Schmidt RA, Schneider CJ, Glomset JA. Evidence for post-translational incorporation of a product of mevalonic acid into Swiss 3T3 cell proteins. J Biol Chem 1984;259:10175—80.

5

Membrane Polar Lipids

OUTLINE

A. Bonds Connecting to Acyl Chains   57

B. Phospholipids   60
    Fatty Acids   61
    Glycerol   61
    Phosphate   62

C. Sphingolipids   71

D. Sterols   74

E. Membrane Lipid Distribution   76

F. Plant Lipids   76

G. Membrane Lipids Found in Low Abundance   78
    Free Fatty Acids and Lysolipids   78
    Lipid-Soluble Vitamins   79
    Ubiquinone (Coenzyme Q)   80

Summary   81

References   82

Free fatty acids are at best a minor component of most membranes, as large quantities are known to disrupt membranes through a 'detergent effect'. When the concentration of free fatty acids in solution reaches a threshold value, known as the critical micelle concentration (CMC), the fatty acids aggregate into mostly spherical structures known as micelles (Figure 5.1), although other structures including ellipsoids are also possible. A typical fatty acid micelle is an aggregate with the hydrophilic carboxylic acid head in contact with the surrounding water and the hydrophobic tail sequestered away from water in the dry micelle center. Micelles can accommodate membrane phospholipids, thus destroying the membrane through the 'detergent effect'. CMCs for fatty acids decrease with increasing chain length and increase with increasing double bonds.

A. BONDS CONNECTING TO ACYL CHAINS

By definition, complex lipids are molecules that, when hydrolyzed, release a fatty acid. The fatty acids are primarily attached to the rest of the lipid via an oxygen ester, although

FIGURE 5.1 Amphipathic molecules form micelles when their critical micelle concentration is exceeded.

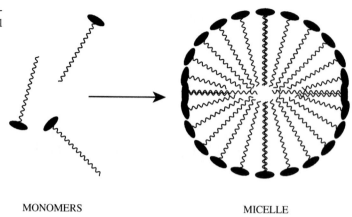

MONOMERS   MICELLE

sphingolipids employ a nitrogen ester (amide). Even sulfur esters (thioesters) are not uncommon (Figure 5.2).

One important example of an ester-containing lipid is triacylglycerol, a non-polar lipid called either a fat (if solid at room temperature) or an oil (if liquid at room temperature). A triacylglycerol has 3 fatty acids attached to a glycerol via ester bonds. A fat has predominantly high T_m saturated fatty acids while an oil has predominantly low T_m unsaturated fatty acids. Esters can be hydrolyzed by boiling in NaOH (soda ash) or KOH (caustic potash). Boiling animal fat in KOH produces glycerol (also called glycerine) and 3 K^+ salts of fatty acids (Figure 5.3). Boiling animal fat with caustic potash is the original method for making soap.

The term used for making soap by hydrolyzing a fat or oil in a strong base is saponification. It is clear that saponification is a process that has been used for a very long time [1], at

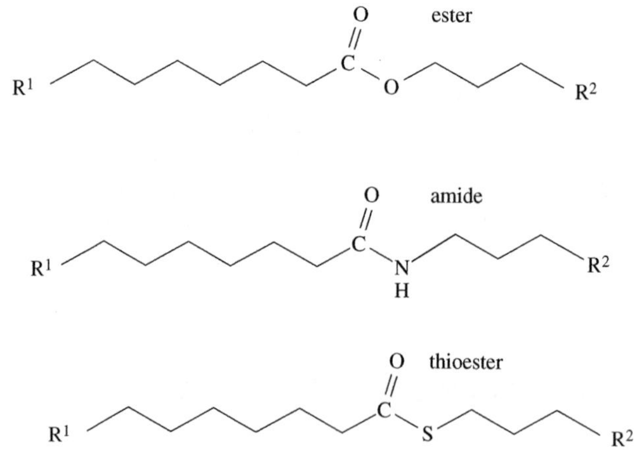

FIGURE 5.2 Types of bonds attaching fatty acids in complex lipids.

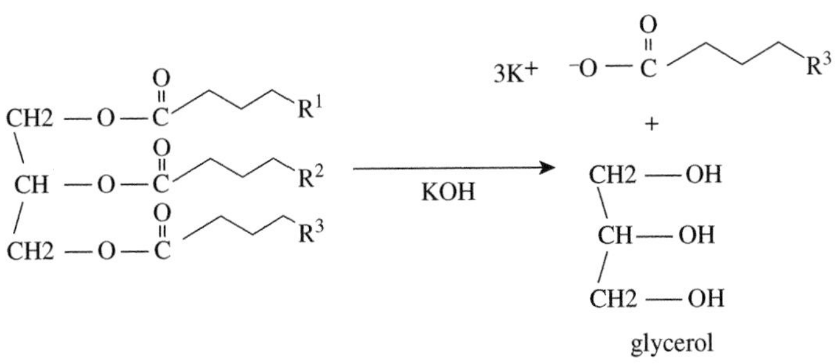

FIGURE 5.3 Hydrolysis of triacylglycerol (saponification) produces glycerol and 3 K^+ salts of fatty acids.

least since the ancient Roman times, as a soap making factory was found in the ruins at Pompeii. Also, the word *sapo*, Latin for soap, was first reported by Pliny the Elder in his great work *Historia Naturalis* [2]. Pliny described the making of soap from tallow and ashes. The origin of the word saponification is most interesting. It has been suggested that it comes from Sapo Mountain near Rome. According to Roman legend from about 1,000 BC, women washing clothes in the Tiber River noticed that their clothes were cleaner if they washed them in a particular area. On the mountain above the special spot in the river was found a place where animals were sacrificed on an altar. The animal fat was inadvertently mixed with heated ashes and lye, producing a crude soap that washed into the Tiber. Hence the term saponification! Although this makes for a great story that for many years has been propagated by soap manufacturers, it is probably only a myth, as Sapo Mountain is in fact a fictitious place. Also, the first archeological evidence of soap manufacturing came from ancient Babylon (2,800 BC), and a formula for soap consisting of water, alkali, and cassia oil was written on a Babylonian clay tablet around 2,200 BC [3].

While many fatty acyl chains are attached to lipids and proteins by hydrolyzable ester or amide bonds, others are attached by non-hydrolyzable ether linkages. Ether bonds are particularly prevalent in very primitive organisms including Archaea. Archaea live in extreme conditions where membrane ester linkages would be highly susceptible to hydrolysis. They characteristically have long chain (32 carbons), branched hydrocarbons linked at each end via ether linkages to glycerols that have either phosphate or sugar residues attached. These unusual lipids are twice the width of regular membrane phospholipids and so can span the bilayer while being anchored in both internal and external aqueous spaces. Another curiosity of these membrane lipids is that the glycerol central carbon isomer is the opposite of that found in phospholipids (discussed below). It is interesting to note that while primitive organisms have stable ether linkages, higher organisms have emphasized less stable ester linkages. Esters can be cleaved by saponification (chemical hydrolysis) or by the action of phospholipases (enzymatic cleavage, discussed below), while ethers are resistant to both processes.

In man one unusual phospholipid class called plasmalogens [4] have both an ester-linked acyl chain as well as an ether linked chain (for the structure of plasmalogens see Figure 5.4). This raises the intriguing possibility that the ether linkage found in plasmalogens perhaps are a vestige held over from an ancient past. For many years plasmalogens and other ether lipids were considered little more than biological curiosities. However, their large abundance in nature indicates ether-lipid bonds must play significant roles in life processes, particularly

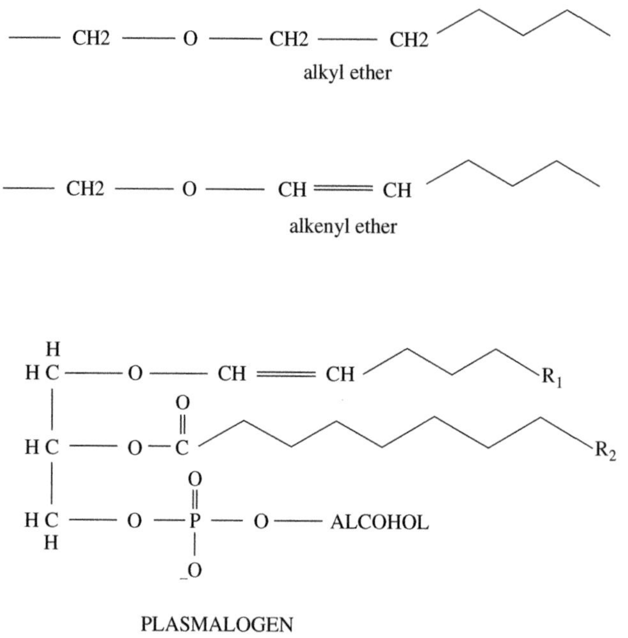

—— CH2 —— O —— CH2 —— CH2 ⌒⌒⌒

alkyl ether

—— CH2 —— O —— CH ══ CH ⌒⌒

alkenyl ether

$$
\begin{array}{l}
\text{H} \\
\text{H C} \!—\! \text{O} \!—\! \text{CH} \!==\! \text{CH} \diagup\diagdown\diagup R_1 \\
\quad| \qquad\qquad \text{O} \\
\qquad\qquad\qquad\;\; \| \\
\text{H C} \!—\! \text{O}\!—\!\text{C} \diagup\diagdown\diagup\diagdown R_2 \\
\quad| \qquad\qquad \text{O} \\
\qquad\qquad\qquad\;\; \| \\
\text{H C} \!—\! \text{O} \!—\! \text{P} \!—\! \text{O} \!—\! \text{ALCOHOL} \\
\;\;\text{H} \qquad\qquad | \\
\qquad\qquad\qquad\;\; \underline{\text{O}}
\end{array}
$$

PLASMALOGEN

FIGURE 5.4   Alkyl ether and alkenyl ether linkages and a plasmalogen. Plasmalogens have an ether link at the *sn*-1 position and an ester link at the *sn*-2 position.

those related to membranes. Vertebrate heart is particularly enriched in ether-linked lipids where about half of the phosphatidylcholine is plasmalogen. Elevated levels of ether-linked lipids in cancer tissues and their role as a platelet activating factor (important in molecular signaling) have greatly stimulated interest in these compounds. Although their molecular function remains unknown, it seems possible it is probably related in some way to its resistance to hydrolysis or its ability to protect cells against the damaging effects of singlet oxygen. Two main types of ether bonds exist in natural lipids, alkyl and alkenyl (vinyl) ethers (Figure 5.4). The double bond adjacent to the oxygen atom in the alkenyl ether has a *cis* or Z configuration.

B. PHOSPHOLIPIDS

The major type of animal membrane polar lipids is commonly referred to as phospholipids (also referred to as glycerophospholipids, GP, in the Comprehensive Classification System of Fahy et al., 2005, [5]). Phospholipids [6–9] are composed of 1 glycerol, 1 phosphate, 2 fatty acids, and an alcohol. The molecule is constructed by removing 4 waters, creating 4 esters (Figure 5.5). The phospholipid class name is derived from the alcohol. In animals there are 7 major classes of phospholipids (see Figure 5.6).

A phospholipid is amphipathic, having both a polar head group and two very hydrophobic, apolar acyl tails. Often a phospholipid is represented by a simple cartoon drawing

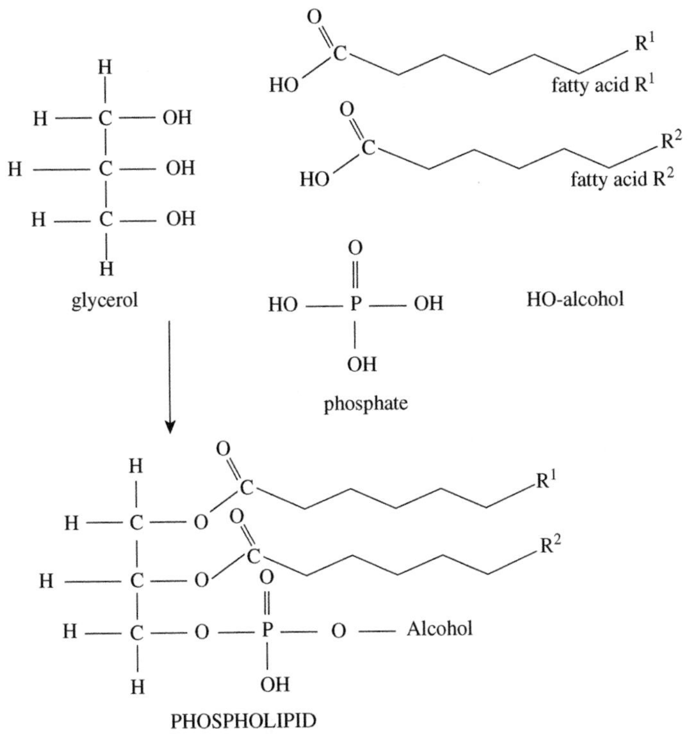

FIGURE 5.5 A phospholipid is formed from the condensation (dehydration) of one glycerol, two fatty acids, one phosphate, and one alcohol.

shown in Figure 5.7. The structure has an oval head (glycerol, phosphate, and the alcohol) and the two acyl chains. The acyl chains are fatty acids esterified to the glycerol.

Fatty Acids

The fatty acids are esterified to the glycerol at carbon positions C-1 (called the *sn*-1 position) and at C-2 (the *sn*-2 position). As a general rule the *sn*-1 chain is saturated while the *sn*-2 chain is unsaturated. The phosphate is esterified to the glycerol C-3 position. The types and structures of membrane fatty acids were discussed in Chapter 4.

Glycerol

Although the molecule glycerol is not optically active (it does not have an asymmetric carbon with 4 different groups attached), once the other phospholipid components are attached, the *sn*-2 position of glycerol becomes asymmetric (it now has 4 different groups attached) and hence is optically active. Consider the case for glycerol-3-phosphate. It comes in two optical isomers, D and L, differing only by the orientation of the same groups about the *sn*-2 carbon (Figure 5.8). The biologically correct form is the D isomer. While generally unappreciated, biological phospholipids are similar to amino acids and sugars in being optically active.

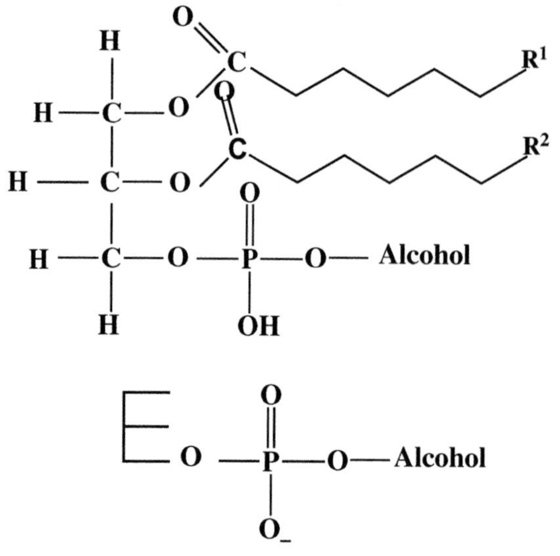

Phospholipid Name	Abbreviation	Alcohol
Phosphatidic Acid	PA	None
Phosphatidylethanolamine	PE	Ethanolamine
Phosphatidylcholine	PC	Choline
Phosphatidylserine	PS	Serine
Phosphatidylinositol	PI	Inositol
Phosphatidylglycerol	PG	Glycerol
Cardiolipin (Diphosphatidylglycerol)	CL	Phosphatidylglycerol

FIGURE 5.6 Simplified structure for phospholipids. Also included are the names and associated alcohol for each of the 7 major phospholipid classes in mammalian membranes.

Phosphate

Phosphoric acid is a tri acid with 3 different dissociable groups having pKas of 12.8, 7.2 and 2.1 (Figure 5.9). As a first alcohol (glycerol) is esterified to phosphate, the highest pKa, 12.8, is eliminated. Attachment of a second alcohol removes the next highest pKa, 7.2, leaving only the lowest pKa of 2.1. Since 6 of the 7 common phospholipids have 2 phosphate oxygens esterified, the only dissociable phosphate group left has a pKa of 2.1. At physiological pH this group is almost totally dissociated (approximately 99.999% dissociated) resulting in a full negative charge on the phosphate.

All 7 classes of phospholipids are amphipathic molecules that contribute to lipid bilayer structure. They vary from no net charge (they are zwitterions) to anions. None have a net positive charge at physiological pH. Therefore one function of most phospholipids is to provide negative charge density to membranes. In addition, each phospholipid class undoubtedly provides other, perhaps unique functions to membranes. However, it is clear that phospholipids have no inherent catalytic activity. Catalysis is provided by resident

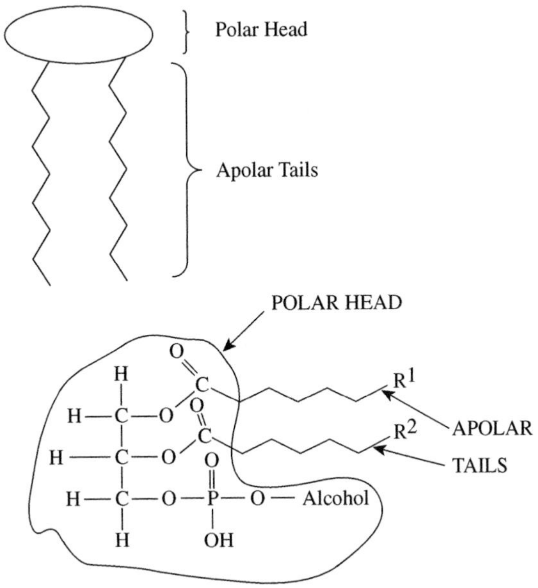

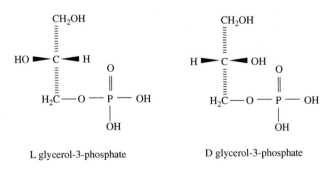

FIGURE 5.7   Cartoon depiction of an amphipathic phospholipid.

membrane proteins. Currently, only a few of the more obvious unique phospholipid functions have been discovered. Unraveling these functions is complex, requiring a combination of *in vivo* and *in vitro* studies. Genetic mutations do not directly affect the phospholipids, but rather their biosynthetic pathways that may indirectly alter many membrane events. Other more subtle functions for each class of phospholipids will undoubtedly be found in the future.

Figure 5.6 depicts a simplified chemical structure that will be employed in the discussion of the individual phospholipid classes. Also included are the names and associated alcohol for each of the 7 major phospholipids.

PA (Figure 5.10) is the smallest and simplest phospholipid and so is the precursor for other more complex, alcohol-containing phospholipids [10]. As with all other

L glycerol-3-phosphate   D glycerol-3-phosphate

FIGURE 5.8   Optical isomers of glycerol-3-phosphate.

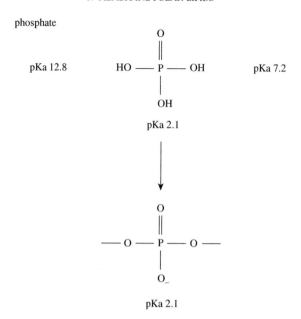

phosphate

pKa 12.8   HO — P — OH   pKa 7.2

pKa 2.1

pKa 2.1

FIGURE 5.9   pKas of dissociable groups on phosphate.

phospholipids, PA comprises part of the membrane lipid bilayer and contributes to the membrane's physical properties. At physiological pH, PA has a net charge of ~ -1.5 and so is a strong anion that contributes negative charge density to the membrane. PA has a unique shape, possessing a small highly-charged head group whose charge lies very close to the glycerol backbone. PA's shape is probably related to its significant role in membrane vesicle fission and fusion events. Membrane PAs can be aggregated by Ca^{2+}, resulting in the formation of PA-rich microdomains that are highly sensitive to pH, temperature and cation concentration. PA also induces negative curvature stress on adjacent proteins, affecting their activity. The activity of more than 10 proteins including PKC and PLC have been shown to be altered by PA (for example, [11]).

PA is a vital cell lipid that is maintained at extremely low levels in the cell by the activity of potent LPPs (lipid phosphate phosphohydrolases). LPPs rapidly convert PA to DAG (diacyl glycerol), a metabolic precursor for many lipids. PA has emerged as a new class of lipid mediators involved in diverse cellular functions in plants, animals, and microorganisms [12].

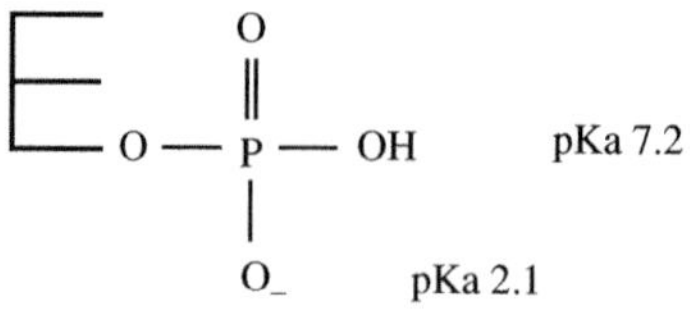

O
‖
O — P — OH   pKa 7.2
|
O_   pKa 2.1

FIGURE 5.10   PA – Phosphatidic acid.

PE (Figure 5.11), infrequently referred to as cephalin, from the word cephalic, meaning pertaining to the head, is the second most prevalent phospholipid in man, but is the principal phospholipid in bacteria. This is due to the ability of man to convert PE to PC while bacteria cannot and so bacteria are relegated to having PE as their major membrane phospholipid. PE, PC, and cholesterol are also the major components of egg yolk, a ready and cheap commercial source of these lipids. In humans, PE is found particularly in nervous tissue including white matter of brain, nerves, and in spinal cord. PE is a primary amine-containing phospholipid and therefore has a highly reactive chemical handle that can be easily derivatized. This property will be exploited in Chapter 9 for various membrane biochemical studies. Since the pKa of the PE amine is ~8.5, at physiological pH the amine is mostly, but not completely, protonated. With the fully dissociated phosphate, PEs have a slightly negative net charge. PEs are therefore not quite fully zwitterions. The PE's primary role in membranes is as a major structural lipid. The PE's head group is not only chemically reactive but is small and poorly hydrated. With unsaturated chains, PEs have a pyramid shape, with a wide base and narrow head. Phospholipid shapes are discussed in Chapter 10. Unlike PCs, PEs can form hydrogen bonds with neighboring polar groups. In the absence of other phospholipids, PEs prefer non-lamellar phase including reverse hexagonal H_{II} phase, an unusual membrane structure that is discussed in Chapter 10.

Natural PCs (Figure 5.12) are often referred to by an old term, lecithin. Actually, lecithin is a complex mixture of many lipids, the predominant one being mixed acyl chain PC. PC is the major phospholipid found in animal membranes and is therefore the most important membrane structural lipid component in man. PC is ideally suited to fit comfortably into membranes. The molecule is can-shaped with its polar head group being about the same width as the sum of its apolar tails (Chapter 10). In addition, PC is a true zwitterion, possessing a formal positive and negative charge, but no net charge. Therefore PC does not exhibit negative charge repulsion, a potential problem with most other phospholipids. In man PC can be made by methylating the nitrogen of PE three times, a process missing in prokaryotes. In addition, dietary lecithin is the major source of the essential biochemical choline.

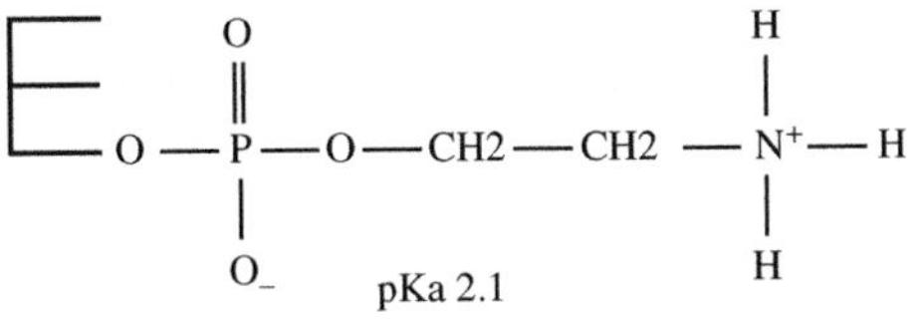

FIGURE 5.11   PE − Phosphatidylethanolamine.

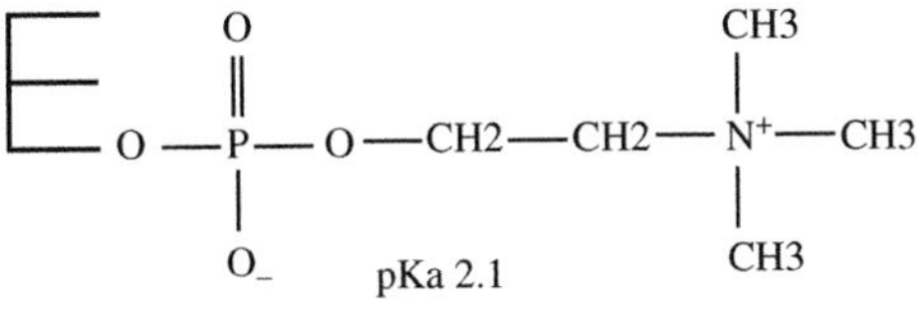

FIGURE 5.12   PC − Phosphatidylcholine.

The major structural phospholipids in animal membranes are PC > PE (Table 5.1). These lipids will predominantly dictate many membrane properties including 'fluidity'. Although 'fluidity' is to a large extent dictated by the length and degree of unsaturation of the acyl chains, the nature of the polar head group is also important. As demonstrated below, PEs with identical chains generally have higher T_ms than PCs by ~20°C to 40°C and therefore at a fixed temperature are less 'fluid' than PCs (Table 5.2).

While PC and PE are predominantly structural lipids responsible for maintaining the lipid bilayer matrix of membranes, PS (Figure 5.13) has many additional functions [13,14]. In fact PS, along with PI, are the most diverse and dynamic of all of the phospholipids. Although PS is an essential lipid in all human cells, its major functions are associated with the brain where it is far more abundant than in other organs. Many clinical trials have suggested that PS supports a variety of brain functions that diminish with age [15,16]. Dietary supplementation with PS can alleviate, ameliorate, or sometimes even reverse age-related decline of memory, learning, concentration, word skills, and mood and may also improve capacities to cope with stress and maintain internal circadian rhythms. From this very brief list of general physiological processes, it is clear that PS must serve a number of distinct molecular functions in addition to just being a component of the lipid bilayer.

PS is involved in both activating and anchoring a variety of proteins to membranes, including ATPases, kinases, receptors, and the crucial signal transducing proteins, protein kinase C and adenylate cyclase [17,18]. PS is also involved in cell—cell recognition and communication processes.

Normally PS is located almost exclusively in the membrane inner leaflet. It is quite unusual for any other phospholipid to be ~100% asymmetrically distributed across a membrane (Chapter 9). This condition is maintained by ATP-dependent amino-PL translocases (also known as flipases). If PS accumulates in the outer membrane leaflet, as it does with age, it signals that the cell should be recycled. In addition, under physiological conditions, PS is an anion and so, similar to PA, interacts strongly with divalent metals like Ca^{2+}.

TABLE 5.1 Average Lipid Composition of
a Mammalian Liver Cell.

Lipid	Mol (%)
PC	45—55
PE	15—25
PI	10—15
PS	5—10
PA	1—2
CL	2—5
SM	5—10
Cholesterol	10—20

TABLE 5.2   Melting Points (T_ms) for PCs and PEs where both the *sn*-1 and *sn*-2
Acyl Chains are Identical.

Acyl chain	diPC (°C)	diPE (°C)
Myristic	23.6	51
Palmitic	41.3	63
Stearic	58	82
Oleic	−22	15
Elaidic	5	41

PEs have higher T_ms by more than 20°C than do homologous PCs.

If we examine the alcohol structures of PE, PC, and PS, it is clear how closely related the three phospholipids are and how easy it might be to interconvert one to the other (Figure 5.14). PE can be converted to PC by simply methylating the ethanolamine nitrogen three times. This step is not possible in bacteria (as the enzyme is missing) and so bacteria lack PC. PS can be converted to PE by simple decarboxylation.

PI (Figure 5.15) is another anionic lipid at physiological pH and so contributes to the lipid bilayer and to the surface negative charge density of membranes. In fact PIs are considered to be one of the most acidic of the phospholipids. Its low concentration (PI is generally even less abundant than PS, Table 5.1) indicates that this phospholipid plays only a small structural role in membranes. However, it has been appreciated for some time that PI does have several crucial roles in cell function [19]. Although PI is present in all tissues and cell types, it is especially abundant in brain where it can account for 10% of total phospholipids. Consistent with the other phospholipids, PI has a saturated fatty acid in the *sn*-1 chain and an unsaturated fatty acid in the *sn*-2 chain. What is unusual about PI is the highly specific nature of its chains. Most PIs are the single molecular species, 18:0,20:4 PI.

Phosphoinositides hold a central role in cell signaling and regulation [20,21]. Often PI is phosphorylated at various positions on the inositol chain, most commonly positions 4 and 5, although position 3 can also be phosphorylated (Figure 5.16). Positions 2 and 6 are not phosphorylated due to steric hindrance. Phosphorylated PIs are referred to as phosphatidylinositol phosphates or polyphosphoinositides. They are usually present at low levels, typically at about 0.5 to 1% of the total lipids of the inner leaflet of the plasma membrane. The major phosphoinositide is PI 4,5-bisphosphate (PIP$_2$).

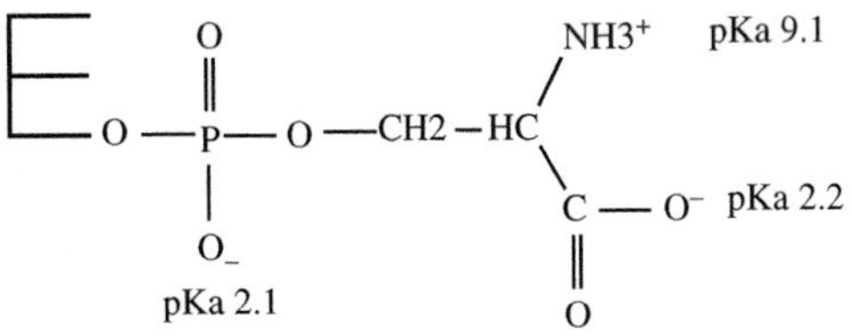

FIGURE 5.13   PS − Phosphatidylserine.

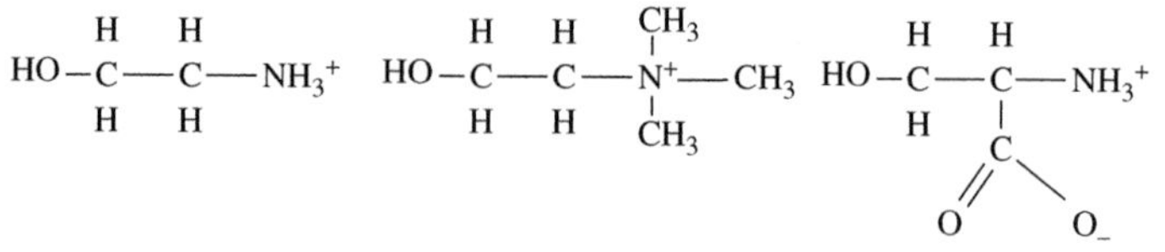

FIGURE 5.14 Close relationship of the head group alcohols of PE, PC, and PS.

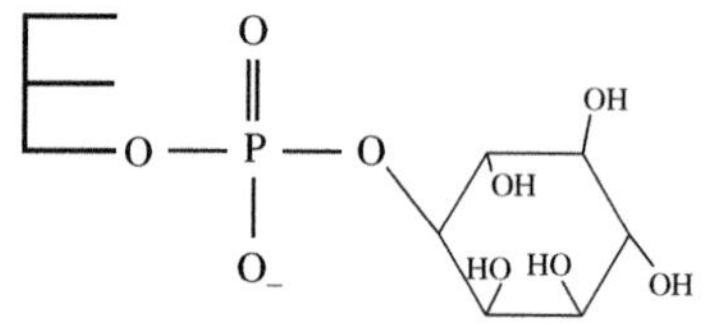

ETHANOLAMINE   CHOLINE   SERINE

FIGURE 5.15   PI — Phosphatidylinositol.

Since the phosphoinositides are highly anionic they are effective at enhancing non-specific electrostatic interactions with cellular proteins, particularly to protein 'pH domains', at inner leaflet plasma membrane interfaces. PIs are in essence an inert storage form of the potent bio-logical signaling second messenger molecules, membrane-bound diacylglycerol (DAG) and water-soluble inositol phosphates. These molecules are released upon hydrolysis of, for example, PIP_2 by phospholipase C (Figure 5.16). In fact, PIs are the major source of DAG in cells. DAG in turn affects the activity of a large, important family of signaling enzymes, The PKCs. The water-soluble inositol phosphates affect the opening of channels for K^+, Na^+, Ca^{2+} and other ions. Stimulation of Ca^{2+} release from the endoplasmic reticulum is an important signaling event. PIP_2 has also been shown to affect cell shape, motility and other processes by interacting with actin on the cytoskeleton. PIP_2 is also a cofactor for phospho-lipase D that produces PA, yet another signaling molecule. As a component of the nucleus, the phosphoinositides affect DNA repair, transcription and RNA dynamics. PI is also involved in binding a variety of proteins to the membrane surface via a glycosyl bridge [22]. Proteins bound in this way are referred to as being glycosyl-phosphatidylinositol (GPI)-anchored proteins (see Chapter 6). It is clear that the phospholipid PI plays an impor-tant role in many aspects of cellular biochemistry in addition to being a component of the membrane lipid bilayer.

PG (Figure 5.17) [23] is another anionic lipid found in mammalian membranes in low amounts (1 to 2% of total phospholipids). PG therefore is also not a major membrane struc-tural lipid. However, PG is the second most abundant lipid in lung surfactant, comprising up to 11% of the surfactant lipids and its level is used to determine the maturity of a baby's lung. PG differs from the other phospholipids already discussed in possessing 2 optically active carbon centers (the sn-2 position in the phosphatidyl group and the central carbon of the alcohol glycerol) and in often having a more unsaturated chain occupying the sn-1 position. It appears that the major membrane role for PG in animals is as a precursor for mitochondrial cardiolipin (CL, Figure 5.18). CL (discussed below) is essential for normal

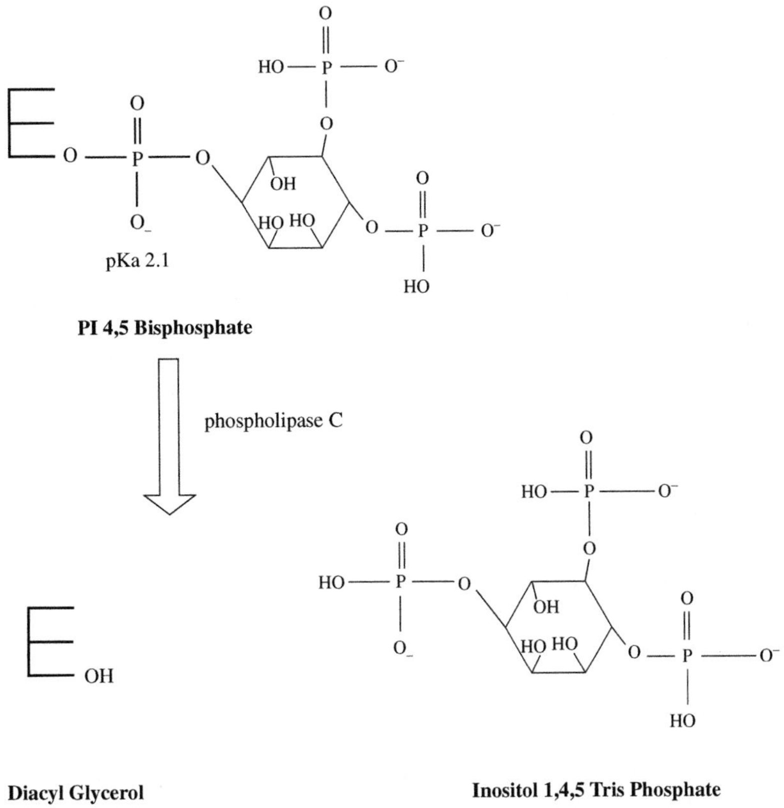

PI 4,5 Bisphosphate

phospholipase C

Diacyl Glycerol   **Inositol 1,4,5 Tris Phosphate**

FIGURE 5.16   Hydrolysis of PI 4,5-bisphosphate (PIP2) by phospholipase C.

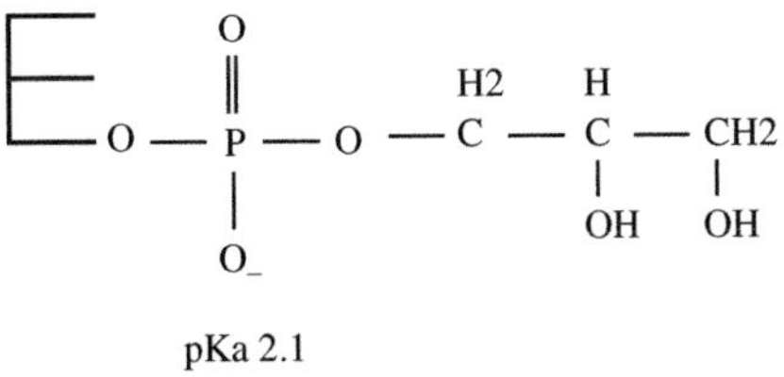

pKa 2.1

FIGURE 5.17   PG – Phosphatidylglycerol.

electron transport and oxidative phosphorylation. While only a minor component of animal membranes, PG can be the main component of some bacterial membranes and is abundant in plant membranes. In plants PG comprises about 10% of the thylakoid (photosynthetic) membrane lipids. In higher plants, PG is unique in that it contains a high proportion of *trans*-3-hexadecenoic acid. This *trans* fatty acid is located exclusively in the *sn*-2 position. However, its role in photosynthesis is not known.

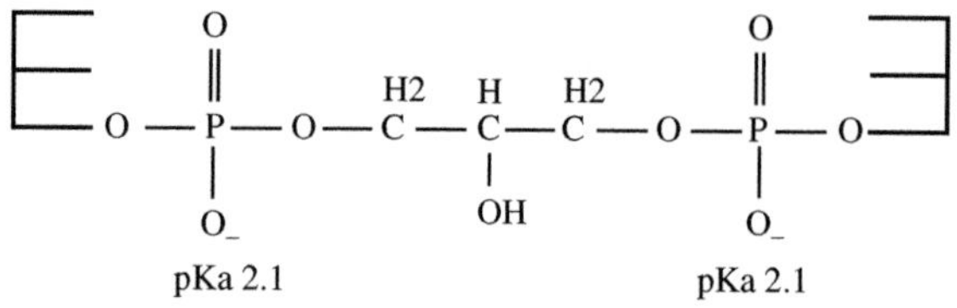

$$\left[\begin{array}{c} O \\ \parallel \\ O-P-O-\overset{H2}{C}-\overset{H}{C}-\overset{H2}{C}-O-P-O- \\ | \\ O_- \\ pKa\ 2.1 \end{array}\right.\quad\begin{array}{c} OH \end{array}\quad\left.\begin{array}{c} O \\ \parallel \\ -P-O- \\ | \\ O_- \\ pKa\ 2.1 \end{array}\right]$$

FIGURE 5.18   CL — Cardiolipin (Diphosphatidylglycerol).

CL (Figure 5.18) is the most unusual of the 7 common mammalian phospholipids. It is essentially 2 PAs glued together by a glycerol. It has an oversized head containing 2 negative charges due to the 2 dissociated phosphates and 4 acyl chains (Figure 5.19). CL has 3 potential optically active carbons, one on each of the two *sn*-2 phosphatidyl groups and the third on the central carbon of the linking glycerol. Saturated chain CLs form normal lamellar phase (the structure is can-shaped) while the more biologically relevant CLs containing highly unsaturated chains prefer non-lamellar phases (the structure is a truncated pyramid with a wider base than head). Under biological conditions where CL is dispersed with other phospholipids (primarily PCs and PEs) the mixture prefers the lamellar (normal bilayer) phase. Non-lamellar phases are discussed in Chapter 10.

CL is found primarily in the mitochondrial inner membrane where it constitutes about 20% of the total lipid and is closely linked with electron transport and oxidative phosphorylation [24]. Particularly effective in supporting the bioenergetic processes are mitochondrial CL with a high content of linoleic acid ($18:2^{\Delta 9,12}$). CL aggregates and stabilizes important electron transport protein complexes including cytochrome c oxidase, NADH dehydrogenase, ATP synthase and various mitochondrial carrier proteins. While most lipids are made in the endoplasmic reticulum, CL is synthesized on the matrix side of the inner mitochondrial membrane.

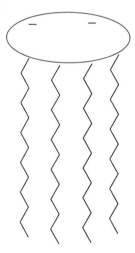

FIGURE 5.19   Cartoon depiction of the structure of cardiolipin (CL).

Cardiolipin was first isolated by M. Pangborn in 1942 [25] and since has been linked to a bewildering array of human metabolic disorders including Tangier's disease, Alzheimer's disease, Parkinson's disease and problems associated with apoptosis. CL from a cow heart is even used as an antigen in the Wassermann test for syphilis. A rare genetic disorder, Barth Syndrome, is associated with abnormal CL biosynthesis. Sufferers of this condition are always male, have mitochondria that are abnormal, and are incapable of sustaining adequate production of ATP. Indeed CL touches many aspects of membrane function and human health.

C. SPHINGOLIPIDS

The family of sphingolipids [26–28] all contain the parent C-18 amino alcohol, sphingosine (Figure 5.20). More than 60 different sphingolipids have been found in human

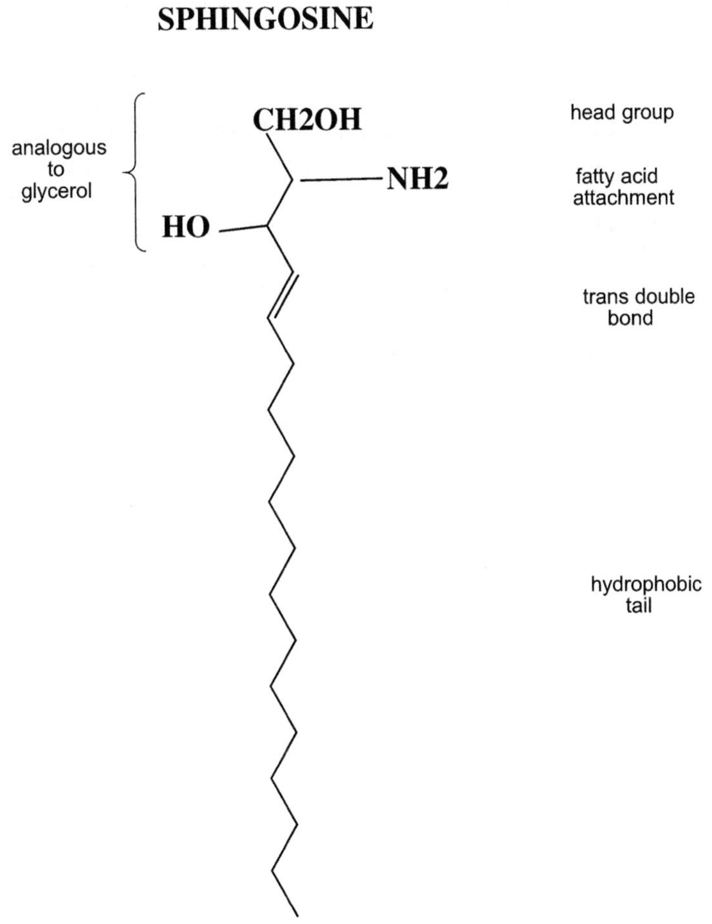

FIGURE 5.20   Structure of sphingosine.

membranes, primarily differing by the components attached to the C-1 alcohol. A variety of fatty acids are also esterified to the nitrogen attached at C-2 forming a saponifiable amide linkage. The first three carbons in sphingosine are structurally analogous to the glycerol in phospholipids. Therefore, like phospholipids, sphingolipids have a polar head group that interacts with water and two hydrophobic tails. However, while both phospholipid acyl chains are variable and can be hydrolyzed, one hydrophobic chain in sphingolipids is neither variable nor hydrolyzable (it is sphingosine) and the other, the variable chain, is attached via a nitrogen ester (amide), not an oxygen ester. The sphingosine chain also contains a permanent *trans* double bond between carbons 4 and 5 while the variable acyl chain can be either saturated or contain *cis* double bonds.

A wide variety of polar head groups may replace the simple proton attached to the −OH at C-1 resulting in the different sphingolipid families listed in Table 5.3.

Most abundant of the human sphingolipids is sphingomyelin (SM, Figure 5.21) where the head group is phosphocholine (or less commonly phosphoethanolamine). SM is rarely found in plants and bacteria. The variable chain in most SMs is longer and more saturated than most phospholipid acyl chains. For example, the most common acyl chains in myelin SM are lignoceric acid (24:0) and nervonic acid (24:1). The general structure of SM therefore resembles PC (or PE). Ceramide with a simple −OH polar head is structurally similar to diacylglycerol. Cerebrosides contain a single sugar as its head group. Cerebrosides with galactose are found in the plasma membrane of neuronal cells while those with glucose are in the plasma membrane of non-neuronal cells. Globosides have 2 or more sugars, usually D-glucose, D-galactose or N-acetyl-D-galactosamine. Gangliosides have several sugars (they are oligosaccharides) and have at least one unusual, and characteristic, 9-carbon anionic sugar, sialic acid (N-acetylneuraminic acid). Sialic acid is discussed with the other membrane sugars in Chapter 7. Gangliosides are further subdivided into families or series that are characterized by the number of sialic acid residues on the sphingolipid:

1 sialic acid   GM series
2 sialic acids   GD series
3 sialic acids   GT series
4 sialic acids   GQ series

TABLE 5.3   Head Groups Attached to the C-1 Position of Sphingosine.

Name	Attached group
Ceramide	−H
Sphingomyelin	phosphocholine, phosphoethanolamine
Cerebroside	glucose or galactose
Globoside	di, tri, or tetra saccharide
Ganglioside	complex oligosaccharide (always with at least one sialic acid)

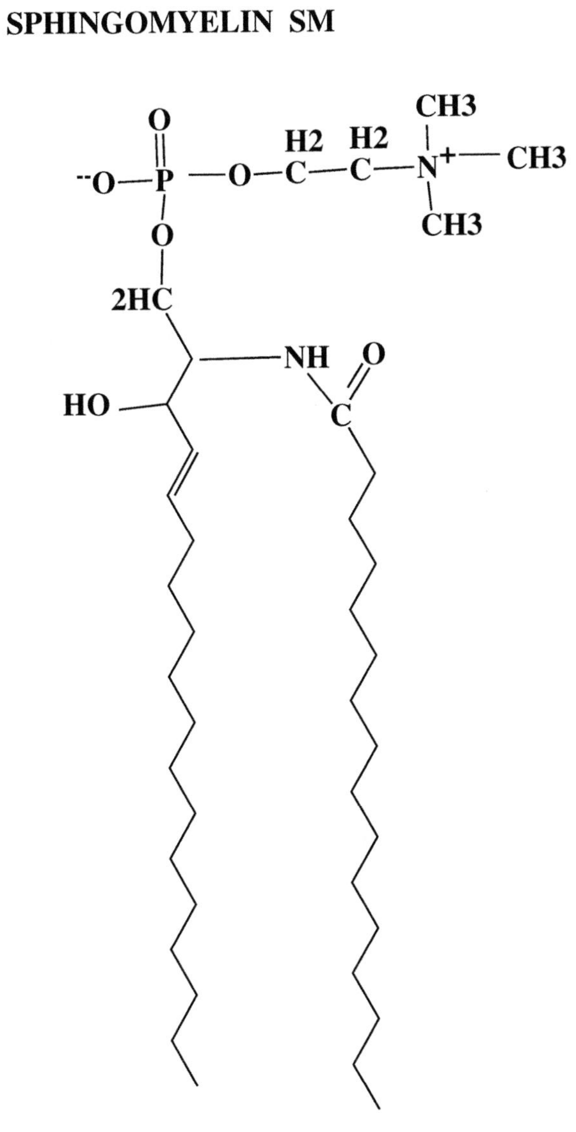

SPHINGOMYELIN SM

FIGURE 5.21   Structure of sphingomyelin (SM).

The specific function of only a few sphingolipids is known. Carbohydrates of some sphingolipids define human blood groups and determine blood type. Gangliosides are found on the outer surface of the plasma membrane where they present points of recognition for extracellular molecules or surfaces of neighboring cells. The absence of specific sphingolipid degrading enzymes are known to be the cause of a number of deadly, incurable genetic lysosomal storage diseases including Tay-Sach's Disease, Gaucher's Disease and Niemann-Pick Disease. Some of these diseases are discussed later in Chapter 7.

D. STEROLS

Sterols [29,30] are major components of many animal, plant, and fungal membranes, but are absent in prokaryotes. In bacterial membranes sterol function is replaced by hopanoids, sterol-like lipids. In animals the major sterol is the much maligned cholesterol, in fungi ergosterol and in plants β-sitosterol (Figure 5.22). It is evident from their structures that these three sterols are very similar, differing only by slight modifications to their floppy tails (Figure 5.22).

Sterols are very water insoluble and so readily partition into membranes. Membrane sterols can be thought of as having three distinct parts, each with a different function (Figure 5.23). Sterols are anchored to the aqueous interface via a polar −OH group. The rest of the molecule is hydrophobic. The 4 rigid sterol rings are responsible for the sterol's major function, controlling membrane 'fluidity' and packing free volume (breathing space for protein function). The function of the floppy tail is far less obvious and is uncertain.

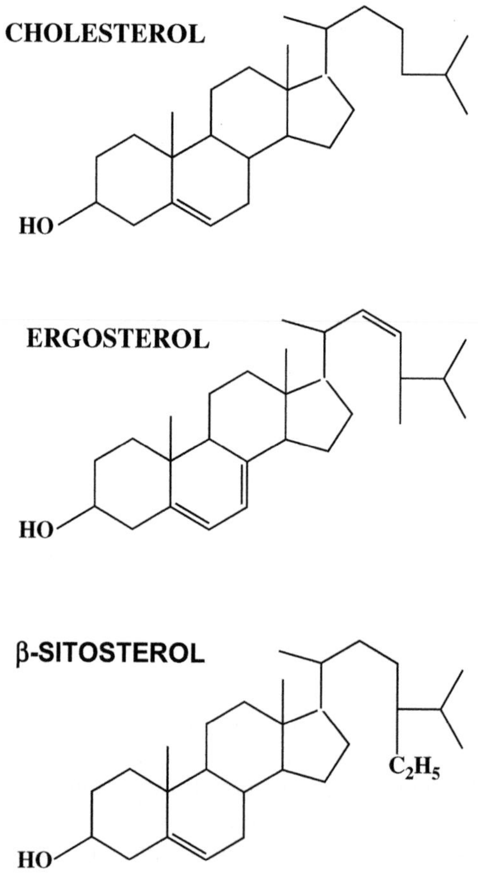

FIGURE 5.22   Structure of the sterols cholesterol (animal), ergosterol (fungi), and β-sitosterol (plant).

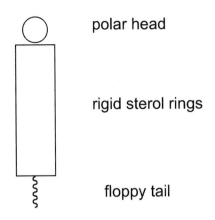

polar head

rigid sterol rings

floppy tail

FIGURE 5.23   Cartoon depiction of the three sections of a sterol.

Overall, the molecule is quite flat. The molecular mode of action of the best understood sterol, cholesterol [31–33], will be discussed in detail in Chapters 9 and 10.

Interest in cholesterol is primarily due to its strong link to high blood pressure and heart disease in man. But most aspects of cholesterol are not evil. Cholesterol was first isolated from gallstones around 1758 by de La Salle. In 1815 Chevreul isolated cholesterol from the unsaponifiable fraction of animal fat and coined the term cholesterine (from the Greek word khole for bile). The major location of cholesterol in vertebrates is the brain, which accounts for one quarter of the body's total cholesterol. In the plasma membrane of animal cells cholesterol is usually the major polar lipid, often comprising more than 50% of the total membrane lipids. It has even been reported that the ocular lens membrane is about 80 mol% cholesterol! Cholesterol is therefore a major structural component of at least the plasma membrane where it is involved in providing mechanical strength, controlling phase behavior and supporting membrane lateral organization and stability. Cholesterol has been proposed to be the molecular 'glue' that holds essential cell signaling lipid rafts together (Chapter 8). In addition cholesterol is a precursor for all steroid hormones (including progesterone, testosterone and estradiol), bile salts and vitamin D. Therefore, cholesterol has two distinct roles: as a structural component of membranes and as a precursor for other essential biochemicals.

In addition to free cholesterol, the sterol can also be found esterified to various fatty acids allowing it to be harmlessly stored as cytosolic droplets or in lipoprotein particles (LDLs). Cholesterol can also be found esterified to a type of secreted polypeptide signaling molecule encoded by the hedgehog gene family. In addition, cholesterol can be sulfated (cholesterol sulfate) or glycosylated (cholesterol glucoside). Indeed, it seems cholesterol is one of those molecules that nature really likes!

The cholesterol story offers an interesting conundrum. Cholesterol provides such a fundamentally essential role in the plasma membrane of animals that it would be easy to conclude that cholesterol or a structurally similar molecule (eg. β-sitosterol or ergosterol) should be essential for life, yet sterols are absent in prokaryotes and they are perfectly functional. Also, in an animal cell, while the plasma membrane is often essentially saturated in cholesterol, the mitochondrial inner membrane is almost completely devoid of cholesterol! Without any cholesterol, how does the mitochondrial inner membrane accomplish the same basic functions as the cholesterol-rich plasma membrane? Cholesterol plays an important role in preventing leakiness to many solutes across the plasma membrane, yet the cholesterol-free mitochondrial inner membrane must be very impermeable to the smallest solute, a proton, to function.

For many years it was believed that cholesterol was nearly absent from plants. However, its presence in higher plants is now universally accepted. Still plants produce a staggering number of plant sterols collectively referred to as phytosterols. More than 250 different phytosterols have been discovered to date. With such a large number of sterols commonly found in plants, it is obvious that large quantities must enter the human diet [34]. Yet photosterols are systematically excluded from the human body, primarily at the level of the intestinal epithelium. Interestingly, it is well documented that plant sterols, particularly β-sitosterol, do compete with cholesterol uptake, thus providing a potentially beneficial role in reducing human cholesterol levels.

E. MEMBRANE LIPID DISTRIBUTION

Table 5.1 lists an 'average' mammalian cell polar lipid composition (these numbers are representative of many examples, for example see [35]). Note the values are reported as a range and not a precise number. Unlike protein and nucleic acid compositions that are defined, lipid compositions can vary considerably with different organisms, different tissues of the same organism, different organelles within a cell, different membrane domains within a single membrane and even different acyl chains of phospholipids in the same cell. Lipid composition can vary with diet, environmental conditions and age of the organism. However there are some general conclusions that are usually observed for mammalian cells. The major structural lipids are usually PC > PE. There are no cationic lipids. The anionic lipids, PA, PS, PI, PG and CL, are all found at lower levels than the structural lipids PC and PE. At physiological pH, PC is a true zwitterion and so has no net charge, while PE is almost a full zwitterion that has a slight net negative charge. Cholesterol can vary from trace amounts (inner mitochondrial membrane) to being the major polar lipid (over 50 mol% in the plasma membrane) depending on the particular membrane. CL is found almost exclusively in the mitochondrial inner membrane. Free, unesterified fatty acids and lysophospholipids (Figure 5.24) are found in very low amounts where they adversely affect membrane structure. High enough amounts of free fatty acids or lysophospholipids will kill a cell through the 'detergent effect'. Fatty acids are distributed unevenly between the two phospholipid acyl chains. The *sn*-1 chains are predominantly, but not exclusively, saturated (often 16:0 or 18:0), while the *sn*-2 chains are mostly unsaturated (18, 20 or 22 carbons long). The large number of unsaturated acyl chains means that, on average, at physiological temperature, most, if not all, of the membrane is in the melted, fluid state.

F. PLANT LIPIDS

The most characteristic feature of plants is photosynthesis and photosynthetic organisms have evolved unique membrane lipids [36] to support this process. Photosynthetic organisms

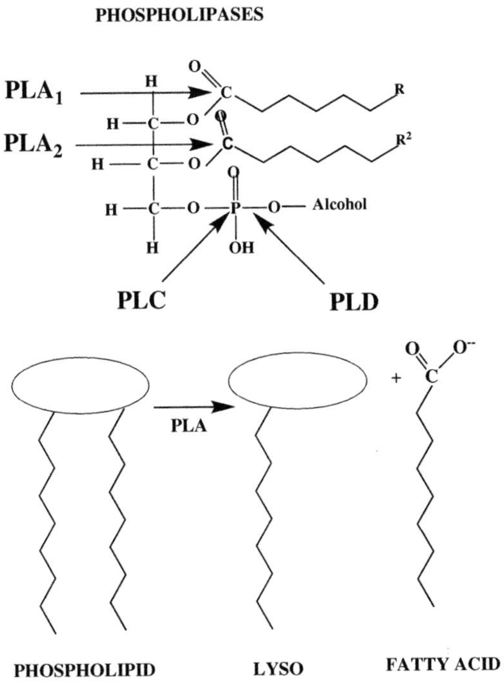

PHOSPHOLIPASES

FIGURE 5.24 Hydrolysis of phospholipids by phospholipases.

including plants, bluegreen algae and some bacteria contain a large amount of glycerolipids (GL) in their thylakoids that must somehow be indispensable to photosynthesis. These lipids often contain the sugar galactose (and so are galactolipids) or sulfur (and so may be sulfolipids) [37]. Since photosynthesis is so prevalent, the major glycerolipid, monogalactosyldiacylglycerol (MGDG, Figure 5.25) is the most abundant membrane lipid on Earth. The glycerolipids are totally devoid of phosphate and, as a result, most have no net charge. An exception is sulfoquinovosyldiacylglycerol (SQDG, Figure 5.25) that is an anion at physiological pH. Phosphate-free plant lipids may have evolved to cope with a soil environment that often has very low phosphate levels and, in fact, phosphate is usually the limiting factor in plant growth.

Despite their unwieldy names, there is a very simple structural relationship between MGDG, DGDG and SQDG (Figure 5.25). MGDG has a relatively small polar head and, with its normal contingent of unsaturated acyl chains, a wide hydrophobic base. Therefore, like PEs with unsaturated acyl chains, MGDG is pyramid-shaped and prefers non-lamellar phase (reverse H_{II} phase, see Chapter 10). The addition of a second galactose in DGDG widens the head enough that the molecule becomes can-shaped and so, like PC, prefers lamellar phase. SQDG is anionic and like PA, PS, PI, PG, and CL in animals, is found in much lower amounts than MGDG and DGDG, the thylakoid membrane structural lipids. The major thylakoid lipids are listed in Table 5.4. Note that the major animal membrane phospholipid, PC, is found at only 2 membrane mol% in plant thylakoids.

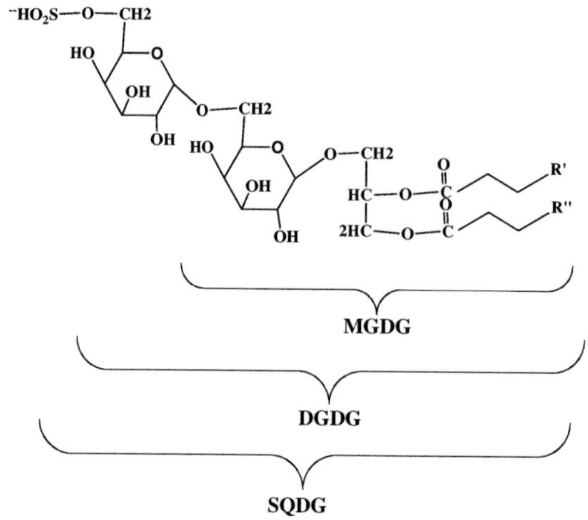

FIGURE 5.25   Structure of three closely related galactosyl diacylglycerols commonly found in plant thylakoid (photosynthetic) membranes.

G. MEMBRANE LIPIDS FOUND IN LOW ABUNDANCE

There are countless numbers of unusual lipids found in membranes in almost undetectable amounts. For example, the oxidation of polyunsaturated fatty acids produces countless byproducts whose structures and biological functions are largely unknown. A few of the more common lipids whose structures and function are appreciated are discussed below.

Free Fatty Acids and Lysolipids

Under physiological conditions phospholipids are degraded (hydrolyzed) by a variety of phospholipases (Figure 5.24) [38]. The phospholipases A_1 and A_2 cleave the hydrophobic

TABLE 5.4   Polar Lipid Composition of the Plant Thylakoid (Photosynthetic) Membrane.

Lipid	Mol %
MGDG	48
DGDG	25
PG	13
SQDG	8
PC	2
Other	4

acyl chains located at positions *sn*-1 and *sn*-2, respectively. Both products of the hydrolysis, free fatty acids and lyso-phospholipids, are harmful to membranes and so are normally present in trace amounts. Significant quantities of a free fatty acid produce a detergent effect that can result in membrane phospholipids dissolving into fatty acid micelles (Figure 5.1). The remaining portion of the phospholipid is referred to as a lyso-phospholipid. Lyso refers to the molecule's ability to disrupt or lyse a membrane. Lyso-phospholipids are cone shaped with a wide polar head and a narrow apolar tail. Molecules with this shape do not readily fit into regular lamellar phase packing, thus distorting the membrane (see Chapter 10). High levels of free fatty acids and lyso-phospholipids are usually the result of artifacts arising from the membrane isolation procedure or arise from membranes isolated from an aged or diseased cell. An exception to this is found in chromaffin granule membranes that appear to be naturally high in free fatty acids. Chromaffin granules are organelles in chromaffin cells of the adrenal medulla, where epinephrine and norepinephrine are synthesized, stored, and released.

The action of phospholipase C on phospholipids produces diacylglycerol (DAG) and the water-soluble phospho-alcohol characteristic of the phospholipid type. As discussed above, DAG is a potent activator of several membrane proteins.

Lipid-Soluble Vitamins

A vitamin [39] is an organic compound that is essential in the diet in small amounts because it cannot be synthesized in sufficient quantities by an organism. Vitamins are classi-fied as either being water-soluble or lipid-soluble. In humans there are 4 lipid-soluble (A, D, E, and K) and 9 water-soluble (8 B vitamins and vitamin C) vitamins. Here we are interested in the major lipid-soluble vitamins A, D, and E (Figure 5.26) due to their favorable partition-ing into membranes.

The family of vitamin As [40] exist in three oxidation states; alcohol (retinol), aldehyde (retinal) and acid (retinoic acid) (Figure 5.26). Vitamin As are involved in complex roles as mediators of cell signaling and regulators of cell and tissue growth and differentiation. The first recognized vitamin A, retinol, was discovered around 1913 by Elmer McCollum and Marguerite Davis in butterfat and cod liver oil. The vitamin was shown to prevent and reverse night blindness. Retinol is now added to skin creams where it is advertised as preventing acne. It also functions as a membrane antioxidant. In the eye, retinal is known to attach to the protein opsin via a Schiff base linkage to lysine, creating rhodopsin or visual purple. Retinal is therefore essential for color vision. Gradients of retinoic acid control the activation of many developmentally important genes.

The structure of Vitamin D (Figure 5.26) indicates it is a metabolic product of choles-terol that results from the opening of ring B in the sterol nucleus. Although vitamin D_3 itself serves no biological function, it is converted to 1,25-dihydroxycholecalciferol, a hormone that regulates calcium uptake in the intestine and calcium levels in the kidney and bone. Vitamin D deficiency results in bone defects and the disease rickets. For this reason it has often been said that vitamin D has 'hormonal' activity. Vitamin D was discovered by Edward Mellanby in 1921 and was shown to have anti-rickets activity.

Vitamin E (Figure 5.26) is a family of related compounds composed of eight different forms, the most biologically important being α-tocopherol [41]. The term α-tocopherol is

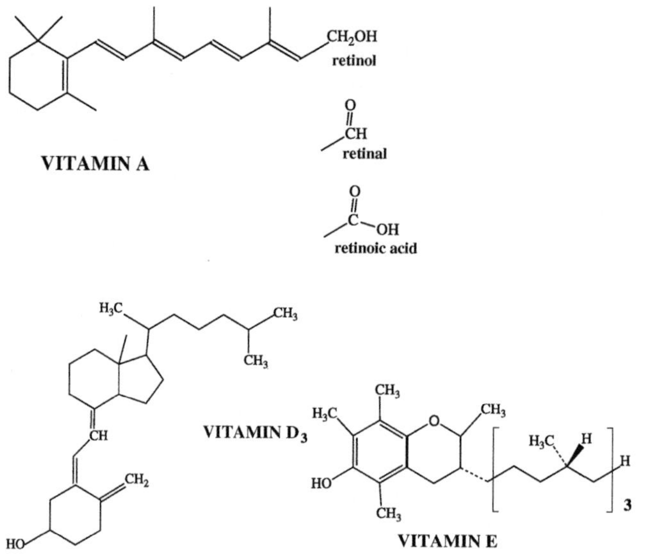

FIGURE 5.26   Structures of the lipid-soluble vitamins: The family of vitamin As, retinol, retinal and retinoic acid; vitamin D; and vitamin E.

derived from a Greek word meaning 'to carry a pregnancy' and was discovered in wheat germ oil in 1922 by Herbert Evans and Katherine Bishop. The vitamin's primary function is to prevent lipid peroxidation in biological membranes. It is the major anti-oxidant in the human body where it has been linked to prevention of a variety of human afflictions including cellular aging and cardiovascular disease.

Superficially α-tocopherol resembles cholesterol, having a polar OH anchoring the molecule to the aqueous interface, a rigid chromanol ring and a floppy hydrophobic tail extending into the membrane interior. A major problem in understanding how α-tocopherol functions in a membrane is deducing how such low levels of the vitamin (estimated to be between ~0.1 to 1 mol% of phospholipids) can protect the large excess of polyunsaturated fatty acids from being peroxidized. This is the equivalent of one molecule of α-tocopherol for every 100—1,000 phospholipids. The anti-oxidant role of α-tocopherol exemplifies many biological processes. While there is a basic understanding of the process itself, it is not at all clear why the process is so efficient.

The lipid-soluble vitamins A, D, and E were all discovered in less than a ten year period in the early 20th century.

Ubiquinone (Coenzyme Q)

Coenzyme Q (also known as CoQ or ubiquinone, Figure 5.27) is a major electron and proton carrying component of the mitochondrial electron transport system [42]. CoQ was discovered in 1957 by Frederick Crane from an extract of beef heart. The molecule has a water-soluble quinone anchored to the hydrophobic membrane interior by a very long methyl branched (isoprenoid) chain. One of the most common ubiquinones has a 50-carbon

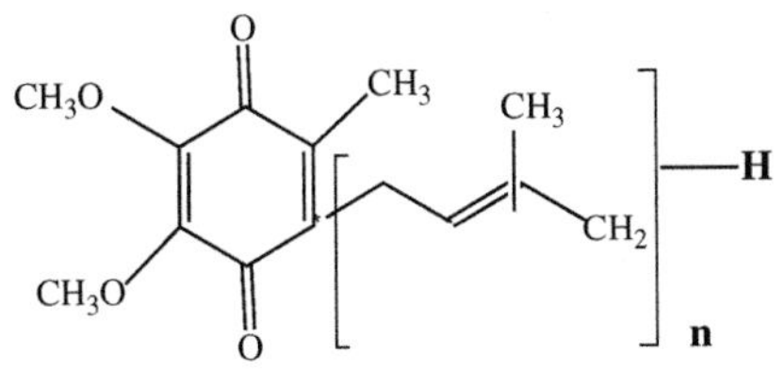

UBIQUINONE (COENZYME Q)

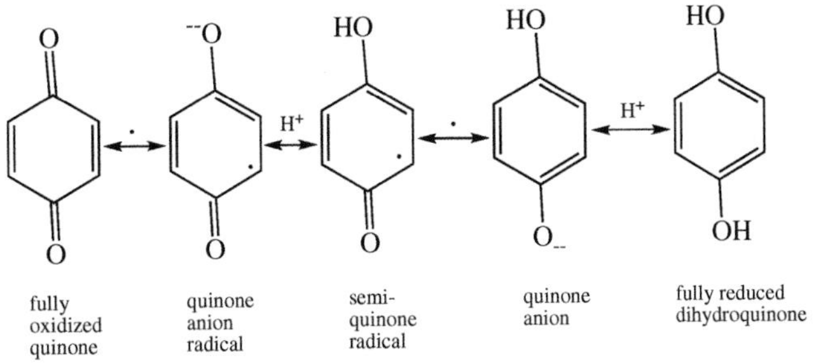

fully	quinone	semi-	quinone	fully reduced
oxidized	anion	quinone	anion	dihydroquinone
quinone	radical	radical		

FIGURE 5.27   Structure (top) and function (bottom) of coenzyme Q (ubiquinone, CoQ). CoQ is versatile as it can carry 1 or 2 electrons and 1 or 2 protons and is a major component of the mitochondrial electron transport chain.

(10 isoprene unit) anchoring chain. This molecule is often referred to as coenzyme Q10 or just CoQ10. CoQ10 has recently been shown to be effective against Parkinson's Disease. Ubiquinone is the most abundant of all the mitochondrial electron carriers, being about 10X more abundant than any other carrier. Ubiquinone is also the most versatile of all of the electron transport components having the ability to carry either 1 or 2 electrons and 1 or 2 protons as depicted in Figure 5.27.

SUMMARY

A major component of biological membranes is complex polar lipids, lipids that when hydrolyzed release free fatty acids. Free fatty acids themselves are a minor component of most membranes, as they can disrupt membrane structure. As a result, fatty acids are usually esterified. Major animal membrane structural lipids can be classified into three basic types; phospholipids, sphingolipids, and sterols. The major type of membrane lipids, phospholipids, are divided into 7 structural classes. Each phospholipid has two esterified fatty acyl chains, one saturated (*sn*-1 chain) and the other unsaturated (*sn*-2 chain), that control properties of the membrane interior. The second membrane lipid type is sphingolipids, so designated because they all contain the parent C-18 amino alcohol sphingosine. More than 60 different sphingolipids have been found in human membranes. The third major membrane lipid type is sterols, particularly cholesterol. In mammalian plasma membranes cholesterol usually comprises more than 50 mol% of all lipids where it controls membrane "fluidity", lipid packing and permeability. Cholesterol has been proposed to be the molecular "glue" that holds membrane lipid rafts together. Lipid composition can vary with diet, environmental conditions, and age of the organism.

By weight, the major component of most membranes is protein, and proteins are responsible for all membrane biochemical activity. Chapter 6 will discuss the many types of structural proteins that are found in membranes.

References

[1]   Cavitch SM. The Natural Soap Book. Storey Publishing; 1994.
[2]   Pliny the Elder. AD 77—79. Historia Naturalis, XXVIII, 191.
[3]   Good Scents Candles & Soap. History and use of soap, www.goodscentscandles.us/soaphistory.php.
[4]   Kanno S, Nakagawa K, Eitsuka T, Miyazawa T. Plasmalogen: A short review and newly-discovered functions. Dietary Fats and Risk of Chronic Diseases. In: Yanagita T, Knapp H, editors. Lipidat: AOCS Press; 2006 [Chapter 14].
[5]   Fahy E, Subramaniam S, Brown HA, et al. A comprehensive classification system for lipids. J Lipid Res 2005;46:839—62.
[6]   Cevc G, editor. Phospholipids Handbook. New York: Marcel Dekker; 1997.
[7]   Hanahan DJ. A Guide to Phospholipid Chemistry. Oxford University Press; 1997.
[8]   Hawthorn JN, Ansell GB, editors. Phospholipids, Vol 4. Elsevier; 1982.
[9]   AOCS The Lipid Library. Complex Glycerolipids, lipidlibrary.aocs.org; 2011.
[10] Leray C. Cyber Lipid Center. Resource Site for Lipid Studies, www.cyberlipid.org/phlip/pgly02.htm; 2011.
[11] Moritz A, De Graan PN, Gispen WH, Wirtz KW. Phosphatidic acid is a specific activator of phosphatidylinositol-4-phosphate kinase. J Biol Chem 1992;267:7207—10.
[12] Moolenaar WH, Kruijer W, Tilly BC, Veerlaan I, Bierman AJ, de Laat SW. Growth factor-like action of phosphatidic acid. Nature 1986;323:171—3.
[13] AOCS The Lipid Library. Phosphatidylserine and Related Lipids Structure, Occurrence, Biochemistry and Analysis. Phosphatidylserine — Structure and Occurrence.
[14] Vance JE, Steenbergen R. Metabolism and functions of phosphatidylserine. Prog Lipid Res 2005; 44:207—34.
[15] Almada A. Phosphatidylserine Boosts Brain Function. Nutrition Science News; 2001.
[16] Kidd PM. Phosphatidyl Serine and Aging—Can This Remarkable Brain Nutrient Slow Mental Decline? ImmuneSupport.com.
[17] Abe T, Lu X, Jiang Y, et al. Site-directed mutagenesis of the active site of diacylglycerol kinase α: calcium and phosphatidylserine stimulate enzyme activity via distinct mechanisms. Biochem J 2003;375:673—80.
[18] Newton AC. Protein kinase C: Structure, function, and regulation. J Biol Chem 1995;270:28495—8.
[19] AOCS The Lipid Library. Phosphatidylinositol and Related Lipids Structure, Occurrence, Composition and Analysis 1. Phosphatidylinositol.
[20] Vivanco I, Sawyers CL. The phosphatidylinositol 3-kinase—AKT pathway in human cancer. Nature Reviews Cancer 2002;2:489—501.
[21] Kuksis A. Inositol Phospholipid Metabolism and Phosphatidyl Inositol Kinases. Elsevier; 2003.
[22] Ferguson MAJ, Williams AF. Cell-surface anchoring of proteins via glycosyl-phosphatidylinositol structures. Annu Rev Biochem 1988;57:285—320.
[23] AOCS The Lipid Library. Phosphatidylglycerol and Related Lipids. Structure, Occurrence, Composition and Analysis.
[24] Houtkooper RH, Vaz FM. Cardiolipin, the heart of mitochondrial metabolism. Cell Mol Life Sci 2008;65:2493—506.
[25] Pangborn M. Isolation and purification of a serologically active phospholipid from beef heart. J Biol Chem 1942;143:247—56.
[26] AOCS The Lipid Library. Sphingolipids. An Introduction to Sphingolipids and Membrane Rafts.
[27] Dickson RC. Sphingolipid functions in *Sacromyces cerevisiae*. Annu Rev Biochem 1998;67:27—48.

[28] Gunstone F. Fatty Acid and Lipid Chemistry. Blackie Academic and Professional; 1996. p. 43–44.

[29] Parish EJ, Nes WD, editors. Biochemistry and Function of Sterols. CRC Press; 1997.

[30] Patterson GW, Nes WD, editors. Physiology and Biochemistry of Sterols. Am Oil Chem Soc. Champaign, IL; 1991.

[31] Yeagle P. The roles of cholesterol in biology of cells. In: Yeagle P, editor. The Structure of Biological Membranes. Boca Raton, FL: CRC Press; 1992 [Chapter 7].

[32] Yeagle PL. The roles of cholesterol in the biology of cells. In: The Structure of Biological Membranes. 2nd ed. Boca Raton, FL: CRC Press; 2005 [Chapter 7].

[33] Finegold LX, editor. Cholesterol in Membrane Models. Boca Raton, FL: CRC Press; 1993.

[34] Dutta PC, editor. Phytosterols as Functional Food Components and Nutraceuticals. Nutraceutical Science and Technology. New York: Marcel Dekker; 2004.

[35] White DA. The phospholipids composition of mammalian tissues. In: Ansell GB, Hawthorne JN, Dawson RMC, editors. Form and Function of Phospholipids. 2nd ed. New York: Elsevier; 1973. p. 441–82.

[36] Williams JP, Khan MU, Lem-Kluwer NW. Physiology, Biochemistry, and Molecular Biology of Plant Lipids. New York: Academic Press; 1997.

[37] AOCS The Lipid Library. Mono- and Digalactosyldiacylglycerols and Related Lipids from Plants and Microorganisms. Structure, Occurrence, Biosynthesis and Analysis.

[38] Gelb MH, Lambeau G. PLA_2: A Short Phospholipase Review. Cayman Chemical, Issue 14; 2003.

[39] McDonald A, Natow A, Heslin J-A, Smith SM, editors. Complete Book of Vitamins and Minerals. Publications International; 2000.

[40] Higdon J. 2003. Updated by Drake, VJ. 2007. Micronutrient Information Center. Vitamin A. Linus Pauling Institute Oregon State University.

[41] Atkinson J, Harroun T, Wassall SR, Stillwell W, Katsaras J. The location and behavior of α–tocopherol in membranes. Mol Nutr Food Res 2010;54:641–51.

[42] Ernster L, Dallner G. Biochemical, physiological and medical aspects of ubiquinone function. Biochim Biophys Acta 1995;1271:105–204.

OUTLINE

A. Introduction   85

B. The Amino Acids   86

C. How Many Membrane Protein Types
are there?   91
  I. Peripheral Proteins   91
  II. Amphitropic Proteins   93
  III. Integral Proteins   93

  a. Endo and Ecto Proteins   94
  b. Trans-membrane Proteins   94
  c. Prenylated Lipid-Anchored
  Proteins   102
  d. GPI-Anchored Proteins   103

Summary   103

References   104

A. INTRODUCTION

Membranes are composed primarily of lipids (Chapters 4 and 5) and proteins (Chapter 6) with a variable amount of carbohydrates (Chapter 7) attached to the surface [1]. By number, lipids are the major component of all membranes where they exceed proteins by ~40 to 1 up to ~200 to 1 and are responsible for the basic membrane structure and environment. Lipids, however, do not directly express any biochemical catalytic activity. The amount of proteins in a membrane varies over a wide range depending on how biochemically active the membrane is. It has been estimated that more than half of all proteins have some association with a membrane. At the low extreme is the myelin sheath whose primary function is to insulate nerve cells. Only ~20% of the myelin sheath by weight is protein. Myelin lipids outnumber the proteins by ~200 to 1. At the other extreme is the mitochondrial inner membrane that supports many complex functions including electron transport, oxidative phosphorylation and numerous trans-membrane transport systems. By weight this membrane is ~75% protein. But even here lipids outnumber the proteins by more than 40 to 1. Since different types of membranes have very different functions, but have similar basic physical properties,

An Introduction to Biological Membranes
http://dx.doi.org/10.1016/B978-0-444-52153-8.00006-4

there is far more functional diversity observed in membrane proteins than is found in membrane lipids.

B. THE AMINO ACIDS

The first basic question is why some proteins are water-soluble while others partition into the oily interior of a membrane. Since all proteins are a linear string of amino acids, it must be that the types and arrangement of the amino acids ultimately determine the eventual location of the protein. The term amino acid [2,3] pertains to any small molecule that possesses both an amine and a carboxylic acid group. There are about 500 known types of amino acids of biological origin. The first amino acid was discovered in 1806 by the French chemists Louis Nicolas Vauquelin and Pierre Jean Robiquet from asparagus juice, and so it was named asparagine [4]. Vauquelin was prolific, authoring some 376 papers covering many topics. He is best known for his discovery of chromium (1797) and beryllium (1798) and for the production of liquid ammonia at atmospheric pressure (Figure 6.1).

There are 20 common amino acids, so-called because they are coded for by the genetic code. It is a natural tendency of scientists to categorize everything and this includes the amino acids. Although there have been several proposed amino acid classifications, they are all artificial and so have advantages and disadvantages. Since this book is primarily interested in membrane proteins, a first approach would be to simply divide the amino acids into water-soluble (polar) and water-insoluble (non-polar) based on the structure of the amino acid side chains. While this is easy to do with most amino acids, some present a problem. For example, what do you do with glycine that has no real side chain (only a −H), and tyrosine that has a polar −OH attached to a non-polar benzene ring?

FIGURE 6.1   L.N. Vauquelin (1763−1829).

TABLE 6.1   List of Non-Polar and Polar Amino Acids.

Non-polar	Polar
Alanine (ala, A)	Glycine (gly, G)
Valine (val, V)	Serine (ser, S)
Leucine (leu, L)	Threonine (the, T)
Isoleucine (ile, I)	Cystine (cys, C)
Proline (pro, P)	Glutamine (gln, Q)
Phenylalanine (phe, F)	Tyrosine (tyr, Y)
Tryptophan (trp, W)	Asparagine (asn, N)
Methionine (met, M)	Lysine (lys, K)
	Arginine (arg, R)
	Histidine (his, H)
	Glutamic Acid (glu, E)
	Aspartic Acid (asp, D)

Table 6.1 presents a simple classification system that was first used to compare water-soluble and membrane proteins. By this analysis an initial comparison was made between two mitochondrial proteins, the very water-soluble cytochrome C and the very water-insoluble mitochondrial 'structural protein.' Surprisingly, both proteins had almost identical ratios of polar/non-polar amino acids. Additional analysis of many proteins indicated that almost all proteins fall within the same range of having between 45—55% polar and 45—55% non-polar amino acids.

The two basic types of classifiers are 'lumpers' and 'splitters.' Dividing the 20 common amino acids into just two classes, non-polar and polar amino acids, is an example of lumping (Table 6.1). There have been other systems that further sub-divide (splitting) the amino acids into multiple classes. Table 6.2 presents a commonly used (splitter) system that is based on the chemical structure of the side chains.

TABLE 6.2   A Classification System Based on the Chemical Structure of the Amino Acid Side Chains.

Aliphatic	Aromatic	Polar uncharged	Sulfur containing	Cation	Anion	Immino
Gly	Phe	Ser	Cys	Lys	Asp	Pro
Ala	Tyr	Thr	Met	His	Glu	
Val	Trp	Asn		Arg		
Leu		Gln				
Ile						

It is now clear that neither a protein's net amino acid composition, nor its distribution into totally artificial classifications, dictate whether the protein will reside in a membrane or not. If the net amino acid composition has little if anything to do with locating a protein into a membrane, the driving force must reside within the amino acid *sequence*. In a protein's three dimensional structure, the more hydrophobic amino acids should reside together, forcing that portion of the protein into the membrane, while the more hydrophilic amino acids should organize into protein domains that will interact favorably with water. Therefore, membrane proteins, analogous to membrane lipids (see Figure 5.7), must also be amphipathic (Figure 6.2), having both hydrophobic and hydrophilic regions.

The outstanding question is how the amino acid sequence relates to a protein's location either into or out of the membrane. One direct possibility would be to determine the precise location of each amino acid in a protein by X-ray crystallography. Unfortunately, however, this has proven to be a very difficult task for most membrane proteins. There are now more than 30,000 known water-soluble globular proteins and more than 20,000 membrane proteins. However, it is so much easier to obtain crystals from globular proteins that >6,000 crystallography structures of these proteins have been determined compared to a few dozen structures of integral membrane proteins. So, are there other methods of identifying protein domains that prefer a location in the hydrophobic core of a membrane? To answer this question, the amino acids cannot simply be lumped into classes, but instead must be characterized individually. This approach involves hydropathy plots. By this method each amino acid is assigned a unique Hydropathy Index value that expresses its lipid/water partition. These values represent a measure of the free-energy change accompanying the movement of an amino acid from a hydrophobic solvent to water. Some amino acids are very hydrophobic while others are very hydrophilic. Still others are in between. Since the assigned Hydropathy Index values are entirely empirical, a number of very different Hydropathy Index tables have been suggested. Probably the major approach for this type of membrane study comes from Kyte and Doolittle [5]. Their method is based on a hydropathy scale that measures the relative hydrophilicity/hydrophobicity of each of the 20 common amino acids. The values are an amalgam of experimentally derived observations taken out of the

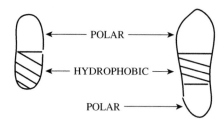

FIGURE 6.2   Amphipathic nature of membrane integral proteins. The protein on the left would sit on one side of the membrane (it would be either an ecto or endo protein) while the protein on the right would span the membrane (it would be a trans-membrane protein).

TABLE 6.3 Kyte-Doolittle Hydropathy Index Values
for the 20 Common Amino Acids [5].

Amino acid	Hydropathy index
Isoleucine	4.5
Valine	4.2
Leucine	3.8
Phenylalanine	2.8
Cysteine	2.5
Methionine	1.9
Alanine	1.8
Glycine	−0.4
Threonine	−0.7
Tryptophan	−0.9
Serine	−0.8
Tyrosine	−1.3
Proline	−1.6
Histidine	−3.2
Glutamic Acid	−3.5
Glutamine	−3.5
Aspartic Acid	−3.5
Asparagine	−3.5
Lysine	−3.9
Arginine	−4.5

literature. Table 6.3 lists the amino acids and their Kyte-Doolittle Hydropathy Index values from the most hydrophobic (isoleucine) to the most hydrophilic amino acid (arginine).

It is obvious that the Kyte Doolittle Hydropathy Index values are for the most part consistent with the first crude classification of the amino acids into polar and non-polar. By Kyte Doolittle, most of the non-polar amino acids have a positive Index while most of the polar amino acids have a negative Index. A wide variety of hydropathy value tables have now been proposed, and all are empirical amalgamations of experimentally derived values. Although all of the hydropathy methods have similar amino acid rankings, all have slight variations in some of the relative amino acid placings. Their empirical numbers also vary over different ranges than do the Kyte Doolittle Hydropathy Index of +4.5 to −4.5.

Once the Hydropathy Index for the amino acids is established, the next step is to plot the Index value for each amino acid in the protein's sequence from N-terminal to C-terminal. Also a mid-point line that is the Hydropathy Index average of all the amino acids is drawn. Unfortunately the first pass through the amino acid sequence assigning the actual Hydropathy Index value for each amino acid in sequence is extremely 'noisy' and cannot be interpreted. Therefore a second and even a third pass are required where the Hydropathy Index of a window of amino acids surrounding a central amino acid is averaged and this value is assigned to the central amino acid in the window. This is referred to as a 'rolling average'. For example using a window of 7 amino acids, amino acid 15 would be assigned the average Hydropathy Index of amino acids 12 through 18. The window would then be shifted to amino acid 16 and the average Hydropathy Index for amino acids 13 through 19 assigned to this central amino acid (amino acid 16). This continues down the entire amino acid chain. A third smoothing pass ('rolling average') may be required to further diminish noise. Readily available computer programs allow the user to easily change the window size from 5 to 25 amino acids and does all of the calculations automatically. Kyte and Doolittle applied this methodology to several proteins whose structure had been determined by crystallography. They found for water-soluble globular proteins a 'remarkable correspondence between the interior positions of their sequence and the regions on the hydrophobic side of the midpoint line, as well as the exterior portions and the regions on the hydrophilic side.' With membrane proteins whose crystallography structures are far less available, possible trans-membrane amino acid stretches can be readily determined. For example, a stretch of 21 hydrophobic amino acids (~33Å) is enough to cross the lipid bilayer. According to the Kyte Doolittle Index, the hydrophobic stretch of amino acids that may cross the membrane must have averaged Index values of >1.25. Figure 6.3 is a Kyte Doolittle Hydropathy Index plot of the single membrane span protein glycophorin [5]. This plot was made from a single pass with a window set at 7 amino acids. The membrane spanning α-helix can be readily picked out between amino acids 75 to 94.

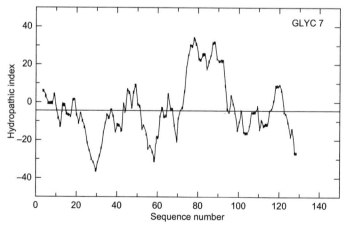

FIGURE 6.3 Hydropathy plot for the single membrane span protein glycophorin. The membrane spanning α-helix can be found between amino acids 75 to 94 [5].

C. HOW MANY MEMBRANE PROTEIN TYPES ARE THERE?

This is a very difficult question to answer with any certainty as it is subject to interpretation. One membrane type gradually blends into another. A first crude approximation could classify proteins as either peripheral (extrinsic) or integral (intrinsic). Peripheral proteins are essentially water-soluble globular proteins that are attached to the membrane surface through electrostatic interactions. They do not significantly penetrate the membrane hydrophobic interior although they may exhibit some weak hydrophobic interaction. Peripheral proteins can be readily removed from the membrane by simply altering the ionic strength of the media or adjusting the pH. The removed protein is freely soluble in water. Integral proteins penetrate the membrane hydrophobic interior to varying extents and so can only be removed from the membrane by far more drastic measures. Integral proteins can only be removed by destroying the membrane and the removed protein is isolated as water-insoluble aggregates. While classifying proteins as either peripheral or integral is easy and not subject to much controversy, it is not very satisfying as most membrane proteins fit into the integral class. In this book, membrane proteins will be classified primarily by the extent and nature of their hydrophobic interactions with membranes. As more is discovered about membrane proteins, devising a meaningful classification system becomes ever more challenging. Overlaps between classes are all too common. Although certainly not perfect, the membrane classification system outlined in Table 6.4 will be employed. This system is an expansion of that proposed for membrane proteins by Robert Gennis in his 1989 book *Biomembranes: Molecular Structure and Function* [1].

I. Peripheral Proteins

Peripheral proteins are held to the membrane surface primarily through electrostatic or hydrogen bonds [6]. They can be divided into two basic types; those that are attached

TABLE 6.4   Membrane Protein Classification System.

I. Peripheral
   A. Bound to Protein
   B. Bound to Lipid

II. Amphitropic

III. Integral
   A. Endo and Ecto
   B. Trans-membrane
      Type I: Single trans-membrane α-helix
      Type II: Multiple trans-membrane spans by α-helices
      Type III: Membrane domains of several different polypeptides assemble to form a channel through the membrane — β-barrels

IV. Lipid-Linked
   A. Myristoylated
   B. Palmitoylated
   C. Prenylated
   D. GPI-linked

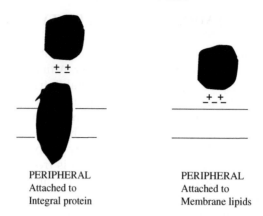

PERIPHERAL
Attached to
Integral protein

PERIPHERAL
Attached to
Membrane lipids

FIGURE 6.4   Peripheral protein attached to an integral protein (left) or to anionic phospholipids (right).

through charges on the surface of integral proteins and those that are attached to the anionic head groups of phospholipids (Figure 6.4).

Peripheral proteins can be removed easily and cleanly from the surface of membranes by either altering the pH or the media salt concentration. One commonly used method that in fact often defines peripheral proteins employs release with alkaline carbonate. If the peripheral protein is attached via Ca^{2+}, a chelating agent will release the protein from the membrane. By definition, a true peripheral protein exhibits little or no interaction with the membrane hydrocarbon interior and is removed from the membrane without attached lipids. There are many well-known examples of peripheral proteins, many of which are equally at home classified as amphitropic proteins (discussed below). Classic examples of peripheral proteins include cytochrome c (binds to an integral protein [7,8]) and the myelin basic protein (binds to membrane phospholipids [9]).

Cytochrome c is probably the best studied peripheral protein. Its function as an essential component in the mitochondrial electron transport system where it links Complex III to Complex IV has been known for many decades. More recently, this very small, primitive protein has been found to be an example of a 'moonlighting' protein, where it performs an entirely different function, that of promoting apoptosis [7]. Cytochrome c is weakly bound to cytochrome c oxidase on the outer side (intermembrane space) of the inner mitochondrial membrane [8]. Cytochrome c can be detached from the mitochondrial membrane by simply washing in 0.15 M KCl. In fact, the protein is often inadvertently lost from the mitochondria during routine isolation and must be replaced. The facile loss of cytochrome c is what makes this peripheral protein such a good trigger for apoptosis.

The best example of a peripheral protein that attaches primarily through electrostatic forces to anionic phospholipids on the surface of a membrane is the myelin basic protein (MBP) [9]. The myelin sheath is considered to be the least dynamic of all membranes and so has very few proteins. Function of the major myelin proteins, myelin proteolipid protein, myelin basic proteins, and myelin-associated glycoprotein, is probably structural. MBP is involved in myelination of nerves in the central nervous system and defects in this protein are believed to be important in multiple sclerosis.

II. Amphitropic Proteins

Amphitropic proteins [10] are a relatively new class whose importance is rapidly becoming more apparent. Unfortunately there is a lot of overlap between the peripheral proteins that are bound to membrane surface phospholipids, ecto and endo integral proteins and amphitropic proteins. Amphitropic proteins can alternate between conformations that are either water-soluble or weakly bound to the membrane surface through a combination of electrostatic and hydrophobic forces. Since the membrane interaction is reversible, an amphitropic protein co-exists as both a free globular protein and a membrane bound protein. The water-soluble globular form is converted into the membrane-bound form following a conformation change induced by phosphorylation, acylation or ligand binding that exposes a previously inaccessible membrane binding site. Membrane interaction is often based on production of a hydrophobic loop of amino acids, sometimes an α-helix, that can penetrate into the lipid bilayer.

There are many important amphitropic proteins that are integrally involved in cell signaling events. Examples include Src kinase, protein kinase C and phospholipase C. Upon binding with the appropriate substrate, conformation of the globular form of the enzyme is altered, exposing a hydrophobic segment on the protein that inserts into the membrane where it comes into close association with its membrane target. For example, in *E. coli* globular puruvate oxidase binds to its substrate pyruvate and cofactor TPP (thiomine pyrophosphate), exposing a hydrophobic helix in the protein. The modified enzyme then binds to the membrane transferring electrons from the cytoplasm to the electrom transport chain in the membrane. The amphitropic protein category also includes water-soluble channel-forming polypeptide toxins including colicin A and α-hemolysin.

III. Integral Proteins

It is estimated that 20–30% of all cellular proteins are integral and these proteins are tightly and permanently bound to the membrane [11]. While peripheral and amphitropic proteins are weakly bound to the surface of membranes and can be readily dissociated producing lipid-free globular proteins, integral proteins are integrated into the membrane hydrophobic interior and can only be removed by more drastic means. Historically, integral proteins have been separated from membranes through acetone and other non-polar organic solvent extractions, chaotropic agents or detergents.

The earliest attempts to isolate integral proteins involved the brute force method of dissolving away the membrane lipids by use of cold acetone. Proteins precipitate into a white powder called an acetone powder. Of course the proteins are completely denatured and virtually worthless for biochemical analysis. Chaotropic agents have been successfully employed to isolate some still functional integral proteins. Chaotropic agents function by disrupting water structure, thus eliminating the hydrophobic effect that is the major force stabilizing membranes. Although not as harmful as acetone, chaotropic agents may also denature membrane integral proteins. Chaotropic agents are highly water-soluble solutes that essentially pull all of the water away from a membrane, destabilizing it. The best example of a chaotropic agent is 6 to 8 M urea. Other examples include high concentrations of guanidinium chloride, lithium perchlorate and thiocyanate. At present the best method for isolating integral proteins involves the use of detergents. This large topic is discussed in Chapter 13 describing protein reconstitution into membranes.

Although thousands of globular proteins have been crystallized and their 3-D structures accurately resolved by crystallography, only a few dozen integral membrane proteins have been resolved [12]. The problem of obtaining 3-D structures for the integral proteins is in obtaining pure crystals in the first place [13]. While peripheral proteins are obtained as monomers that are free of membrane lipids, integral proteins are isolated as aggregates with lipids (or detergents) still attached. The aggregates are almost impossible to crystallize. Therefore structural information about integral proteins is often obtained by indirect methods like enzyme degradation studies and hydropathy plots.

a. Endo and Ecto Proteins

Endo and ecto proteins do not traverse the entire membrane. Instead they are inserted into the membrane's hydrophobic interior from either the cytoplasmic side (endo protein) or the extracellular side (ecto protein). An example of such a protein is cytochrome b_5. Cytochrome b_5 is an electron transport protein associated with the endoplasmic reticulum (ER). It is a component of the mixed function oxidase system involved in the desaturation of fatty acids. Intact cytochrome b_5 can only be removed from the ER membrane by detergents. It is therefore by definition an integral protein, but since it has none of its 152 amino acids exposed on one side of the membrane, it is not a trans-membrane protein. When exposed to trypsin the cytochrome is cleaved at the aqueous interface producing a 104 amino acid water-soluble N-terminal peptide and a 48 amino acid membrane soluble C-terminal peptide (Figure 6.5) [14]. The C-terminal membrane-bound segment has many hydrophobic amino acids and is depleted in hydrophilic amino acids.

b. Trans-membrane Proteins

A trans-membrane protein must span the entire membrane with substantial segments exposed on both the outside and inside aqueous spaces. The membrane that must be

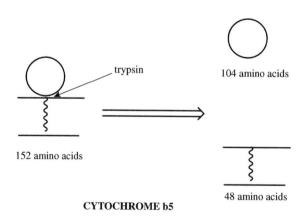

CYTOCHROME b5

FIGURE 6.5 Cleavage of cytochrome b_5 by trypsin producing a 104 amino acid water-soluble protein and a membrane-bound 48 amino acid peptide.

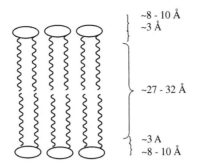

~8 - 10 Å
~3 Å

~27 - 32 Å

~3 A
~8 - 10 Å

FIGURE 6.6   Three sections of the membrane lipid bilayer. The inner hydrocarbon region is ~27 to 32 Å thick. The very narrow boundary region between the hydrophobic inner core and the hydrophilic interfacial regions is ~3 Å. Finally, the polar head group region is ~8 to 10 Å thick.

spanned is composed of a lipid bilayer that can be divided into 3 sections (Figure 6.6). The inner hydrocarbon region is ~27 to 32 Å thick. The very narrow boundary region between the hydrophobic inner core and the hydrophilic interfacial regions is ~ 3 Å. Finally, the polar head group region is ~8 to 10 Å, although this may be wider in membranes that include large amounts of carbohydrate-rich components.

All membrane proteins are 100% asymmetrically distributed with respect to the membrane and do not flip-flop across the membrane. Therefore, trans-membrane proteins are locked in place. A characteristic of many trans-membrane proteins is the presence of tyrosines and tryptophans at the aqueous interface [15]. These amino acids serve as interfacial anchors that can interact simultaneously with the membrane hydrophobic interior and the aqueous exterior. Also, for reasons not entirely clear, the cationic amino acids lysine, histidine, and arginine occur more commonly on the cytoplasmic side of membranes (positive-inside rule [16]). The inner (cytoplasmic) leaflet is also the location of the majority of membrane primary amine lipids, PE and PS [17]. Virtually all integral proteins have at least one stretch of ~20 hydrophobic amino acids that implies the existence of a trans-membrane α-helix. α-Helical domains are found in all types of biological membranes and so represent a common membrane structural motif [18]. Since each amino acid in an α-helix spans ~1.5 Å, it takes approximately 20 amino acids to cross the ~30 Å of the hydrophobic interior. As expected, the trans-membrane α-helices are generally devoid of polar amino acids while non-membrane spanning portions of the protein are usually enriched in these amino acids. α-Helices can cross the membrane a single time or multiple times.

TYPE I. SINGLE TRANS-MEMBRANE α-HELIX: GLYCOPHORIN

Since erythrocytes, unlike other human cells, do not have any complicating internal membranes, and are easily obtained in pure form from blood, they have served for decades as the 'laboratory for membrane studies'. Many of the classic membrane biochemical and biophysical studies discussed in later chapters were first worked out on erythrocytes, and many of the membrane protein studies were first done on resident glycophorin. However, despite the fact that glycophorin is one of the most studied of all membrane proteins, its biological function remains elusive. Glycophorins a and b are the major glycoproteins in the erythrocyte plasma membrane and bear the antigenic determinants for the MN and Ss blood groups. By weight, 60% of glycophorin is carbohydrate and the abundant sialic acid (see Chapter 7) imparts a negative charge to the erythrocyte outer surface. The negative charges on glycophorin cause the erythrocytes to repel one another preventing them from clumping in the blood. With age, the loss of sugars triggers destruction of old erythrocytes.

Glycophorin a is a single span integral protein of 131 amino acids whose cartoon structure is depicted in Figure 6.7 [19]. By weight, 60% of glycophorin's mass is carbohydrate, all of which are attached to amino acids on the N-terminal, exterior segment. All of the carbohydrates are attached to only a few of the first 50 amino acids of glycophorin's 80 amino acid exterior sequence. As with cytochrome b₅, this extracellular segment can be cleaved by trypsin. The attached sugar tetrasaccharides are O-linked to serine or threonine while the larger oligosaccharides are N-linked to asparagines. O-linked serine and threonine and N-linked asparagine are the usual way sugars are linked to membrane proteins (see Chapter 7). Glycophorin a is held in the membrane via a 22-amino acid single span trans-membrane α-helix [20]. Finally, glycophorin a has a 29 amino acid C-terminal segment that extends into the aqueous space in the erythrocyte interior. This segment is totally devoid of sugars. Both the N-terminal and C-terminal non-membrane domains have many hydrophilic amino acids that are not found in the trans-membrane α-helix segment. Glycophorin a is locked in place by an α-helix that is flanked by positively charged amino acids lysine and arginine. These cationic amino acids interact with the negatively charged phospholipid head groups. In the erythrocyte membrane glycophorin a probably exists as dimers that form a coiled structure involving the α-helices.

TYPE II. MULTIPLE TRANS-MEMBRANE SPAN BY α-HELICES: BACTERIORHODOPSIN

Proteins span membranes via two major structures, α-helices or β-barrels. The first of these structures that was shown to be located inside membranes is the α-helix, a secondary structure first formulated in 1948 by Linus Pauling while lying in bed, sick from a cold. The legend states that boredom led Pauling to start playing with paper and scissors and

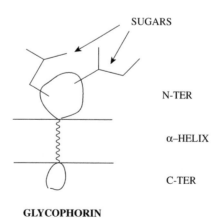

GLYCOPHORIN

FIGURE 6.7 Cartoon structure of glycophorin showing extra-cellular (N-terminal), trans-membrane (α-helix) and intracellular (C-terminal) domains [19].

voila — the α-helix was born! A single span α-helix integral protein, glycophorin a, was discussed above. More common than the single span proteins are the multiple span α-helix proteins. Of particular importance is a very large family of proteins that have 7 membrane-spanning α-helices. These proteins have been referred to as the 'magnificent seven' due to their biochemical importance [21]. Examples of proteins containing the 7 α-helix motif include the large family of G-protein-coupled receptors (GPCR) including the visual receptor rhodopsin and the olfactory receptor, and a variety of channels including voltage gated potassium channels, mechanosensitive channels, aquaporin, chloride channels, and polysaccharide transporters. Included in the G-protein coupled 7 α-helix family are receptors for most of the important hormones and signaling molecules in man including γ-aminobutyric acid (GABA), adenosine, bradykinin, opioid peptides, somatostatin, vasopressin, dopamine, epinephrine, histamine, glucagons, acetylcholine, serotonin, prostaglandins, platelet activating factor, leukotrienes, calcitonin and follicle stimulating hormone (FSH). The 7 α-helix motif is so important that it has been estimated that more than half of all commercial pharmaceuticals are modulators of the GPCR.

The best studied of the 7 α-helix motif proteins is bacteriorhodopsin, a product of the salt loving archaea *Halobacterium* [22]. Under anaerobic conditions this bacteria switches its metabolism to produce enormous quantities of bacteriorhodopsin, a 7 α-helix trans-membrane protein that accumulates in two-dimensional crystalline patches known as a 'purple membrane'. The 'purple membrane' may occupy up to 50% of the cell surface. This highly unusual feature allows for the easy isolation and crystallization of bacteriorhodopsin. Bacteriorhodopsin's function is bioenergetic. Under anaerobic conditions, it captures light to generate a trans-membrane proton gradient that is used to drive ATP synthesis. The light absorbing entity is a retinal connected by a Schiff base to a lysine buried deep in the membrane hydrophobic interior.

The crystallographic structure of bacteriorhodopsin has been determined to 1.55 Å resolution [23]. The protein was shown to have seven trans-membrane α-helices accounting for 80% of the protein's mass. The helices are connected by short, extra-membrane, non-helical loops (Figure 6.8a and 6.8b) [25]. In agreement with the Kyte Doolittle hydropathy plot analysis (Figure 6.8a and 6.8b) [24], each α-helix is composed of about 20 amino acids, the vast majority of which are hydrophobic. However, there are several amino acids found in the α-helices that would not be expected. These include lysine-216 that binds to the retinal, 2 prolines that normally terminate α-helices, as well as the charged amino acids lysine, arginine, aspartic acid and glutamic acid. Since each of the 7 α-helices has at least one of these unfavorable amino acids, application of a Kyte Doolittle hydropathy plot *without* employing a rolling average to smooth the curves would not detect any 20 or more amino acid α-helix spans in the protein. The 7 α-helices are clustered together and are oriented a little off perpendicular to the membrane plane. Since this structural motif is typical of the entire family of GPCRs, bacteriorhodopsin was then used to model the other 7 α-helix family members until some additional crystallography data finally became available in 2007.

TYPE III. MULTIPLE TRANS-MEMBRANE SPANS BY β-BARRELS

The second major type of trans-membrane structure is the β-barrel [26]. This structure is stabilized inside the membrane by the same hydrophobic forces that stabilize α-helix integral proteins and, furthermore, the protein's folding is facilitated by water-soluble chaperones.

(a)

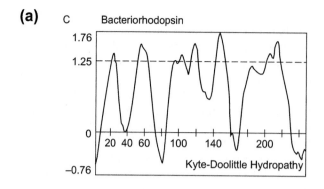

(b)

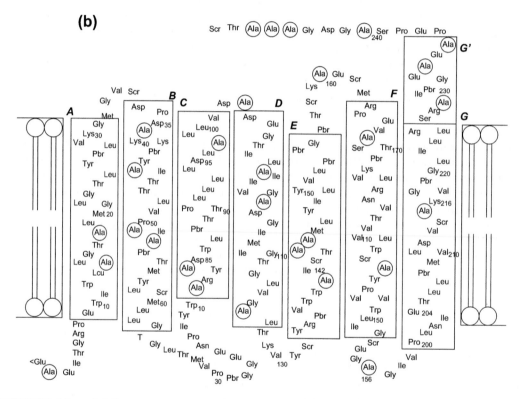

FIGURES 6.8a and 6.8b   Bacteriorhodopsin Kyte Doolittle Hydropathy plot (Top Panel) and trans-membrane sequence (Bottom Panel) demonstrating the 7 trans-membrane α-helices [24,25].

β-Barrels are very common in the outer membrane of gram-negative bacteria (they account for 2–3% of the bacterial genes), but are also found in the leaky outer membranes of chloroplasts and mitochondria. β-Barrels are composed of 20 or more β-sheet segments that line a cylinder that spans the membrane. It is through this cylinder that various polar solutes diffuse. Two examples of typical β-barrel trans-membrane protein is depicted in Figure 6.9

(a)

(b)

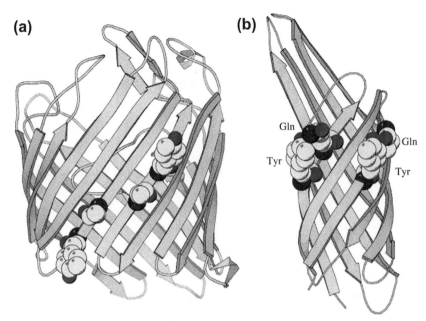

Gln

Tyr

Gln

Tyr

FIGURE 6.9 Two 'typical' examples of trans-membrane β-barrel proteins, (a) a AY2 motif in OmpF and (b) a YQ2 motif in OmpX. Omp stands for outer membrane protein of which there are several distinct types. *Reproduced with permission [27].*

[27]. Shown are side views of two bacterial outer membrane β-barrel proteins [27]. In β-barrels the trans-membrane segments are comprised of only 7 to 9 amino acids [28] compared to the 20 plus amino acids in a trans-membrane α-helix. Also, every other amino acid in a β-barrel is hydrophobic, alternating with hydrophilic amino acids. This makes detecting trans-membrane β-barrel segments by hydropathy plots much more difficult than α-helix segments. In a β-barrel, hydrophobic amino acids face the acyl chains of the membrane hydrophobic interior while the hydrophilic (mostly charged) amino acids line the inside of the aqueous trans-membrane channel. The β-barrel strands are arranged in an anti-parallel fashion. Currently, there are more than 50 known β-barrel structures that fall into three categories; up-and-down, Greek key, and jelly roll. Bacterial porins [27] are composed of 16−18 strands that are connected by beta turns on the cytoplasmic side and long loops of amino acids on the extracellular side. The porin channel is partially obstructed by an amino acid loop called the eyelet that gives the channel solute selectivity.

TYPE IV. LIPID-ANCHORED PROTEINS

Another classification involves proteins that have lipid anchors. Many of these proteins, however, are actually trans-membrane proteins that have additional lipid anchors. They could be classified as either a type of integral protein or as a lipid-anchored protein. For example, rhodopsin is the prototypical example of a 7 α-helix membrane spanning protein, but is also anchored to the membrane through covalently attached myristic acid. The choice of classification for this protein is arbitrary. Although there are additional examples of lipid anchors, the most important involve myristic acid (Figure 6.10), palmitic acid (Figure 6.11), prenylated hydrophobic acyl chains (Figure 6.12) and proteins attached through glycosyl-phosphatidylinositols (GPI-anchored) (Figure 6.13). The GPI-anchored proteins are always

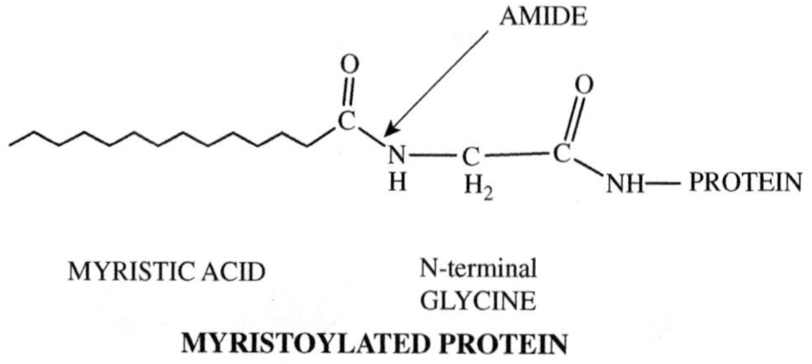

MYRISTOYLATED PROTEIN

FIGURE 6.10   Myristoylated protein. Protein is N-myristoylated through an amide to its N-terminal glycine.

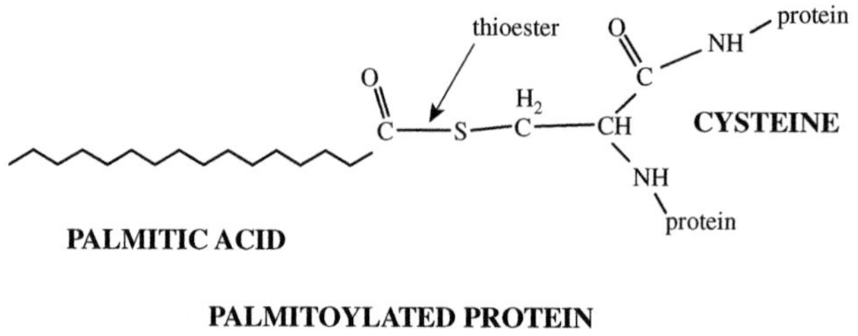

PALMITOYLATED PROTEIN

FIGURE 6.11   Palmitoylated protein. Protein is S-palmitoylated through a thioester to a cysteine near the protein's N-terminal.

FIGURE 6.12   Prenylated proteins. Proteins attached through a thioether to either farnesyl (15-carbons, 3-isoprene units) or geranylgeranyl (20-carbons, 4-isoprene units) at a C-terminal cysteine on the protein.

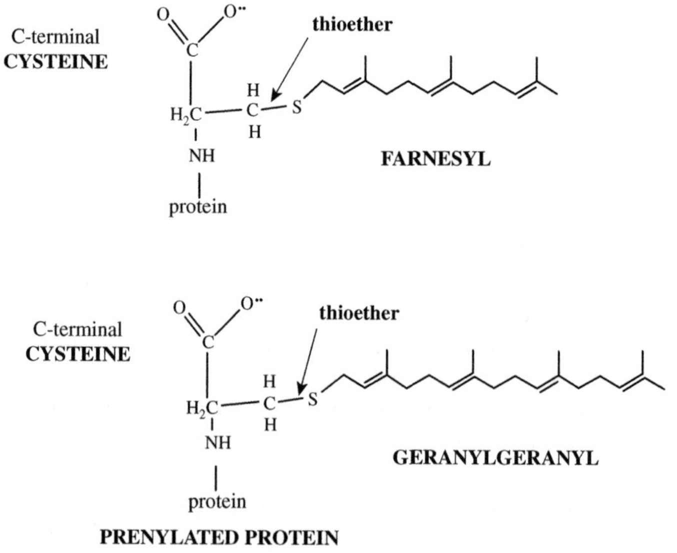

PRENYLATED PROTEIN

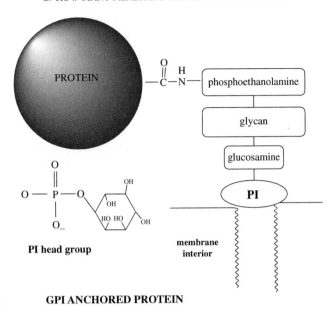

PI head group

membrane interior

GPI ANCHORED PROTEIN

FIGURE 6.13   Cartoon drawing depicting a typical GPI-anchored protein.

attached to the exo-cytoplasmic (outer leaflet) surface, while the others are lipid-anchored to the endo-cytoplasmic inner leaflet surface of cells. Since lipid anchors by themselves are very weak, these proteins are often additionally attached by other, usual interactions (electrostatic and hydrophobic).

1. MYRISTOYLATED LIPID-ANCHORED PROTEINS   The 14-carbon saturated fatty acid, myristic acid (14:0), is often employed as a lipid anchor for proteins [29]. This may seem an odd choice since myristic acid, in sharp contrast to the other commonly employed fatty acid lipid anchor palmitic acid (16:0), is found in membranes only at very low levels. Myristoylation was first reported in 1982 for the proteins calcineurin B [30] and the catalytic subunit of the cyclic AMP-dependent protein kinase [31]. Myristic acid was found to be acylated to the anchored protein via an amide linkage to an N-terminal glycine (Figure 6.10). This linkage is found only in eukaryotes and viral proteins and is established early in translation. Generally, it is an irreversible protein modification. Additional anchoring for myristoylated proteins is provided by electrostatic interactions between positively charged protein side chains and negatively charged membrane phospholipids. Some N-myristoylated proteins undergo additional fatty acyl modifications by attachment of palmitoyl groups to cysteines via reversible thio esters. An example of these dually acylated proteins include members of the Src family of tyrosine kinases (eg. Fyn, Lck). N-Myristoylated proteins have a variety of important functions including roles in several signal transduction cascades. Under unusual circumstances where myristic acid is limiting, some other fatty acids including shorter-chain and unsaturated, can be attached to the N-terminal glycine.

2. PALMITOYLATED LIPID-ANCHORED PROTEINS   Palmitoylated (16:0) proteins are the most extensively studied of the lipid-anchored proteins [32]. The first member of the palmitoylated class was identified by Braun and Rohn in 1969 in *E coli* cell wall [33]. Later, palmitoylation was found in the myelin proteolipid protein (PLP). PLP is the major protein in the myelin sheath, accounting for ~40% of membrane proteins. PLP is a 4 α-helix span trans-membrane protein that contains 6 cysteine residues that are thought to be acylated, primarily with palmitic acid. Other less common acylations involve stearic acid (18:0) and oleic acid ($18:1^{\Delta 9}$). Acylation of a fatty acid to the sulfur side chain of cysteine involves reversible formation of a thioester (Figure 6.11). In fact being a reversible fatty acid attachment is perhaps the most important feature of this lipid linkage. Reversibility is in sharp contrast to the myristoylations discussed above and the prenylations discussed below. Facile palmitoylation and depalmitoylation has the potential to regulate enzyme activity using acyltransferases and thioesterases, respectively. This is highly reminiscent of the better recognized enzyme control mechanisms employing phosphorylation by kinases and dephosphorylation by phosphatases. Among the large and ever-growing family of palmitoylated proteins are rhodopsin, band 3 of erythrocytes, and the lung surfactant proteolipid. Although acylation through a thioester to cystein is the most important palmitoylation, palmitic acid has also been reported to be O-acylated to serines and threonines and N-acylated to the ε-amine side chain of lysine. In Sonic Hedgehog, palmitic acid is reported to be attached to an N-terminal cysteine via an amide linkage.

c. Prenylated Lipid-Anchored Proteins

Membrane integral proteins can also be lipid-anchored through irreversible thioether linkages to very hydrophobic isoprene groups. These proteins are thus termed prenylated proteins [34]. In 1978 Kamiya et al [35] discovered the first prenylated protein (actually a simple 11-amino acid polypeptide), a fungal mating factor. As discussed in Chapter 5, prenylated groups (repeating 5-carbon branched units of 3-methyl-2-buten-1-yl) derive their name from isoprene. Although it appears that the major function of these prenylated chains is to anchor the attached protein to the hydrophobic core of membranes, these groups have an additional, very different and vital role in membranes. Prenyl groups have been shown to be important for protein–protein binding through specialized prenyl-binding domains [36]. Therefore, prenylated groups are both lipid anchors and are intimately involved in localizing certain proteins to specific domains, or to specific proteins within the membrane.

Membrane prenylation involves either 15-carbon (3 isoprene) farnesyl or 20-carbon (4 isoprene) geranylgeranyl (see Chapter 4), covalently attached through a thioether to cysteine residues at or near the C-terminus of proteins (Figure 6.12). Once formed, thioethers cannot be readily removed and so are considered to be permanent attachments. As with myristoylated and palmitoylated proteins, prenylation anchors the protein to the cytoplasmic side (inner leaflet) of the membrane. Known prenylated proteins include nuclear lamins, Ras and Ras-associated G proteins, and protein kinases. Since prenylation is essential for the function of many proteins involved in cellular signaling and trafficking pathways, the process is becoming a major therapeutic target for multiple diseases. Already there are more than 20,000 known prenylated proteins, and this number increases daily. Therefore, it is reasonable to assume that many more prenylated proteins have yet to be discovered.

d. GPI-Anchored Proteins

While myristoylation, palmitoylation and prenylation all anchor proteins to the cytoplasmic side (inside) of membranes, some extracellular proteins are anchored to the outer surface of cells through the acyl chains of glycosylphosphatidylinositol (GPI) (Figure 6.13) [37,38]. GPI-anchored proteins were discovered at about the same time (1976) by Ikazawa [39] and by Low [40]. GPI-anchored proteins are not trans-membrane but also cannot be removed by alkaline carbonate. Therefore, by definition, they are not peripheral proteins. Their major anchor seems to be through the acyl chains of PI that, without additional electrostatic or hydrophobic interactions, would be very weak. GPI anchored proteins only face outside and are normally clustered. The attached protein is indirectly linked from its C-terminal through a phosphoethanolamine, to a glycan (three mannose sugars), to a glucosamine and finally to the inositol sugar of PI (Figure 6.13).

GPI-anchored proteins are found in most eukaryotes, but not bacteria. Although GPI-anchors are not considered to be a common motif of membrane protein structure, there are some 45 GPI-anchored proteins in man. In yeast, where all GPI proteins are known, only 15 of the 5,790 proteins (about 1 in 400 proteins) have GPI-anchors. GPI-anchored proteins have a wide variety of essential functions including enzymatic, antigenic, adhesion, membrane organization, and roles in several receptor-mediated signal transduction pathways. Among the better known GPI-anchored proteins are carbonic anhydrase, alkaline phosphatase, Thy-1, and BP-3. The GPI-anchored protein can be released from the membrane upon hydrolysis by phospholipase C, and in fact this is how they were discovered [39,40]. There are several, mostly rare, human genetic disorders that have been linked to GPI proteins. Perhaps the best known is Marfan syndrome, whose most famous carrier was reputed in a 1962 theory to be Abraham Lincoln. Although it has received a great deal of attention, this interesting story is no longer believed. It is more likely that Lincoln had a different disorder, multiple endocrine neoplasia type 2B, that causes skeletal features almost identical to Marfan syndrome.

SUMMARY

Soon after Gorter and Grendel proved the existence of lipid bilayers, it became obvious that proteins were also a component of membranes. It is a protein's amino acid sequence, not its net content, that locates it to membranes. Hydropathy Plots are used to predict which segments of a protein cross the membrane. It is not clear how many distinct types of membrane proteins exist. A first crude approach identifies peripheral (loosely attached to the membrane surface by electrostatics) and Integral (penetrates into the hydrophobic interior). Integral proteins span membranes via two major structures, a-helices or b-barrels. a-Helices can cross the membrane single or multiple times. A very large family of proteins, including G-protein-coupled receptors, has 7 membrane-spanning a-helices. b-Barrels are composed of 20 or more b-sheet segments that line a membrane-spanning cylinder. Many integral proteins are also anchored to the membrane cytoplasmic surface by lipids, primarily myristyl, palmityl or prenyl groups. Some outer surface proteins are anchored through the acyl chains of glycosylphosphatidylinositol (GPI). Most membrane surface proteins are heavily glycosylated.

The final major membrane component is carbohydrates (sugars) that, if present, are attached to the membrane outer leaflet. Chapter 7 will discuss the carbohydrates that are commonly found attached to membrane lipids (glycolipids) and proteins (glycoproteins).

References

[1] Gennis RB. Biomembranes: Molecular Structure and Function. New York: Springer-Verlag; 1989.

[2] Barrett GC, Elmore DT. Amino Acids and Peptides. London: Cambridge University Press; 1998.

[3] Nelson DL, Cox MM. Lehninger Principles of Biochemistry. 5th ed. New York, NY: W Freeman; 2008.

[4] Bradford H, Vickery C, Schmidt LA. The history of the discovery of the amino acids. Chem Rev 1931;9:169–318.

[5] Kyte J, Doolittle RF. A simple method for displaying the hydropathic character of a protein. J Mol Biol 1982;157:105–32.

[6] Marsh D, Horvath LI, Swamy MJ, Mantripragada S, Kleinschmidt JH. Interaction of membrane-spanning proteins with peripheral and lipid-anchored membrane proteins: perspectives from protein-lipid interactions (Review). Mol Membr Biol 2002;19:247–55.

[7] The New World Encyclopedia. Cytochrome c, www.newworldencyclopedia.org/entry/Cytochrome_c.

[8] Garber EA, Margoliash E. Interaction of cytochrome c with cytochrome c oxidase: an understanding of the high- to low-affinity transition. Biochim Biophys Acta 1990;1015:279–87.

[9] Boggs JM. Myelin Basic Protein. Hauppauge NY: Nova Scientific Publishers; 2008.

[10] Johnson JE, Cornell RB. Amphitropic proteins: regulation by reversible membrane interactions (review). Mol Membr Biol 1999;16:217–35.

[11] Molecular Cell Biology. NCBI (National Center for Biotechnology Information). Section 3.4 Membrane Proteins, www.ncbi.nlm.nih.gov.

[12] Membrane Proteins of Known Structure. From the Stephen White Laboratory, University of California, Irvine.

[13] Caffrey M. Membrane protein crystallization. J Struct Biol 2003;142:108–32.

[14] Ozols J, Gerard C, Nobrega FG. Proteolytic cleavage of horse liver cytochrome b5. Primary structure of the heme-containing moiety. J Biol Chem 1976;251:6767–74.

[15] De Planque MR, Bonev BB, Demmers JA, Greathouse DV, Koeppe 2nd RE, Separovic F, Watts A, Killian JA. Interfacial anchor properties of tryptophan residues in transmembrane peptides can dominate over hydrophobic matching effects in peptide-lipid interactions. Biochemistry 2003;42:5341–8.

[16] Von Heijne G, Gavel Y. Topogenic signals in integral membrane proteins. Eur J Biochem 1988;174:671–8.

[17] Op den Kamp JAF. Lipid Asymmetry in Membranes. Annu Rev Biochem 1979;48:47–71.

[18] General Principles of Membrane Protein Folding and Stability. From the Stephen White Laboratory, University of California, Irvine.

[19] Tomita M, Marchesi VT. Amino-acid sequence and oligosaccharide attachment sites of human erythrocyte glycophorin. Proc Nat Acad Sci USA 1975;72:2964–8.

[20] Furthmayr H. Structural composition of glycophorins and immunochemical analysis of genetic variants. Nature 1978;271:519–24.

[21] Sakmar TP. Twenty years of the magnificent seven. The Scientist 2005;19:22–5.

[22] Lanyi JK. Bacteriorhodopsin Annu Rev Physiol 2004;66:665–88.

[23] Luecke H, Schobert B, Richter H-T, Cartailler J-P, Lanyi JK. Structure of bacteriorhodopsin at 1.55 Å resolution. J Mol Biol 1999;291:899–911.

[24] Engelman DM, Goldman A, Steitz TA. The identification of helical segments in the polypeptide chain of bacteriorhodopsin. Methods Enzymol 1982;88:81–9.

[25] Krystek SR, Metzler WJ, Novotny J. Hydrophobic profiles for protein sequence analysis. Current Protocols in Protein Science 2001; Unit 2.2.

[26] Wimley WC. The versatile β-barrel membrane protein. Curr Opin Struct Biol 2003;13:404–11.

[27] Ronald Jackups Jr, Sarah Cheng, Jie Liang, Sequence motifs and antimotifs in β-barrel membrane proteins from a genome-wide analysis: The Ala-Tyr dichotomy and chaperone binding motifs. J Mol Biol 2006;363:611–23, http://dx.doi.org/10.1016/j.jmb.2006.07.095.

[28] Achouak W, Heulin T, Pages JM. Multiple facets of bacterial porins. FEMS Microbiol Lett 2001;199:1–7.

[29] Farazi TA, Waksman G, Gordon JI. The biology and enzymology of protein N-myristoylation. J Biol Chem 2001;276:39501—4.

[30] Aitken A, Cohen P, Santikarn S, et al. Identification of the NH2-terminal blocking group of calcineurin B as myristic acid. FEBS Lett 1982;150:314—8.

[31] Carr SA, Biemann K, Shoji S, Parmelee DC, Titani K. n-Tetradecanoyl is the NH2-terminal blocking group of the catalytic subunit of cyclic AMP-dependent protein kinase from bovine cardiac muscle. Proc Natl Acad Sci USA 1982;79:6128—31.

[32] Bijlmakers M-J, Marsh M. The on—off story of protein palmitoylation. Trends in Cell Biology 2003;13:32—42.

[33] Braun V, Rohn K. Chemical characterization, special distribution and function of lipoprotein (murein lipoprotein) of the E. coli cell wall. The specific effect of trypsin on the membrane structure. Eur J Biochem 1969;10:426—38.

[34] Gelb MH, Scholten JD, Sebolt-Leopold JS. Protein prenylation: from discovery to prospects for cancer treatment. Curr Opin Chem Biol 1998;2:40—8.

[35] Kamiya Y, Sakurai A, Tamura S, Takahashi N. Structure of rhodotorucine 4 a novel lipopeptide. inducing mating tube formation in Rhodosporidium toruloides. Biochem Biophys Res Commun 1978;83:1077—83.

[36] Kloog Y, Cox AD. Prenyl-binding domains: potential targets for Ras inhibitors and anti-cancer drugs. Semin Canur Biol 2004;14:253—61.

[37] Ferguson MAJ, Williams AF. Cell-surface anchoring of proteins via glycosyl-phosphatidylinositol structures. Annu Rev Biochem 1988;57:285—320.

[38] Ikezawa H. Glycosylphosphatidylinositol (GPI)-anchored proteins. Biol Pharm Bull 2002;25:409.

[39] Ikezawa H, Yamanegi M, Taguchi R, Miyashita T, Ohyabu T. Studies on phosphatidylinositol phosphodiesterase (phospholipase C type) of Bacillus cereus. I. purification, properties and phosphatase-releasing activity. Biochim Biophys Acta 1976;450:154—64.

[40] Low MG, Finean JB. Non-lytic release of acetylcholinesterase from erythrocytes by a phosphatidylinositol-specific phospholipase C. FEBS Lett 1977;82:143—6.

C H A P T E R

7

Membrane Sugars

O U T L I N E

A. Introduction	107	D. GPI-Anchored Proteins	114
B. Glycolipids	111	Summary	115
C. Glycoproteins	112	References	115

A. INTRODUCTION

In addition to lipids and proteins, a third major component exists, primarily on the outer surface of the cell plasma membrane. This component is a collection of very water-soluble carbohydrates (sugars) [1−4]. The word 'carbohydrate' is a contraction of 'hydrated carbon', $(C\ H_2O)_n$, the empirical formulation of most sugars. For example, the most common sugar, glucose, is $C_6H_{12}O_6$ or $C_6\ (H_2O)_6$, and is thus a 'hydrated carbon' (Figure 7.1).

Unlike membrane lipids and proteins, carbohydrates have no hydrophobic segments and so are never found integrated into the non-polar interior of membranes. Instead, they sit on the membrane surface, where they are favorably surrounded by water. Their location, almost exclusively on the outer surface of the plasma membrane, strongly indicates that their main function is to interact with the cell's external environment. It is well established that, like lipids, carbohydrates have no enzymatic functions and so must influence membrane biochemistry through indirect methods. General roles for membrane carbohydrates include: biosynthetic sorting, protecting membrane proteins from proteolysis, assisting in protein folding (serves as chaperones), enhancing protein stability, roles in cell identification, recognition and cell−cell adhesion, providing immunological properties (they are antigens), and serving as receptors. Carbohydrates clearly have a large impact on plasma membrane function.

An Introduction to Biological Membranes
http://dx.doi.org/10.1016/B978-0-444-52153-8.00007-6

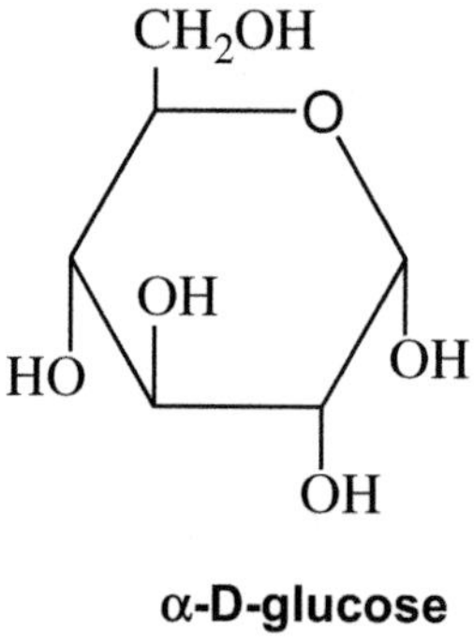

CH₂OH

α-D-glucose

FIGURE 7.1   α-D-Glucose.

The amount of carbohydrate on a membrane protein varies tremendously from 0 in some multiple span trans-membrane proteins to 60% by weight for glycophorin and >85% for some blood group substances. In fact some plasma membranes have so many surface carbohydrates that in cross-section they resemble a hairbrush, with a lot of sugar on the outer surface and none on the cytoplasmic inner surface. The term used to describe these membranes is glycocalyx or 'sugar coat' [5]. A glycocalyx is commonly found on the outer surface of bacteria and mammalian epithelial cells (Figure 7.2). Membrane sugars are attached to either lipids (glycolipids [6–8]) or proteins (glycoproteins, [9,10]). There are nine basic sugars commonly found attached to membranes: α-D-glucose; α-D-galactose; α-D-mannose; α-L-fucose; α-L-arabinose; α-D-xylose; N-acetyl glucosamine; N-acetyl galactosamine; and N-acetyl neuraminic acid (sialic acid) (Figure 7.3). The carbohydrates can be either single sugars or short, <15 unit, polysaccharides, and are always located on the extra-cellular membrane leaflet.

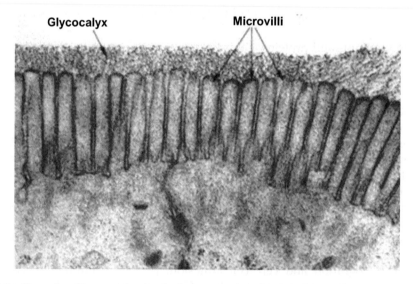

FIGURE 7.2   Glycocalyx. Electron micrograph of a cross section of microvilli in the human digestive track. The glycocalyx can be seen at the top of the columnar microvilli. *Reproduced with permission* [23].

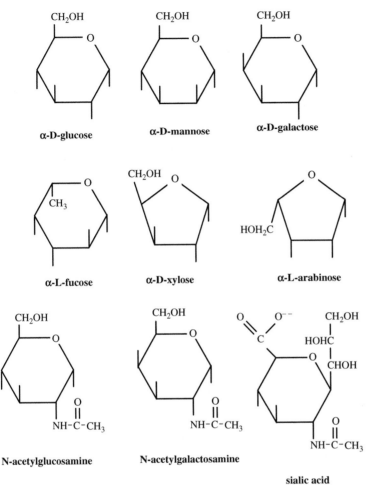

FIGURE 7.3 The 9 carbohydrates found on membrane glycolipids and glycoproteins.

One under-appreciated aspect of carbohydrates is their enormous potential for information storage. Carbohydrates exist in either an aldehyde or a ketone series. In addition, each carbon location that has a carbon with four different groups on it (positions 2 through 5 for glucose) comes in two orientations or isomers. An aldehyde 6-carbon (hexose) sugar therefore has 4 asymmetric carbons, each with two isomers or 2^4 or 16 different possible isomers. The potential for information storage is greatly enhanced when two or more sugars are connected through what is known as a glycosidic bond. For example, linking two glucoses together by removing water between carbon #1 of one glucose (referred to as the anomeric carbon) to the alcohol at position number 4 of the second glucose produces the disaccharide maltose (Figure 7.4).

It is generally assumed that biological information storage resides within the primary sequence of proteins and nucleic acids. However it is becoming more evident that polysaccharides have far more information storage capacity than these other biopolymers. This

α-D-glucose   β-D-glucose

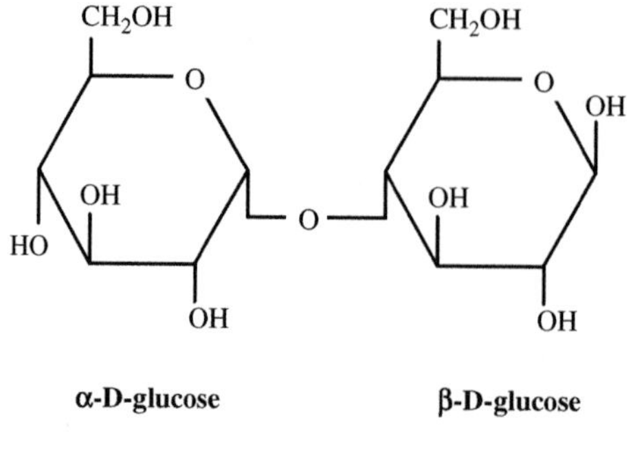

MALTOSE

FIGURE 7.4   The di-saccharide maltose. Two α-D-glucose sugars are attached through a glycosidic bond from the anomeric carbon (C-1) of the left glucose to C-4 of the right glucose.

has led to a search for the elusive 'carbohydrate code'. Many different sugar links are possible, even for disaccharides, and this number increases logarithmically upon each additional linked sugar in the polysaccharide. In a 1997 paper entitled *Information capacity of the carbohydrate code*, R.A. Laine calculated the possible number of ways that sugars could be linked [11]. For a linear tri-saccharide, there are 12 possible linkage sites (4 on each sugar) that can be arranged into 6,000,000 different structures. This is in sharp contrast to a tri-peptide composed of the 20 different common amino acids that can be arranged in only 8,000 (20^3) different ways. Therefore, in this simple example, carbohydrates have a greater information storage capacity, by an order of magnitude of three, compared to proteins.

Of the nine sugars commonly found in membranes (Figure 7.3), one is highly unusual and supports a vast array of functions. Sialic acid [12,13] is a 9-carbon carboxylic acid that is an anion under physiological conditions. Actually sialic acid is a family of similar sugars, the most common being N-acetylneuraminic acid. Sialic acid was named by Gunnar Blix in 1952 after the Greek word for saliva. By the time Blix gave the name to sialic acid, he had been working on this family of compounds for more than 15 years. Sialic acid is widely found in gangliosides and glycoproteins of animal plasma membranes, but is also found in most other organisms including plants, fungi, and bacteria. Glycoproteins of cancer cells that can metastasize are particularly rich in sialic acid. The negative charge on sialic acid has many diverse functions, some advantageous, some not. For example, anionic sialic acid helps keep erythrocytes from clumping in the blood stream but is also the binding site and entry port for the Human Influenza Virus. Sialic acid-rich oligosaccharides help retain waters close to the cell surface and thus are involved in water uptake. Sialic acid also 'hides' mannose units from incorrectly reacting with components of complement. Erythrocytes have a 120-day life span after which they are targeted for destruction. Sialic acid plays an important role in this process. Aged erythrocytes are known to lose sialic acid from their membranes, exposing the remaining neutral sugars to host antibodies. After antibody binding, the senescent cell is removed from circulation by the reticuloendothelial (RES) system, primarily within the spleen. This mechanism is responsible for removing one million old erythrocytes per second from the blood.

B. GLYCOLIPIDS

In animals the major glycolipids are sphingolipids (discussed in Chapter 5 and [14−16]). The glycolipids include cerebrosides that have only one sugar attached (either glucose or galactose), globosides that have either a di-, tri-, or tetra-saccharide attached, but do not contain sialic acid and gangliosides that have complex oligosaccharides attached that includes at least one sialic acid residue.

The name ganglioside was coined by E. Klenk in 1942 for lipids isolated from ganglion cells of the brain. In fact gangliosides can amount to 6% of brain lipid weight where they constitute 10−12% of the total lipid content of neuronal membranes (20−25% of the membrane outer leaflet lipids). More than 60 gangliosides that differ in number and arrangement of the sugars, particularly sialic acid, have now been identified. In the plasma membrane outer leaflet, glycolipids accumulate into lipid rafts (see Chapter 8) and therefore are involved in cell signaling. They serve as receptors of signaling proteins including interferon, epidermal growth factor, nerve growth factor, and insulin. In addition, gangliosides bind specifically to viruses and to various bacterial toxins, such as those from botulism, tetanus, and cholera. It has been proposed that toxins utilize the gangliosides to hijack an existing retrograde transport pathway from the plasma membrane to the endoplasmic reticulum. Since binding to cholera toxin is the primary experimental marker used to identify lipid rafts on the surface of plasma membranes, it may be through lipid rafts that these toxins function. Due to information stored within the 'carbohydrate code' and important functions of sialic acid, gangliosides are also intimately involved in complex cell−cell recognition and serve as essential antigens in immunology. Gangliosides are not found outside of the animal kingdom.

Malfunctions in the enzymatic degradation of sphingosine-containing glycolipids, particularly the gangliosides, have been linked to a family of devastating, incurable genetic diseases that are grouped together as lysosomal storage diseases (LSDs) [17]. These diseases all result when a specific lytic enzyme is defective or missing, resulting in abnormal accumulation of the enzyme's substrate in the lysosomal membrane. One series of these diseases results from the failure to degrade the ganglioside: $GM_1 \rightarrow GM_2 \rightarrow GM_3$ (Figure 7.5)

The best known disease involving failure to degrade a ganglioside (generally known as gangliosidoses or sphingolipidoses) is Tay-Sachs Disease [18]. Tay-Sachs was the first gangliosidosis identified (in 1881) and results from a genetic mutation that fails to produce the enzyme hexoseaminidase A that normally would convert $GM_2 \rightarrow GM_3$ + galactosamine (Figure 7.5). As a result, GM_2 accumulates in the lysosomal membrane producing a plethora of eventually fatal complications. In GM_2, G stands for glycoside, M for mono (or one) sialic acid and 2 indicates it is the second mono-sialo ganglioside ever characterized. Accumulation of GM_2 occurs primarily in the lysosomes of tissues that are normally enriched in gangliosides (neurons and brain) resulting in severe neurologic symptoms and death before age 5. Symptoms include seizures, mental retardation, paralysis, blindness, extreme sensitivity to noise and the appearance of an unusual cherry-red macular spot. The trait is carried by

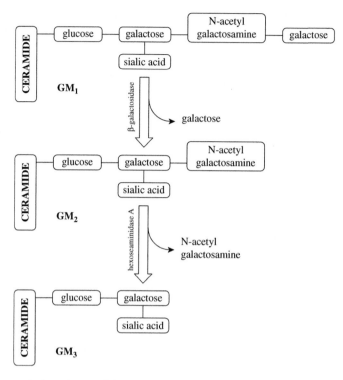

FIGURE 7.5   Enzymatic degradation of the ganglioside GM$_1$ → GM$_2$ → GM$_3$.

1/27 of normal adults of Eastern European (Ashkenazi) Jewish origin and is 100 times more prevalent in Jews than other populations.

Approximately 50 lysosomal storage diseases have now been identified, all resulting in the accumulation of unhydrolyzed membrane sphingolipids. These relatively rare, incurable genetic diseases occur in less than 1 in 100,000 births. They primarily affect children, resulting in premature deaths, often within a few months or years of birth. Table 7.1 lists a few of the better known LSDs.

C. GLYCOPROTEINS

Most plasma membrane proteins have attached sugars, all of which face the outside. On the plasma membrane outer surface there is a higher percentage of proteins with attached sugars (glycoproteins) than lipids with attached sugars (glycolipids). Glycoproteins [9,10] are ubiquitous in nature, although they are relatively rare in bacteria. The sugars are attached co- or post-translationally to a protein through either the nitrogen of asparagine or the oxygen of serine, threonine, hydroxylysine, or hydroxyproline via a process known as glycosylation (Figure 7.6). Attachment to asparagine is known as N-glycosylation and to serine, threonine, hydroxylysine, or hydroxyproline as O-glycosylation. N- and O-glycosylations

TABLE 7.1   Some Better Known Lysosomal Storage Diseases (LSDs) with the Missing
Enzyme and Accumulated Sphingolipid.

Disease	Missing enzyme	Accumulating sphingolipid
Fabrey	α-Galactosidase A	Gb3
Gauchers	β-Glucosidase	Glucoceramide
GM$_1$ Gangliosidoses	β-Galactosidase	GM$_1$
Hunter	Iduronate-2-sulfatase	Mucopolysaccharides
Krabbe	Galactosylceramidase	Galactoceramide
Niemann-Pick	Sphingomyelinase	Sphingomyelin
Pompe	α-Glucosidase A	Gb3
Sandhoff	Hexoseaminidase A & B	GM$_2$
Tay-Sachs	Hexoseaminidase A	GM$_2$

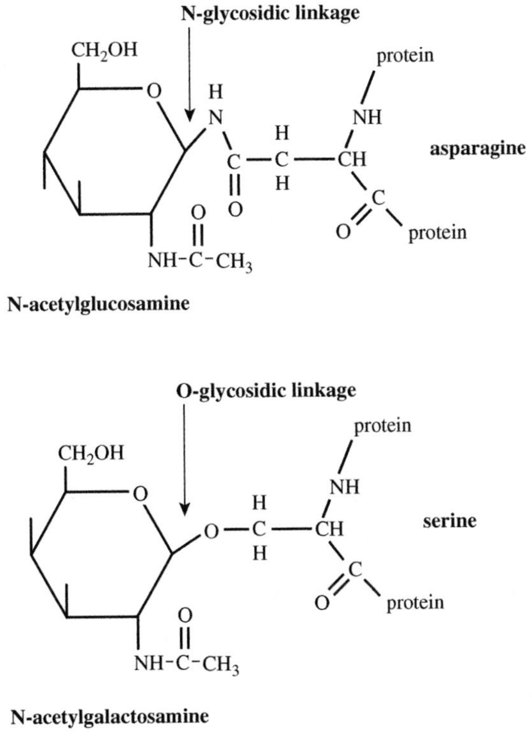

FIGURE 7.6   Attachment of sugars to proteins occurs at either an asparagine (the sugar is N-linked) or to serine, threonine, hydroxylysine, or hydroxyproline (the sugar is O-linked).

are very different. The sugar directly bound to asparagine is N-acetylglucosamine. O-linked sugars tend to be shorter (mono-, di- or tri-saccharides) and more diverse than are the N-linked sugars that contain much longer polysaccharides. For most membrane glycoproteins, N-linked sugars are more common on dynamic enzymatic proteins while O-linked sugars are predominant on structural proteins. Since a large protein will normally have numerous possible amino acids that could theoretically have sugar attachments, there are a number of different locations where sugars will be found. In addition there may be a variety of different sugars that can be attached at a single location. This complexity leads to what is known as *microheterogeneity*. For example, ovalbumin has only one site of glycosylation, yet more than a dozen different attached sugars have been identified at this site.

In the late 1990s, computer-based sequencing searches were applied to membrane glycosylation by Nathan Sharon and colleagues [19,20]. In one study of known membrane proteins, the question was addressed of how common glycoproteins are and how the sugars are attached to the proteins. Complete sequences of 1823 integral membrane proteins with extracellular features were identified. Of these, 1676 glycoproteins (92% of the total integral proteins) were identified, and of these 1630 or 97.3% were N-glycosylated. This leaves only 46 (2.8%) possible O-glycosylated proteins. Not glycosylated were 116 multiple span proteins, leaving only 14 possible single span proteins, at most, that were not glycosylated. From this study, two major conclusions can be drawn. The vast majority of plasma membrane integral proteins are glycosylated and, of these, there are many times more N-glycosylations than O-glycosylations.

D. GPI-ANCHORED PROTEINS

A highly unusual family of membrane integral proteins is GPI (glycosylphosphatidylinositol)-anchored (Chapter 6, and [21,22]). These proteins are all anchored to the membrane outer leaflet and have no trans-membrane segment. Therefore GPI-anchored proteins are also ecto-proteins. Membrane anchorage is primarily through the two acyl chains of PI and so binding is much weaker than trans-membrane proteins. GPI-anchored proteins are linked at their C-terminus to the membrane PI via a phosphoethanolamine to a tetrasaccharide composed of three mannoses and one glucosamine. The structure is shown in Figure 6.13.

GPI-anchored proteins are found in most organisms, but not bacteria. Although widely distributed, GPI-anchored proteins are not abundant. In yeast, where all GPI proteins are known, only 1 in 400 proteins have GPI-anchors and even these are very similar to one another. In man, the 45 known GPI-anchored proteins are most abundant in neurons where they appear to be concentrated in lipid rafts (see Chapter 8). Here they have functions in receptor-mediated cell signaling pathways. Being outer surface proteins, they also function in cell-adhesion and as cell surface antigens. GPI-anchored proteins can be released from the cell surface by phospholipase C which also produces diacylglycerol and by phospholipase D which generates PA. Diacylglycerol and PA are both highly reactive, membrane-bound compounds (see Chapter 5).

Several human genetic disorders have been linked to defects in GPI-anchored proteins. Although the most common of these is paroxysmal nocturnal hemoglobinuria (UPNH), the best known is Marfan syndrome, the disorder Abraham Lincoln was for many years believed to have had (see Chapter 6). In addition, it is now believed that all prion proteins are GPI-anchored. Prions are the infectious agent responsible for a number of incurable brain or neuronal diseases including 'mad cow disease' and, in humans, Creutzfeldt–Jakob Disease.

SUMMARY

In addition to lipids and proteins a third major component, carbohydrates, exists primarily on the outer surface of the cell plasma membrane where they can be anchored to lipids (glycolipids) and proteins (glycoproteins). Carbohydrates are highly water-soluble and so are never found in a membrane's hydrophobic interior. Their main function is to interact with the cell's external environment. The amount of carbohydrate on a membrane protein varies from 0 to >85 weight percent. There are 9 basic sugars commonly found attached to membranes as either single sugars or short, <15 unit, polysaccharides and are always located on the extra-cellular membrane leaflet. One sugar, sialic acid, is highly unusual, being a 9-carbon carboxylic acid that is an anion under physiological conditions. In animals the major glycolipids are sphingolipids. Sphingolipid (ganglioside) sugar metabolism disorders constitute the approximately 50 fatal lysosomal storage diseases. Glycosylphosphatidylinositol (GPI)-anchored proteins are a very different type of glycoprotein.

So far we have discussed the major components of membranes - lipids, proteins and carbohydrates. Chapter 8 will assimilate these components into working models for membrane structure, covering lipid bilayers to lipid rafts.

References

[1] Nelson DL, Cox MM. Lehninger Principles of Biochemistry. 5th ed. New York, NY: W Freeman; 2008.

[2] Stick RV, Williams S. Carbohydrates: The Essential Molecules of Life. 2nd ed. Elsevier; 2009.

[3] Wang PG, Bertozzi CR, editors. Glycochemistry – Principles, Synthesis and Applications. Marcel Dekker; 2001.

[4] Fraser-Reid BO, Tatsuta K, Thiem J, et al. editors. Glycoscience – Chemistry and Chemical Biology. 2nd ed. Springer; 2008.

[5] Reitsma S. The endothelial glycocalyx: composition, functions, and visualization. Eur J Physiol 2007;454:345–59.

[6] The AOCS Lipid Library. Complex Glycerolipids, lipidlibrary.aocs.org/lipids/complex.html; 2011.

[7] Hirabayashi Y, Ichikawa S. Roles of glycolipids and sphingolipids in biological membrane. In: Fukuda M, Hindsgaul O, editors. The Frontiers in Molecular Biology Series. IRL Press at Oxford Press; 1999. p. 220–48.

[8] Moss GP. IUPAC-IUB Joint Commission on Biochemical Nomenclature (JCBN). Nomenclature of Glycolipids, http://www.chem.qmul.ac.uk/iupac/misc/glylp.html; 1997.

[9] Kornfeld R, Kornfeld S. Comparative aspects of glycoprotein structure. Ann Rev Biochem 1976;45:217–38.

[10] Tauber R, Reutter W, Gerok W. Role of membrane glycoproteins in mediating trophic responses. Gut 1987;28(Suppl.):71–7.

[11] Laine RA. Information capacity of the carbohydrate code. Pure Appl Chem 1997;69:1867–73.

[12] Schauer R. Chemistry, metabolism, and biological functions of sialic acids. Adv Carbohydr Chem Biochem 1982;40:131–234.

[13] Schauer R. Sialic acids and their role as biological masks. Trends Biochem Sci 1985;10:357–60.

[14] AOCS. The Lipid Library. Sphingolipids. An Introduction to Sphingolipids and Membrane Rafts; 2013.

[15] Dickson RC. Sphingolipid functions in *Sacromyces cerevisiae*. Annu Rev Biochem 1998;67:27–48.

[16] Gunstone F. Fatty Acid and Lipid Chemistry, p. 43–44. Blackie Academic and Professional; 1996.

[17] Futerman AH, van Meer G. The cell biology of lysosomal storage disorders. Nat Rev Mol Cell Biol 2004;5:554–65.

[18] National Institutes of Neurological Disorders and Stroke (NINDS). National Institutes of Health. NINDS Tay-Sachs Disease Information Page, www.ninds.nih.gov/disorders/taysachs/taysachs.htm.

[19] Sharon N, Lis H. Glycoproteins: structure and function. In: Gabius H-J, Gabius S, editors. Glycosciences-Status and Perspectives. Weinheim, Germany: Chapman and Hall; 1997. p. 123.

[20] Apweiler R, Hermjakob H, Sharon N. On the frequency of protein glycosylation, as deduced from analysis of the SWISS-PROT database. Biochim Biophys Acta 1999;1473:4–8.

[21] Ferguson MAJ, Williams AF. Cell-surface anchoring of proteins via glycosyl-phosphatidylinositol structures. Annu Rev Biochem 1988;57:285–320.

[22] Ikezawa H. Glycosylphosphatidylinositol (GPI)-anchored proteins. Biol & Pharm Bull 2002;25:409.

[23] Neu J, Hilton B. Update on host defense and immunonutrients. Clin Perinatol 2002;29(1):41–64.

From Lipid Bilayers to Lipid Rafts

OUTLINE

A. Development of Membrane Models   117

B. The Danielli-Davson Pauci-Molecular Model   118

C. The Robertson Unit Membrane Model 119

D. Benson and Green's Lipoprotein Subunit Models   120

E. The Singer-Nicolson Fluid Mosaic Model   122

F. Simons' Lipid Raft Model   125

Summary   127

References   128

A. DEVELOPMENT OF MEMBRANE MODELS

By 1935 all of the basic elements were in place for someone to propose a realistic model of a membrane (see Chapter 2, also [1]). Comparing cell permeability measurements with solute partition coefficients, Overton had suggested the lipid nature of a membrane and Gorter and Grendel had proposed that there was the precise amount of lipid in the erythrocyte membrane to exactly cover the cell surface twice. Thus, at least the erythrocyte membrane was a lipid bilayer. In fact, the then obscure Gorter-Grendel paper implied that the membrane was entirely lipid! Using trans-membrane electrical measurements, in 1925 Fricke correctly determined that the hydrophobic part of a membrane was ~33 Å thick. Also in 1925, Sumner purified the first protein, urease. The prevailing opinion at that time was that biological catalysts were small, hard to isolate organic molecules that were hidden in all the 'cellular junk'. Although it took almost a full decade for the importance of protein catalysts (enzymes) to be universally accepted, by 1935 it was obvious that proteins were to be a major player in understanding life. Proteins were soon found to be a major component of all membrane compositional analysis. One readily measurable membrane property, surface tension, was severely impacted by both membrane lipids and proteins. The surface tension of pure water is

72.8 dynes/cm at 25°C. Phospholipids reduce this to ~7 to 15 dynes/cm. Proteins further reduce the surface tension down to that of a biological membrane (~0.1 dynes/cm). The conclusion was that proteins must play an important role in membrane structure. However, by 1935 the only proteins that were understood to any extent were water-soluble globular proteins. Nothing was known about membrane proteins.

By 1935, questions about overlap between the related fields of biochemistry, protein structure and function, and membranes were starting to emerge. Although it was clear that different membranes supported different biochemical functions, it was not known whether the structures of different membranes were completely different, or if they had important features in common. In other words, are the various membranes more similar or different? It was clear that membranes were 'semi-permeable', being permeable to some things while impermeable to others. How was semi-permeability accomplished? Any proposed model for membrane structure would have a lot of things to consider [2].

The first realistic model of a biological membrane was that of James Danielli and Hugh Davson, initially proposed in 1935 [3]. Their Pauci-Molecular model was subsequently tweaked and altered as membrane science progressed [2]. The history of membrane models will be discussed around some major milestones, starting with the Danielli-Davson model and progressing through the Robertson Unit Membrane model, to the Singer Fluid Mosaic model, and finally to Simons' Lipid Raft model. Two misdirected models that do not employ a lipid bilayer, and so clearly fall off this major track, the Benson and Green Lipoprotein Subunit models, will also be discussed.

B. THE DANIELLI-DAVSON PAUCI-MOLECULAR MODEL

In 1935 Danielli and Davson proposed their model for a biological membrane [3]. Their Pauci-Molecular model was the first to employ a lipid bilayer. While the seminal experiment of Gorter and Grendel in 1925 [4] was the first to demonstrate the possibility of a lipid bilayer barrier (membrane) surrounding a cell, these authors did not actually propose a membrane model. Instead, ten years later Danielli and Davson used the lipid bilayer as the foundation for a new membrane model. Ironically the original 1935 Danielli-Davson paper did not even mention the Gorter-Grendel work! They realized that a membrane was not just lipid, as suggested by Gorter-Grendel, and somehow proteins had to be incorporated. Since at that time only globular proteins were known, Danielli and Davson simply attached globular proteins loosely to the lipid bilayer surface in the form of a protein–lipid–protein sandwich (Figure 8.1a). Note their original model had no membrane-penetrating protein. Later, Danielli, realizing the membrane had to be semi-permeable, added thin, peptide-lined trans-membrane channels (Figure 8.1b). One important feature of this simple model is that it was proposed to be the basic foundation of all biological membranes. In addition, at the heart of their model was the lipid bilayer. These concepts have remained intact through all subsequent revisions of the Pauci-Molecular model. The major problems with this first model were the failure to incorporate proteins into the membrane bilayer interior and the total lack of dynamics in the model. The Pauci-Molecular model is static. The model predicts that all membranes would be identical and fails to indicate how variety would be achieved, nor does the model take into account membrane asymmetry. Also, the globular protein coat

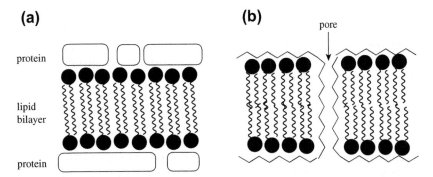

FIGURE 8.1 The Danielli-Davson Pauci-Molecular model. (a) shows the original model consisting of a lipid bilayer with globular proteins covering the surface. (b) shows a later revision of (a) with a trans-membrane peptide pore.

would prevent favorable interactions between the phospholipid head groups and water. While the Danielli-Davson Pauci-Molecular membrane model was a good start, and crucial parts of the model, particularly the lipid bilayer, are still invoked today, significant alterations were needed.

C. THE ROBERTSON UNIT MEMBRANE MODEL

In a 1957 meeting of the American Physiological Society, J. David Robertson proposed a slight modification of the Danielli-Davson model membrane [5,6]. He called his model the Unit Membrane model. Robertson was generally regarded as the premier electron micros-copist of cell membranes and electron micrographs formed the basis of his model. Upon staining membranes with $KMnO_4$, Robertson always observed what is often referred to as 'railroad tracks', two dark bands separated by a light band (Figure 8.2, [7]). Robertson defined the Unit Membrane as 'measuring ~75 Å across, consisting of two parallel dense lines

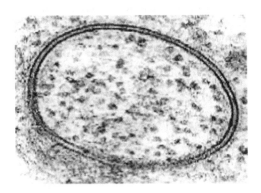

FIGURE 8.2 A typical electron micrograph of a plasma membrane demonstrating the 'railroad tracks' on which Robertson based his 'Unit Membrane' model [7]. The plasma membrane is observed to be two dark lines surrounding a light interior.

25 Å wide separated by a light zone ~25 Å wide'. According to the Unit Membrane model, all membranes have the same basic structure. They are composed of a lipid bilayer with protein monolayers attached covalently to both sides, very similar to the Danielli-Davson model. Whereas the Danielli-Davson model was non-committal concerning the thickness of a membrane, Robertson's model was not.

As with the Danielli-Davson model, Robertson's Unit Membrane model had its advantages and disadvantages. Robertson's EM (electron microscope) pictures allowed for a precise membrane measurement. Indeed, Robertson's EM images were the first direct observations of membrane structure. Robertson found the same tri-laminar structure ('railroad tracks') in the plasma membranes from a wide variety of cell types as well as from internal cellular membranes including those of the mitochondria, endoplasmic reticulum, and the nuclear envelope. Unfortunately the same flaws that eventually doomed the Danielli-Davson model also doomed the Robertson model. The Unit Membrane model is a static model and fails to appreciate either the dynamic aspects of membranes or the asymmetric nature of membrane structure. Membrane dynamics would have to await further advances in technology to address these problems. The eventual replacement of Robertson's use of $KMnO_4$ by gluteraldehyde and OsO_4 indicated that the intra-cellular membranes are not directly connected to one another, as Robertson predicted, but instead are connected by tiny vesicles. The newer methodology also showed that the plasma membrane has ties to the internal cytoskeleton. The Unit Membrane model was so similar to the Pauci-Molecular model that the two are often melded into one Danielli-Davson-Robertson membrane model.

D. BENSON AND GREEN'S LIPOPROTEIN SUBUNIT MODELS

Throughout the 1960s membrane science was still in its infancy and the most important missing aspects involved membrane asymmetry and dynamics. It should not be surprising that even the venerable lipid bilayer was in question by some. Two well-disseminated membrane models replaced the lipid bilayer with lipoprotein subunits (depicted in Figure 8.3). The Lipoprotein Subunit models were proposed by Andrew Benson (Figure 8.4, for photosynthetic thylakoids, [8]) and David Green (for mammalian mitochondria, [9]). These models, depicted in Figure 8.3, received considerable attention due to the stature of their inventors. Andrew Benson was a world-renowned plant physiologist from his early

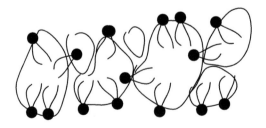

FIGURE 8.3 The Benson-Green Lipoprotein Subunit model membrane. Proteins are depicted as large, open sack-like structures while phospholipids are the small, black spheres with attached acyl chains. Note the complete absence of a lipid bilayer.

FIGURE 8.4   Andrew A. Benson (1917–). *Reproduced with permission [28].*

work on the Calvin-Benson cycle of photosynthetic dark reactions. David Green was a leading expert in the field of mitochondria, whose reputation was based on fracturing the mitochondrial electron transport/oxidative phosphorylation system into five isolatable complexes. Partial or complete mitochondrial oxidative phosphorylation could be reconstituted by simply mixing the appropriate complexes. Therefore both Benson and Green were experts on highly specific and sophisticated membranes, Benson the photosynthetic thylakoid, and Green the bioenergetic inner mitochondrial membrane. Both of these membranes are at the upper range of membrane protein/lipid ratios. Therefore, in these complex membranes the lipid bilayer would likely be reduced. Both the Benson and Green models were essentially a lipoprotein monolayer devoid of a lipid bilayer (Figure 8.3).

The Green model (also called the Protein Crystal model) was particularly well described. In his 1970 *Proceedings of the National Academy of Sciences* paper [9], Green states, 'The essential principle underlying this model is that when membrane proteins polymerize, the points of contact between proteins are few, and cavities lined with predominantly non-polar amino acids are formed. Phospholipid molecules become oriented with the fatty chains inserted into the cavities while the polar heads remain on the surface of the membrane. This orientation applies to both faces of the membrane continuum. All the lipid known to be present in membranes can be accommodated in this manner.'

Therefore, no lipid bilayer was required. While this proposal was quite interesting, it failed to explain membrane structure for most membranes, ignored the rapidly accumulating data on the lipid bilayer-based models (for example, lateral diffusion discussed in Chapter 9), and ignored the likelihood that a single mutation could destroy the entire membrane. These lipoprotein-based membrane models disappeared with the advent of Singer's Fluid Mosaic model. Before they vanished, however, they did clearly indicate a major problem with all membrane models, the lipid bilayer 'sea' must be very limited in scope and must be very crowded (see Chapter 11, [10]).

E. THE SINGER-NICOLSON FLUID MOSAIC MODEL

By 1972 information concerning membrane structure was rapidly accumulating from a wide variety of new approaches. It was becoming clear that the Danielli-Davson-Robertson model for membranes was inadequate. Of particular concern was the failure to properly account for membrane structure and topology (asymmetry), membrane protein diversity, and the lack of appreciation for membrane dynamics. The major positive aspect of the Danielli-Davson-Robertson model was having a lipid bilayer as its centerpiece.

In what was to become an iconic paper on membranes, in 1972 Singer (Figure 8.5) and Nicolson [11] proposed a new membrane model they called the Fluid Mosaic model. With some later modifications, this model is the one still *in vogue* today. In this chapter, basic components of the Fluid Mosaic model will be presented, while experiments on membrane physical properties supporting the model will be discussed in Chapter 9. In their paper, Singer and Nicolson defined a membrane as 'an oriented, two-dimensional viscous solution of amphipathic proteins (or lipoproteins) and lipids in instantaneous thermodynamic equilibrium'. Their model retained the lipid bilayer but substantially altered many other aspects of the static Danielli-Davson-Robertson model.

A major problem with Robertson's Unit Membrane model was assigning the two dark bands he noted in all of his EM pictures to continuous protein coats on both sides of a white lipid bilayer core. We now know that shielding the polar lipid headgroups from water would substantially diminish bilayer stability. Also, these protein coats were symmetrically distributed on each side of the membrane and did not span the hydrophobic core. Therefore the Unit Membrane model could not possibly account for trans-membrane vectorial (directional) biochemical activity. The Fluid Mosaic model took into account the reality of membrane protein structure (discussed in Chapter 6). Both peripheral and integral proteins were proposed to be distributed asymmetrically across the membrane (Figure 8.6). Carbohydrates

FIGURE 8.5   S.J. Singer (1924–). *Courtesy of Oregon State University.*

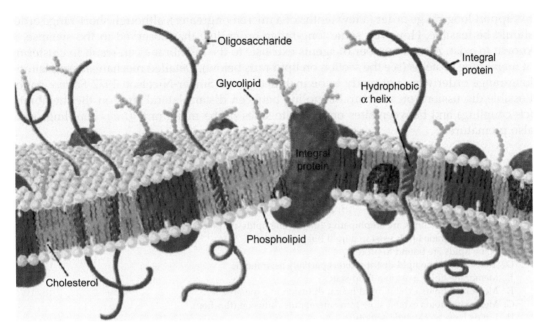

FIGURE 8.6   The Singer-Nicolson Fluid Mosaic model for membrane structure. *Reprinted with permission [29].*

were attached to proteins (glycoproteins) and lipids (glycolipids), but were found only on the outer membrane surface. Integral proteins were described as being 'globular and amphipathic' and 'having a significant fraction of its volume embedded in the membrane'. This early description of protein structure was later modified to include trans-membrane α-helical and β-barrel spans (discussed in Chapter 6). Membrane components were distributed in a 'mosaic structure' where both proteins and lipids were dispersed inhomogeneously resulting in lateral and trans-membrane patches called domains [12].

A second important element Singer and Nicolson added to their model can be loosely described as 'membrane dynamics'. The Danielli-Davson-Robertson model was essentially a frozen picture. The Fluid Mosaic model recognized for the first time that a biological membrane is dynamic, fluid, and ever changing. Although identifying the vast diversity of membrane lipids, proteins, and carbohydrates is by itself a difficult task, it is membrane dynamics that presents the most challenges. Dynamics involves how the various membrane components interact with one another. A variety of techniques are required to cover the vast size and time domains encountered in membrane studies. Various aspects of membrane dynamics are discussed in Chapter 9.

The original Fluid Mosaic model predicted totally free lateral diffusion of both lipids and proteins. This concept was later modified by proposing limited diffusion for some proteins due to their interactions with the essentially immobile cytoplasmic proteinaceous mesh known as the cytoskeleton [13]. The Fluid Mosaic model was the first to recognize both membrane fluidity and lateral mobility. The existence of an extensive lipid bilayer matrix and free lateral diffusion in the bilayer presented a dilemma. A bilayer should not be able to support long-range order (a few tenths of a micron or greater), although short-range order should be feasible. However, some long-range order, like that observed in the synapse, is known to exist. Also, a number of agents extrinsic to the membrane can result in clustering of membrane proteins (see the section on lipid rafts below). Detailed mechanisms explaining long-range order were not ready to be included in the Singer-Nicolson 1972 *Science* paper. Possible discussions on structural coupling between distant lateral sites on the membrane (*cis* coupling) and between sites on opposite sides of the membrane (*trans* coupling) were also premature.

TABLE 8.1   Salient Features of the Fluid Mosaic Model.

I. Membrane Lipids:
  A. Molecular structures are amphipathic (membrane lipids are polar).
  B. Most membrane lipids exist in a lipid bilayer matrix.
  C. Some lipids are bound to protein.
  D. Trans-membrane lipid distribution is partially asymmetric.
  E. Membrane lipids exist in a fluid state.
  F. Membrane lipids exhibit rapid lateral diffusion in a membrane.
  G. Membrane lipids exhibit slow trans-membrane diffusion (flip-flop).
  H. Lipids have no catalytic activity.

II. Membrane Proteins:
  A. Membrane protein molecular structure is amphipathic.
  B. Membrane protein distribution is totally asymmetric.
  C. Membrane protein lateral diffusion rate varies from rapid to immobile.
  D. Absolutely no trans-membrane protein diffusion (flip-flop) occurs.
  E. Membrane proteins exhibit substantial diversity (biochemical activities).

Protein types
1. Peripheral (essentially globular)
  a. bound to lipid (electrostatic)
  b. bound to protein (electrostatic)
2. Amphitropic
3. Integral: Endo, Ecto, and Trans-membrane
  a. α-Helix: Single Span, Multiple Span
  b. β-Barrel
  c. Lipid Linked
    i. Myristoylated, Palmitoylated, Prenylated
    ii. GPI-anchored
4. Glycoproteins (general type of protein that has attached sugars)

III. Membrane Carbohydrates:
  A. Totally water-soluble — no interaction with membrane interior.
  B. Glycolipids.
  C. Glycoproteins.
  D. Membrane distribution is totally asymmetrical.
  E. Always found on membrane outer surface.
  F. Lateral diffusion is determined by the attached lipid or protein.
  G. No trans-membrane diffusion (flip-flop) occurs.
  H. Carbohydrates have no catalytic activity.

Table 8.1 lists the salient features of the Fluid Mosaic model. Some of these features (e.g. β-barrels, lipid-linked proteins) were discovered after the Fluid Mosaic model was first formulated in 1972.

F. SIMONS' LIPID RAFT MODEL

In the original 1972 Fluid Mosaic model, Singer and Nicolson did not postulate lateral heterogeneity into functional membrane domains although it was generally accepted that membrane phospholipids and proteins were probably heterogeneously distributed. However, in the 1970s, reports were starting to emerge that membrane 'lipid clusters' may be common features of biological membranes. In a 1982 paper in the *Journal of Cell Biology*, Karnovsky and Klausner formalized the concept of lipid domains in membranes [14]. Interestingly, one type of lipid cluster that kept arising was that of cholesterol and sphingolipids [15,16]. Indeed, protein-free lipid membrane models showed that cholesterol has a higher affinity for SM (sphingomyelin) than it does for other phospholipids [16]. Although lipid phase separations in membranes were starting to achieve acceptance, their biological relevance awaited a link to a biological function. The link proved to be spectacular when cholesterol- and sphingolipid-rich microdomains were closely associated with essential cellular signaling events. However, the real interest in these microdomains did not come until after they received the very catchy and descriptive name 'Lipid Raft'. This critical link was made by Kai Simons (Figure 8.7) in a 1997 paper in *Nature* [17]. Membrane studies would be changed forever.

Since 1997, lipid rafts have generated countless primary research papers and many excellent reviews (e.g. [18–25]. However, one wonders if these microdomains had been more correctly called 'phase separated cholesterol- and sphingolipid-rich, liquid ordered microdomains' instead of 'Lipid Rafts' whether they would have achieved such notoriety!

Lipid rafts are cholesterol- and sphingolipid-rich, liquid ordered (l_o) state platforms (microdomains) that float in a non-raft liquid disordered matrix (Figure 8.8.). The l_o state has properties midway between those of a liquid crystalline (disordered, fluid) and gel

FIGURE 8.7   Kai Simons (1938–). *Courtesy of BMBF/Unternehmen Region.*

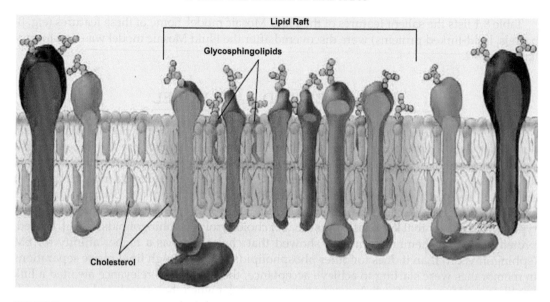

FIGURE 8.8   Simons' Lipid Raft model. *Reprinted with permission [30].*

(ordered, solid) state. These states are discussed in detail in Chapter 9. Briefly, the acyl chains of lipid rafts are more saturated than are the acyl chains of the surrounding membrane matrix and so pack more tightly, producing the l_o state. It is believed that lipid rafts are held together by cholesterol, reflecting the sterol's high affinity for sphingolipids. Typically, raft cholesterol levels are twice that found in membrane non-raft regions, while the SM concentration is ~50% higher and the PC level is reduced in proportion to the increase in SM. As a result of the tight packing, lipid raft portions of the membrane are thicker than the surrounding, more loosely packed membrane, and so protrude above the surrounding membrane surface. Lipid rafts are believed to house a variety of raft-characteristic proteins that are intimately involved in cell signaling events. Often these proteins affect the phosphorylation state of signaling proteins that can be modified by local kinases and phosphatases to affect down-stream signaling.

An original hallmark of lipid rafts is their ability to be isolated from the remaining membrane matrix by extraction in cold (4°C) non-ionic detergents (e.g. Triton X-100 or Brij-98). Under cold, detergent conditions, the membrane matrix is solubilized while the lipid rafts are insoluble (they are detergent resistant). Because of their composition and detergent resistance, lipid rafts are also referred to as detergent-insoluble glycolipid-enriched complexes (GEMs or DIGs) or detergent resistant membranes (DRMs). It is believed, although by no means proven, that lipid rafts are normally extremely small (~10–200nm) but their size can be greatly enhanced by various natural and artificial cross-linking agents.

In a 2006 Keystone Symposium on Lipid Rafts and Cell Function [26], lipid rafts were defined as 'small (10–200nm), heterogeneous, highly dynamic, sterol- and sphingolipid-enriched domains that compartmentalize cellular processes. Small rafts can sometimes be stabilized to form larger platforms through protein–protein interactions'.

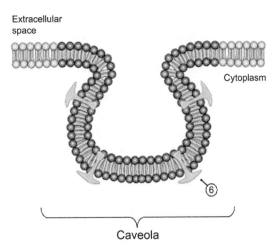

Extracellular
space

Cytoplasm

Caveola

FIGURE 8.9   Caveolae, a type of lipid raft. *Reproduced with permission [31]*

While the lipid raft story is compelling, it is wrought with potential problems. In fact rafts have occasionally been referred to as 'unidentified floating objects' or UFOs. The cold temperature detergent procedure is crude and very slow compared to the likely timescale of lipid raft stability, and is full of potential artifacts. The older raft isolation procedures are being improved and non-detergent methodologies have been developed (Chapter 13) [27]. The very small raft size, being below the classical diffraction limit of a light microscope, has made direct observation of rafts very difficult. Raft size also brings into question their stability. Even if lipid rafts exist, they may only occur on a time scale that is irrelevant to biological, diffusion-controlled processes.

One question that is not close to resolution is how many types of microdomains constitute a functional membrane. While lipid rafts have garnered almost all of the attention, it seems likely that there probably exist many types of undiscovered non-raft microdomains. Even the lipid rafts may be composed of countless sub-raft domains. Currently two basic types of rafts have been identified, planar lipid rafts (also referred to as non-caveolar, or glycolipid rafts) and caveolae. Planar rafts are continuous with the membrane surface, while caveolae are flask-shaped invaginations (Figure 8.9). Caveolae are easily distinguished from planar rafts by their unique flask-shape and the presence of a characteristic protein, caveolin. In contrast, planar rafts contain flotillin proteins. The function of both caveolin and flotillin is to attract and retain signaling molecules into rafts. Other distinct types of raft and non-raft microdomains will undoubtedly be discovered in the future.

SUMMARY

By 1935 basic elements were in place to propose a realistic membrane model. This was accomplished by Danielli and Davson in their Pauci-Molecular model, based on a lipid bilayer that had proteins uncomfortably attached in the form of a protein—lipid—protein sandwich. A slight improvement on this model was proposed by Robertson in his 1957 Unit Membrane model. His model was based on electron micrographs, the first direct observations of membrane structure, and featured structures that resembled 'railroad tracks'. In the 1960s, Benson (for the photosynthetic thylakoid) and Green (for the mitochondrial inner membrane) proposed lipid bilayer-free models, called the Lipoprotein Subunit model. These early membrane models did not appreciate membrane asymmetry or dynamics. In 1972 Singer and Nicholson proposed their Fluid Mosaic model that retained the lipid bilayer, but accounted for previous model shortcomings, recognizing membrane fluidity and lateral mobility. An extension of this model was provided by Kai Simons' 1997 Lipid Raft model that proposed sphingolipid- and cholesterol-rich, liquid ordered, signaling microdomains.

The next three chapters will discuss various membrane physical properties. These properties are essential to understanding how membranes work. Chapter 9 will discuss membrane thickness, bilayer stability, membrane protein, carbohydrate and lipid asymmetry, lipid trans-membrane movement (flip-flop), membrane lateral diffusion, membrane-cytoskeleton interaction, lipid melting behavior, and membrane fluidity.

References

[1]   Eichman P. From the Lipid Bilayer to the Fluid Mosaic: A Brief History of Model Membranes. SHiPS Resource Center for Sociology, History and Philosophy in Science Teaching.
[2]   De Weer P. A century of thinking about cell membranes. Annu Rev Physiol 2000;62:919—26.
[3]   Danielli JF, Davson H. A contribution to the theory of permeability of thin films. J Cell Comp Physiol 1935;5:495—508.
[4]   Gorter E, Grendel F. On bimolecular layers of lipids on the chromocytes of the blood. J Exp Med 1925;41:439—43.
[5]   Robertson JD. The cell membrane concept. J Physiol 1957;140:58P—9P.
[6]   Robertson JD. The molecular structure and contact relationships of cell membranes. Prog Biophys Biophys Chem 1960;10:344—418.
[7]   Sandraamurray: http://en.wikipedia.org/wiki/File:Annular_Gap_Junction_Vesicle.jpg; public domain.
[8]   Benson AA. On the orientation of lipids in chloroplasts and cell membranes. J Am Oil Chem Soc 1966;43:265—70.
[9]   Green DE, Vanderkooi G. Biological membrane structure, I. The protein crystal model for membranes. Proc Natl Acad Sci USA 1970;66:615—21.
[10] Ellis RJ. Macromolecular crowding: obvious but unappreciated. TRENDS Biochem Sci 2001;26:597—604.
[11] Singer SJ, Nicolson GL. The fluid mosaic model of the structure of cell membranes. Science 1972;175:720—31.
[12] Devaux PF, Morris R. Transmembrane asymmetry and lateral domains in biological membranes. Traffic 2004;5:241—6.
[13] Doherty GJ, McMahon HT. Mediation, modulation, and consequences of membrane-cytoskeleton interactions. Annu Rev Biophys 2008;37:65—95.
[14] Karnovsky MJ, Kleinfeld AM, Hoover RL, Klausner RD. The concept of lipid domains in membranes. J Cell Biol 1982;94:1—6.
[15] Estep TN, Mountcastle DB, Barenholz Y, Biltonen RL, Thompson TE. Thermal behavior of synthetic sphingomyelin—cholesterol dispersions. Biochemistry 1979;18:2112—7.
[16] Demel RA, Jansen JW, van Dijck PW, van Deenen LL. The preferential interaction of cholesterol with different classes of phospholipids. Biochim Biophys Acta 1977;465:1—10.
[17] Simons K, Ikonen E. Functional rafts in cell membranes. Nature 1997;387:569—372.
[18] Brown DA, London E. Functions of lipid rafts in biological membranes. Annu Rev Cell Develop Biol 1998;14:111—36.
[19] Simons K, Toomre D. Lipid rafts and signal transduction. Nat Rev Mol Cell Biol 2000;1:31—9.

[20] Edidin M. Shrinking patches and slippery rafts: Scales of domains in the plasma membrane. Trends Cell Biol 2001;11:492−6.

[21] Edidin M. The state of lipid rafts: From model membranes to cells. Annu Rev Biophys Biomol Struct 2003;32:257−83.

[22] Pike LJ. Lipid rafts: heterogeneity on the high seas. Biochem J 2004;378:281−92.

[23] Simons K, Vaz W. Model systems, lipid rafts, and cell membranes. Annu Rev Biophys Biomol Struct 2004;33:269−95.

[24] Pike LJ. The challenge of lipid rafts. J Lipid Res 2009;50:S323−8.

[25] Lingwood D, Simons K. Lipid rafts as a membrane-organizing principle. Science 2010;327:46−50.

[26] Pike LJ. Rafts defined: a report on the Keystone symposium on lipid rafts and cell function. J Lipid Res 2006;47:1597−8.

[27] Smart EJ, Ying Y-S, Mineo C, Anderson RGW. A detergent-free method for purifying caveolae membrane from tissue culture cells. Proc Natl Acad Sci USA 1995;92:10104−8.

[28] Govindjee. Celebrating Andrew Alm Benson's 93rd birthday. Photosynthesis Research, 2010;105(3):201−8.

[29] Lotchi bio: http://lotchibio05.blogspot.com/2010/12/fluid-mosaic-model.html.

[30]   Davis, A. Cells 101: Business Basics in Inside the Cell, National Institute of General Medical Sciences. 2005; 28−9

[31] Sebastião AM, Colino-Oliveira M, Assaife-Lopes N, Dias RB, Ribeiro JA. Lipid rafts, synaptic transmission and plasticity: Impact in age-related neurodegenerative diseases. Neuropharmacology, 2013;64:97−107.

Basic Membrane Properties of the Fluid Mosaic Model

OUTLINE

A. Size and Time Domains   132

B. Membrane Thickness   134
   Planar Bimolecular Lipid Membranes   134
   X-Ray Diffraction   136

C. Membrane Asymmetry   137
   Protein Asymmetry   137
   Carbohydrate Asymmetry   139
     Lectins   139
     Lipid Asymmetry   139
   Chemical Modification   140
     Enzymatic Modification   141
   Erythrocyte Lipid Asymmetry   143

D. Lateral Diffusion   144
   The Frye-Edidin Experiment   145
   Fluorescence Recovery after
   Photobleaching (FRAP)   145
   Single Particle Tracking (SPT)   148
   Membrane Lateral Diffusion Rates:
   Conclusions   150

E. Lipid Trans-membrane Diffusion
(Flip-Flop)   151
   Chemical Method to Determine Flip-Flop
   Rates   152
   Phospholipid Exchange Protein Method
   to Determine Flip-Flop Rates   154
     Flippase, Floppase, and Scramblase   156
     Some General Conclusions from
     Flip-Flop Studies   157

F. Lipid Melting Behavior   157
   Differential Scanning Calorimetry   159
   Fourier Transform Infrared (FT-IR)
   Spectroscopy   163
   Fluorescence Polarization   164

G. Membrane 'Fluidity'   170

Summary   172

References   172

Every aspect of membrane structural studies involves parameters that are very small and very fast. As a result a variety of highly specialized and esoteric biophysical methodologies must be employed, often simultaneously. It is the nature of the highly sophisticated techniques used to investigate membrane structure that unfortunately forms an almost insurmountable impediment between physical and biological approaches to membrane studies. This chapter will investigate several fundamental membrane properties using specific examples taken from the research literature. Emphasis will be placed on how the technique was employed to address a particular question. It is not the objective of this chapter, or this book, to discuss the intricate details of each technique, but rather to understand what questions the technique can address. In other words, the emphasis will be placed on application and not detailed methodology.

A. SIZE AND TIME DOMAINS

Generally parameters of interest to the study of biological membranes will range in size from microns (10^{-6} m) to Angstroms (Å, 10^{-10} m) and in time from milliseconds (10^{-3} sec) to

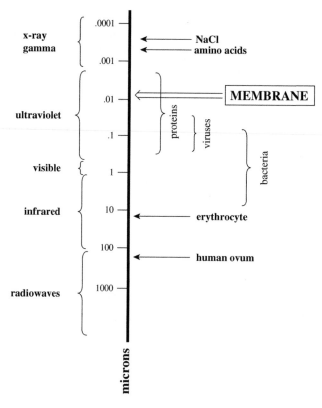

FIGURE 9.1   The electromagnetic spectrum and the size of various important biological components. Note the average thickness of a biological membrane is a little less than 10 nm or 100 Å.

picoseconds (10^{-12} sec). Both size and time ranges are so vast that multiple instrumentations must be employed. Size domains are demonstrated by use of the electromagnetic spectrum shown in Figure 9.1. The electromagnetic spectrum is the range of all possible electromagnetic radiation frequencies. It has no distinct beginning, nor does it have a distinct end. It covers wavelengths from the size of the universe down to a fraction of the size of an atom. The spectrum is continuous and infinite. However, for convenience, the spectrum is classified into several, overlapping ranges. At the long wavelength end are radiowaves, followed by infrared, the visible region, ultraviolet, X-rays, and at the shortest wavelengths, gamma rays. The shorter the wavelength, the higher is its frequency and energy. While short wavelength radiation has better resolving power, it is also a more destructive power. Long wavelength radiowaves can be used for some crude imaging such as ultrasound to safely image a fetus still in the womb, but would be of little use in imaging a membrane. Electromagnetic radiation between 380 nm and 760 nm can be detected by the human eye and so is referred to as visible light. Shorter wavelength blue light has better resolving power than does longer wavelength red light. Note, however, that even the wavelength of blue light is much longer (380 nm) than the thickness of a biological membrane. Even including proteins, membrane thickness is generally <10 nm. Therefore a biological membrane cannot be directly imaged with a light microscope, but must employ much shorter wavelength radiation, primarily destructive X-rays. It is unfortunate that non-harmful, long wavelength radiation has poor resolving power, while good resolution can only be achieved with biologically harmful short wavelength radiation.

Also complicating studies of biological membranes are the very fast times associated with many biological events (micro seconds to picoseconds). Figure 9.2 shows the time ranges

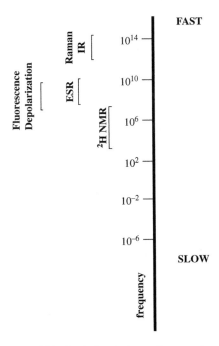

FIGURE 9.2   Time domains for several biophysical techniques. Frequencies are per second.

over which several important biophysical techniques, commonly employed in membrane studies, can be used. Of these techniques, the fastest is Raman = IR > ESR = Fluorescence Polarization > NMR. There are, however, modifications of each of these techniques that can substantially expand the normal range to much faster times. The technique must match the question being addressed. For example, take the case of one well-studied, fundamental membrane problem that is discussed in more detail in Chapter 10. Early ESR studies indicated that lipids in immediate contact with membrane proteins (termed annular lipids) have different properties than the bulk bilayer lipids. However, according to NMR, all lipids are the same and therefore annular lipids do not exist. The reason for this discrepancy is that ESR detects events that are ten times faster than can be observed by the slower NMR.

B. MEMBRANE THICKNESS

Although not fully appreciated at the time, the thickness of a biological membrane was accurately determined by Hugo Fricke in 1925 [1]. Using trans-membrane electrical measurements, Fricke correctly determined that the hydrophobic center of a membrane was ~33 Å thick. This was in close agreement with the early pioneers working on lipid monolayers, Rayleigh, Pokels, and Langmuir, who determined the thickness of a triolein lipid monolayer was ~13—16 Å (Chapter 2). Doubling this value to account for membrane lipids existing in bilayers, reported in 1925 by Gorter and Grendel, yields a correct hydrophobic membrane interior of ~26—32 Å. Fricke's failure to appreciate the importance of his own measurements stemmed from his lack of understanding of the bilayer. It was not until 1957, when Robertson imaged the membrane using electron microscopy (see Chapter 8), that the full membrane span of about 75 Å, including lipid bilayer and protein, was established.

Two other important methods that have been used to determine membrane thickness (planar bimolecular lipid membrane analysis and X-ray diffraction) are considered below.

Planar Bimolecular Lipid Membranes

For the decades between 1925 and 1960, the lipid bilayer was believed to be an essential feature of membranes. However, it was not clear if the lipid bilayer was stable enough to actually exist. Several investigators, including Irwin Langmuir in ~1937, tried but failed to make a lipid bilayer using the reverse process involved in making soap bubbles. Whereas a lipid bilayer is water — lipid bilayer — water, a soap bubble is air — inverse lipid bilayer — air (Figure 9.3). Fascination with soap bubbles is centuries old and its early scientific investigation is summarized in a classic 1911 book by C.V. Boys [2]. In 1961 Mueller and co-workers reported the first preparation of a large, stable lipid bilayer they termed the planar bimolecular lipid membrane or BLM [3]. They found that BLMs could be made from a variety of polar lipids dissolved in organic solvents. Originally these investigators used a chloroform/methanol/brain lipid extract since they were interested in the nerve process of excitability. The solution was spread over a submerged 1—2 mm hole in a Teflon cup (Figure 9.4) by use of a small artist's brush or later by a syringe. The hole was monitored by 90 degree reflected light through a small telescope. After application of the lipid solution, the hole appeared gray, but soon developed many bands of color, the result of partial

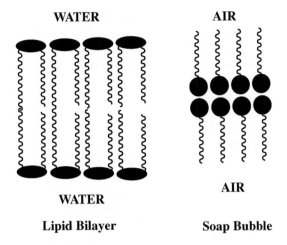

FIGURE 9.3 The lipid bilayer compared to a soap bubble. Both are made from amphipathic molecules, only their orientation is reversed.

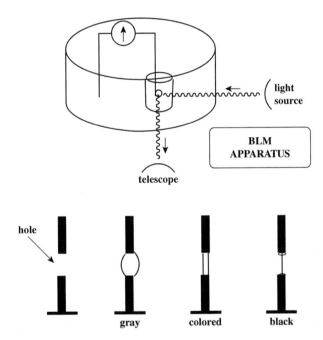

FIGURE 9.4 Apparatus for making the original BLM (top). Appropriate polar lipids are dissolved in an organic solvent and added across a small aperture in a Teflon cup (bottom). The aperture is observed using a small telescope by 90° reflected light. Initially the lipid solution appears gray. As the membrane thins, partial destructive interference of the reflected light makes the developing membrane appear multi-colored. Eventually a black spot appears in the thick, colored membrane identifying a small area that has become bimolecular. The spot rapidly fills almost the entire aperture indicating the entire membrane has become a lipid bilayer. Various membrane electrical properties can be measured using electrodes on either side of the aperture.

destructive interference (Figure 9.4). Eventually a black spot, indicative of total destructive interference, appeared. The black spot rapidly expanded to occupy the entire hole. Using Bragg's law, it was a simple calculation to determine that the BLM was ~40–80 Å thick. This experiment proved that the lipid bilayer was indeed stable!

X-Ray Diffraction

X-Ray diffraction can generate limited high-resolution structural information, but only for membranes that have highly ordered crystalline repeating units. A few biological membranes (e.g. myelin sheath and the rod outer segment) are naturally stacked and so are ideally suited for X-ray diffraction studies. In fact, as early as 1935 X-ray studies on myelin were consistent with the presence of a lipid bilayer. Also, mitochondria and erythrocyte membrane vesicle preparations and phospholipid vesicles (liposomes) can be collapsed by centrifugation, also making them suitable for X-ray diffraction analysis. The electron density profiles of all membranes look very similar by X-ray diffraction. They consist of a low electron density hydrocarbon interior flanked by a high electron density polar group on either side. Figure 9.5 shows an electron density profile of DMPC and DMPC/cholesterol liposomes [4]. This bilayer has a 43 Å separation between polar head groups. The addition of more lipids or even proteins only slightly modifies the profile. Details of membrane proteins,

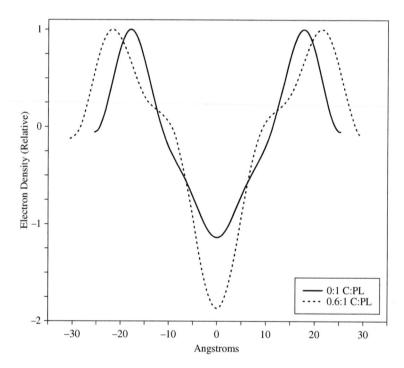

FIGURE 9.5   Electron density profile obtained by X-ray diffraction for lipid vesicles (liposomes) made from DMPC (solid line) or DMPC/cholesterol (0.6:1 mol:mol, dotted line). The most electron-dense part of the spectrum corresponds to the polar head group regions. The least electron-dense region is the bilayer center found between the two leaflets [4].

FIGURE 9.6   Humberto Fernandez-Moran (1924—1999) [56].

for example, cannot be obtained by X-ray diffraction. Note that the addition of cholesterol (dotted line) increases the distance between the electron dense peaks demonstrating that cholesterol increases membrane bilayer thickness. Therefore, X-ray diffraction does not yield molecular detail of membrane structure, but does support the concept of a lipid bilayer.

C. MEMBRANE ASYMMETRY

Even before membrane asymmetry had been experimentally proven, it was universally agreed that proteins and probably carbohydrates should be 100% asymmetrically distributed across the membrane. The situation for membrane lipids, however, was not as certain.

Protein Asymmetry

It seems logical that all functional membrane proteins should be oriented in only a single trans-membrane direction to prevent futile cycles. For example, if the function of a protein was to pump a solute into a cell, it would not make any sense if some of the same solute transport proteins were oriented in the opposite direction, thus pumping the same solute back out of the cell. But logic is not proof. A classic 1964 electron microscopy (EM) paper by Humberto Fernandez-Moran (Figure 9.6) is often cited as the first definitive evidence that membrane proteins are 100% asymmetrically distributed across membranes. Fernandez-Moran was a pioneer in electron microscopy and has been credited for developing the diamond knife and ultra-microtome for EM use. Most amazing is the fact that he was able to obtain his Ph.D. from the University of Munich in 1944! In 1964 he published a paper with David Green who later proposed the Lipoprotein Subunit model for membranes (Chapter 8). In this paper Fernandez-Moran used negative staining with phosphotungstate to obtain EM images of what was then known as a mitochondrial elementary particle (EP) [5]. We now know EP is actually the F_1 ATPase. It was the unique shape of this particle when imaged by EM that facilitated its use as a membrane directional marker. To quote Fernandez-Moran '...the elementary particle (EP), consists of three parts: (1) a spherical or polyhedral head piece (80 to 100 Å in diameter); (2) a cylindrical stalk (about 50 Å long and 30 to 40 Å wide); and (3) a base piece (40 x 110 Å).' All of the stalked EPs faced the same direction (towards the mitochondrial matrix, see Figure 9.7). Later it was shown that all three bioenergetic membranes, the mitochondrial inner membrane, the bacterial plasma membrane, and the thylakoid, all have similar-shaped ATPases. For each membrane all of the ATPases face in the same direction (inward for the mitochondrial inner membrane and the bacterial plasma membrane and outward for the thylakoid). No ATPase has been found to simultaneously face in both directions.

In addition to the classic Fernandez-Moran experiment, there have been countless other indications that membrane proteins are always 100% asymmetrically distributed across membranes. Using patch clamp techniques all ion channels have been shown to orient in only one direction. The activity of many membrane enzymes has been shown to be only associated with one side of the membrane. For example, the Na^+/K^+ ATPase hydrolyzes ATP only from the inside and pumps Na^+ out of the cell while pumping K^+ into the cell (Chapter 14). Hormones and membrane-protein specific antibodies will bind to only one membrane surface.

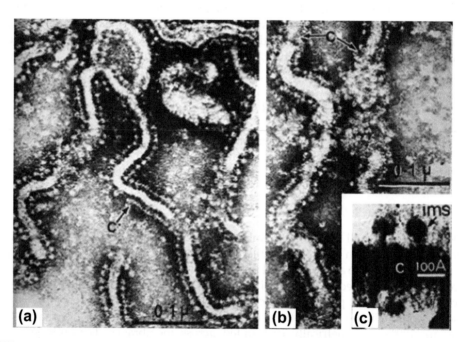

FIGURE 9.7   The Fernandez-Moran experiment. The mitochondrial inner membrane (the Cristae) was imaged by EM using negative staining with phosphotungstate. Three different perspectives of the same mitochondrial inner membrane are shown. The mitochondrial elementary particles (EPs), now known to be the F_1 ATPase, are the white lollipop-shaped objects. Note they are all attached to the same membrane leaflet (the inner leaflet of the mitochondrial inner membrane) and face into the matrix. The thick white tube-like space is the space exterior to the matrix. The matrix is the dark space in which the elementary particles protrude. Reprinted with Permission.©1964, Rockefeller University Press. Originally published in The Journal of Cell Biology 22:63–100. [5]

Carbohydrate Asymmetry

The primary function of membrane-bound carbohydrates is to interact with the extra-cellular environment by providing the cell surface with a unique sugar identity (Chapter 7). Function strongly implies, but does not prove, that carbohydrates should face outside the cell. Fortunately, carbohydrates are relatively easy handles to access and label, and so a variety of methodologies are available. The best and most employed methodology involves lectins.

Lectins

Lectins are small, sugar-binding proteins that are highly specific for their sugar ligand [6]. In fact, the term 'lectin' is derived from the Latin word *legere* meaning 'to select'. Lectins are ubiquitous in living organisms and are often referred to as 'agglutins' due to their ability to aggregate animal cells. Lectins are non-enzymic in action and non-immune in origin. Their most common biological role is assisting in various kinds of cell recognition events. In addition, lectins can target mannose-6-phosphate-containing hydrolytic enzymes to the lysosome, play important roles in the immune system by recognizing carbohydrates that are found exclusively on target pathogens and clear certain glycoproteins from the circulatory system. Lectins are also routinely used for blood typing.

The lectin Ricin, isolated from the seeds of castor bean, is currently employed as a highly toxic biochemical warfare agent. A famous incident involving ricin occurred in 1978 when the Bulgarian dissident writer Georgi Markov was stabbed in his calf by an umbrella while waiting for a bus on the Waterloo Bridge in London. The umbrella, wielded by a member of the Soviet KGB, injected a fatal dose of ricin into Markov's calf [7].

In addition to their many important biological functions, lectins have been employed as biochemical tools to study membranes. Lectins are routinely used in affinity chromatography, affinity electrophoresis, and affinity immunoelectrophoresis to isolate glycoproteins and plasma membrane vesicles. This aspect of using lectins as biochemical tools in isolating plasma membranes is discussed in Chapter 12. There are now hundreds of commercially available lectins, the most common being concanavalin A, or Con A. A brief list of commonly used lectins and their sugar ligands can be found in Table 12.1. Unmodified lectins are, of course, not detectable and so would be useless for membrane imaging studies. Therefore a variety of tags have been successfully attached to lectins and many of these are now commercially available. Included in the modified lectins are fluorescent-labeled (visible light), radio-labeled (autoradiography) and heavy metal-labeled (electron microscopy) lectins. All lectin-based imaging studies confirm that carbohydrates are 100% asymmetrically distributed across the membrane with sugars always facing the outside.

Lipid Asymmetry

It probably surprised no one that proteins and carbohydrates were shown to be 100% asymmetrically distributed across membranes. Their functions would strongly imply this. However, it was not as evident what to expect of lipid asymmetry. The major function of membrane lipids is to provide the fundamental bilayer that serves as the barrier controlling leakiness as well as the environment to support membrane biochemical activity. There is no reason to think that these functions would require lipid asymmetry. This would have to be tested [8,9].

There are now three major and very distinct methods to measure lipid asymmetry in membranes: chemical modification; enzymatic modification; and lipid exchange. Any successful technique has to cause a measurable alteration of a membrane lipid. The method must modify the lipid without destroying the membrane or making the membrane leaky to the reagent. The chemical, enzymatic or lipid exchange agents must only alter the external leaflet lipid without being exposed to the inner leaflet lipid and the procedure must be fast compared to the rate of trans-membrane lipid diffusion (flip-flop).

Chemical Modification

The first report of lipid asymmetry in membranes came from Mark Bretscher's lab at Oxford in 1972 [10]. Bretscher (Figure 9.8) developed the primary amine binding reagent, formylmethionylsulphone methyl phosphate (FMMP, Bretscher's Reagent, Figure 9.9) for his studies. He reported that with intact erythrocytes, most PE and PS (primary amine lipids) could not be labeled by FMMP. However, almost all PE and PS could be labeled in leaky erythrocyte ghosts. Therefore, he concluded that most PE and PS reside on the inner leaflet of erythrocytes. Later FMMP was replaced by the more available trinitrobenzenesulfonate (Figure 9.9). Both compounds are employed identically. The intact cells are exposed to the reagent and after a short incubation the cells are removed from the solution and washed free of the reagent. The membrane lipids are then extracted and separated, usually by thin layer chromatography (TLC). These methods are discussed in Chapter 13. The amounts of labeled (outer leaflet) and non-labeled (inner leaflet) PE and PS are quantified and the outside-to-inside ratio for both lipids determined.

Another interesting pair of primary amine-detecting reagents used in lipid asymmetry determinations is ethylacetimidate (EAI) and isethionylacetimidate (IAI). Both compounds, shown in Figure 9.9, attach to PE and PS by the same mechanism. The difference between

FIGURE 9.8   M. Bretscher (1940–). *Courtesy of Mark S. Bretscher.*

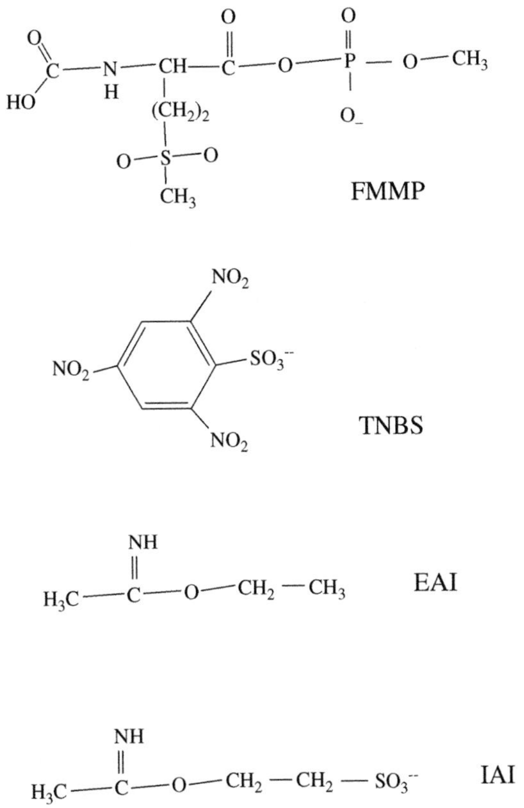

FIGURE 9.9 Types of primary amine tags used to label PE and PS in membrane lipid asymmetry studies: FMMP (formylmethionylsulphone methyl phosphate, Bretscher's Reagent); TNBS (trinitrobenzene sulfonate); EAI (ethylacetimidate); and IAI (isethionylacetimidate).

the two reagents is that IAI is charged and cannot penetrate the membrane and so only labels outer leaflet lipids. In contrast, EAI is uncharged and so can readily cross membranes, labeling both inner and outer leaflet lipids.

Enzymatic Modification

PE and PS both have a highly reactive primary amine on their polar head group. Therefore they lend themselves well to covalent attachment by a number of reagents. However the remaining membrane lipids do not possess such a handle and so other methods had to be sought. Chemical methods do exist that can modify the non-primary amine head groups, but these methods are much harsher and usually result in membrane destruction. Enzymatic alteration of a polar head group can replace chemical modification. Enzymes have a tremendous advantage in being highly specific for the substrate lipid. The large size of an enzyme also makes it impossible to cross an intact membrane. Therefore, the enzyme will only alter the outer leaflet lipid. While a variety of enzymes have been employed, only three will be discussed here: phospholipase D; sphingomyelinase; and cholesterol oxidase.

Phospholipase D (specific for PC, PLD$_{(PC)}$) can be used to measure the asymmetry of PC. The enzyme catalyzes the reaction:

$$PC \xrightarrow{\text{PLD}_{(PC)}} PA + Choline$$

Since the enzyme cannot cross the membrane, it only converts outer leaflet PC to PA. It does not affect the inner leaflet PC. The PC and PA content is determined before and after hydrolysis by PLD$_{(PC)}$. The decrease in PC equals the increase in PA, and this represents the outer leaflet PC content. From this measurement the membrane asymmetry of PC can be determined.

In a similar fashion, sphingomyelinase (SMase) can be used to measure the membrane asymmetry of SM. The enzyme catalyzes the following reaction:

$$SM \xrightarrow{\text{SMase}} ceramide + phosphorylcholine$$

As with PLD$_{(PC)}$, only the outer leaflet SM is hydrolyzed and SM and ceramide content is determined before and after enzyme treatment. The decrease in SM equals the increase in ceramide, and this represents the outer leaflet SM content. From this measurement the membrane asymmetry of SM can be determined.

Some other types of phospholipases are less useful since they produce products that destabilize the membrane. For example phospholipase A$_1$ or A$_2$ produce lysophospholipids and free fatty acids, both of which adversely affect membrane structure. Phospholipase C produces the non-lamellar phase-preferring compound diacylglycerol (Chapter 5).

Cholesterol is not hydrolysable and so its asymmetry must be determined using a very different method. Cholesterol also presents an additional problem since its inherent rate of trans-membrane diffusion (flip-flop) is so much faster than phospholipids. Accurate measurement of cholesterol asymmetry is difficult and no method is without its flaws. One common method employs the use of cyclodextrin (Chapter 10) [11]. A less commonly used method is outlined in Figure 9.10. It presents a theoretically easy UV method that would be available to

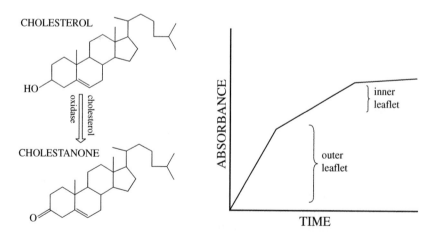

FIGURE 9.10   Conversion of cholesterol to cholestanone by cholesterol oxidase. The initial rapid increase in absorbance is due to oxidation of outer leaflet cholesterol to UV light-absorbing cholestanone. The slower oxidation occurs after inner leaflet cholesterol has flipped to the outer leaflet. Cholesterol asymmetry can be determined from the ratio of inner to outer leaflet cholesterol.

most laboratories. As the C-3 hydroxyl group of cholesterol is oxidized to a ketone by the enzyme cholesterol oxidase, its UV absorbance increases. Since the enzyme is too large to cross a membrane, it can only oxidize cholesterol exposed on the outer surface of a cell. Therefore if cells are mixed with cholesterol oxidase in a quartz cuvette, the UV absorbance will increase rapidly until all of the outer leaflet cholesterol is oxidized. Subsequent, slower oxidation can be measured as cholesterol flips from the inner to the outer leaflet. From these measurements, cholesterol's membrane asymmetry and flip-flop rate can be determined.

Erythrocyte Lipid Asymmetry

Since the 1925 Gorter and Grendel experiment establishing the lipid bilayer, erythrocytes have been 'the laboratory for membrane studies'. It was in the erythrocyte that lipid asymmetry was first established. Figure 9.11 depicts asymmetry of the basic erythrocyte lipids. The choline-containing lipids (PC and SM) are predominantly in the membrane outer leaflet, while the primary amine lipids (PE and PS) are found predominantly in the membrane inner leaflet. PS is unusual in being 100% asymmetrically distributed to the inner leaflet. This is attributed to the presence of an ATP-dependent flipase [12−14] whose function it is to take any outer leaflet PS and immediately flip it back to the inner leaflet, preventing accumulation of outer leaflet PS. PA, PI, and PI 4,5 *bis* phosphate are predominantly located in the inner leaflet, while cholesterol is mainly associated with the outer leaflet, although its rapid flip-flop rate would indicate it moves back and forth across the membrane spending more time in the outer leaflet. The sugar-containing gangliosides are always (100% asymmetric) in the outer leaflet. CL is evenly distributed across the membrane.

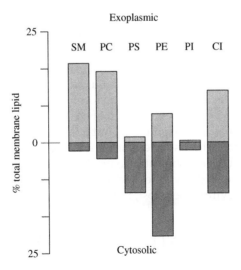

FIGURE 9.11 Distribution of common membrane lipids between the inner and outer leaflets of erythrocytes. Note the choline-containing lipids (SM and PC) are predominantly found in the outer (exoplasmic) leaflet, while the primary amine lipids (PE and PS) are associated with the inner (cytoplasmic) leaflet. Also, PI is found primarily on the inner leaflet while CL is equally distributed between the leaflets. *Courtesy of TISTORY*

It is not an easy task to understand the variety of forces driving and maintaining partial lipid asymmetry [15–17]. For example, lipid asymmetry must be severely altered at membrane segments that are tightly curved. In order to fit the enormous quantity of internal membranes into the tiny intra-cellular space (discussed in Chapter 1), there must be countless places in internal membranes that are under tremendous curvature strain. These locations are difficult to study since the strain is instantly released upon breaking open the cell. Therefore membrane curvature is best studied by making spherical lipid vesicles of known diameter and hence strain. The tightest vesicle that is achievable with phospholipids is ~200 Å in diameter. These vesicles, referred to as SUVs (small unilamellar vesicles, see Chapter 13), are characterized by having two-thirds of their lipids in the outer leaflet and one-third crunched into the inner leaflet. In mixed-lipid SUVs, the larger head group lipids (PC, SM, glycolipids) tend to accumulate in the less densely packed outer leaflet while lipids with smaller head groups (e.g. PE, cholesterol) are forced to accumulate in the more compact inner leaflet. If SUVs undergo a series of fusions, the curvature strain is released and the membrane becomes essentially planar in the new large unilamellar vesicles (LUVs, see Chapter 13). In the LUVs, lipid asymmetry is lost.

As one example, if a mixed lipid PC/PS SUV is made at pH 7.0, PC is predominantly found in the outer leaflet, while PS is more prevalent in the inner leaflet. At pH 7.0, PC is larger than PS. However, if the pH is increased, PS flips to the outside and PC moves to the inside. At high pH PS becomes larger than PC, due to negative charge repulsion of the PS head group. At pH 7.0, PS has a net charge of -1, while at high pH the charge of PS is -2. The problem with phospholipid size is far more complex than this. One must keep in mind that the size of a phospholipid is also dependent on the nature of the attached acyl chains. The phospholipid base diameter increases with the number of double bonds (see Chapter 11). One would therefore expect that lipids with more double bonds would be forced into the outer leaflet in membrane segments with high curvature. But there is an additional complication, as lipids with more double bonds are more 'compressible' than lipids with saturated chains (discussed in Chapter 11). In sharp contrast, cholesterol is almost totally non-compressible. The net effect of all of these forces is unknown and at present unpredictable.

Lipid partial asymmetry was first worked out in the erythrocyte and the results are depicted in Figure 9.11. If the erythrocyte is transformed into a spherical 'ghost' upon emulsion in hypotonic media (the erythrocyte is 'hemolyzed'), much of the intracellular cytoskeleton is lost. The ghost still has lipid asymmetry, but it is much reduced compared to the intact erythrocyte. If the erythrocyte lipids are extracted and used to make an LUV, no lipid asymmetry is observed. If the LUVs are sonicated to produce SUVs, lipid asymmetry is once again established. However, this lipid asymmetry does not resemble either that observed for the intact erythrocyte or the erythrocyte ghost, but instead reflects curvature strain.

D. LATERAL DIFFUSION

A key aspect of the Fluid Mosaic model is 'fluidity'. The model predicts that the membrane is in constant flux (it is fluid) with components always in the process of exchanging lateral positions with other components. The first proof of membrane lateral mobility was the iconic

1970 immunofluorescence paper by Frye and Edidin [18]. Subsequent, improved methodologies to follow lateral diffusion include fluorescence recovery after photobleaching (FRAP) and single particle tracking (SPT), discussed below.

The Frye-Edidin Experiment

By 1970 it was evident that cells were not rigid structures, but changed shape as the cell moved or developed pseudopodia. The objective of the Frye-Edidin experiment was to prove that plasma membrane (surface) proteins could move laterally in the membrane. These investigators used two very different cell lines, one a human line, the other a mouse line. The experiment is depicted in Figure 9.12. Each cell type had different surface antigens and so could be readily distinguished by fluorescent-labeled antibodies. The technique is referred to as immunofluorescence. The antibody for the mouse antigen was labeled with fluorescein and so fluoresced green, while the antibody against the human antigen was labeled with rhodamine and appeared red. The cells were fused at 37°C with a Sendai virus producing a heterokaryon that initially had one half of the fused cell green and one half red. After 40 minutes the red and green colors were totally mixed. One interpretation of this observation was that the plasma membrane proteins were free to diffuse laterally in the plane of the membrane. However, Frye and Edidin did realize that other explanations were possible and they tested these. Protein synthesis inhibitors (puromycin, cycloheximide, and chloramphenicol) and the uncoupler 2,4-dinitrophenol had no effect on the process. Therefore, intermixing of the surface antigens was not the result of newly synthesized antigens at various locations in the plasma membrane, nor was the intermixing due to an energy-dependent translocation. However, if the temperature was reduced from 37°C to 15°C, the process was greatly slowed. The dependence on temperature was consistent with a change in membrane viscosity that would affect diffusion rates. This paper concluded that a membrane 'is not a rigid structure, but is fluid enough to allow free diffusion of surface antigens resulting in their intermingling within minutes after the initiation of fusion'.

This seminal paper in membrane studies was one of the very first in Michael Edidin's (Figure 9.13) long and distinguished career at Johns Hopkins. He is also well respected in an area far removed from membranes. He is a world expert on the history of watch making and repair!

Fluorescence Recovery after Photobleaching (FRAP)

While the Frye-Edidin experiment clearly established that membrane proteins can diffuse laterally, it proved to be a difficult method to obtain accurate rates of diffusion. The Sendai virus is not a reliable fusogen. Its fusion agent is likely a component picked up from the host cell as the virus escapes and so varies tremendously in effectiveness from preparation to preparation. In addition, while the time that fusion begins (T = 0) can be accurately determined, the time at which complete mixing is achieved is imprecise. Therefore other, more reliable and accurate methods to measure lateral diffusion in membranes were sought. Advances in laser technology led to the accelerated development of what is now the major technique in determining lateral diffusion of lipids and proteins in

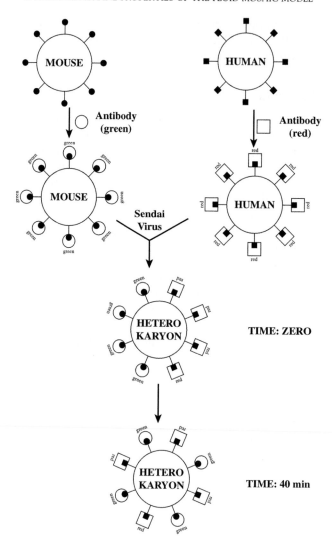

FIGURE 9.12   The Frye-Edidin experiment. Cell cultures were obtained from a mouse and a human cell line. Each cell type had different surface antigens. Fluorescent-labeled antibodies were obtained for each cell line. The antibody for the mouse antigen was labeled with fluorescein and so appeared green, while the antibody for the human antigen was labeled with rhodamine and so appeared red. The cells were fused with a Sendai virus producing a heterokaryon that initially had one half of the mega-cell green and one half red. After 40 minutes the red and green colors were totally mixed indicating membrane protein lateral diffusion.

membranes. This technique is referred to as fluorescence recovery after photobleaching or FRAP (Figure 9.14).

FRAP had its beginnings in a 1976 paper by Axelrod et al. [19]. The technique requires a high quality light microscope, a general low intensity light source, and a highly focused, high intensity light, normally a laser. FRAP begins by uniformly labeling the surface of

FIGURE 9.13 Michael Edidin. *Courtesy of Michael Edidin*

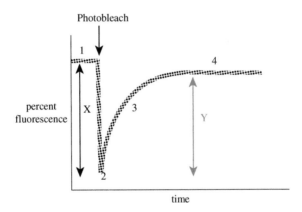

FIGURE 9.14 Data obtained for a typical FRAP experiment. The initial fluorescence background for the unbleached cell is indicated by line (1). Photobleaching (arrow) decreases fluorescence in the spot (2). The net loss in fluorescence is X. Brownian motion allows for recovery of fluorescence in the spot (3). Eventually fluorescence returns in the spot to a new baseline (4) that is lower than the pre-bleached fluorescence (1). The fluorescence recovery is Y (Y < X). The % recovery is therefore Y/X x 100 and represents the percent of the membrane component that is diffusible. The percent of the membrane component that is non-diffusible is 100 − (Y/X x 100). Lateral mobility is directly related to the slope of the recovery process. A steeper slope indicates a faster recovery and higher diffusion rate. [20]

a cell with a fluorescent-labeled antibody or other lipid or protein-specific fluorescent tag. For example, a fluorescent lectin could be bound to a specific glycolipid or glycoprotein. The next step involves acquiring a basic fluorescence background image of the cell surface that will be photobleached. A small diameter, intense light pulse is then used to rapidly photobleach a small spot on the cell surface. Originally the light source was a broad spectrum mercury or xenon lamp and was used in conjunction with a color filter to assure the spot would receive light that matched the absorbance of the fluorescent probe. Tunable lasers have now replaced the older light sources and filters. The intense laser pulse photobleaches a spot of predetermined size and shape. This spot appears dark on a highly fluorescent, bright, unbleached background. The laser destroys the fluorescence of the probe but has no effect on the membrane component (lipid or protein) to which the tag is associated. With time, the still fluorescent-labeled membrane components diffuse into the photobleached spot, while the non-fluorescent (bleached) components diffuse out of the spot. Eventually the once bleached spot recovers most of its fluorescence, although never fully recovering.

A detailed plot of data obtained from a typical FRAP experiment is shown in Figure 9.14 [20]. The initial fluorescence background for the unbleached cell is indicated by line (1). Photobleaching (arrow) decreases fluorescence in the spot (2). The net loss in fluorescence is X. Brownian motion allows for recovery of fluorescence into the spot (3). Eventually fluorescence returns in the spot to a new baseline (4) that is lower than the pre-bleached fluorescence (1). The fluorescence recovery is Y ($Y < X$). The percent recovery is therefore $Y/X \times 100$ and represents the percent of the fluorescent-tagged membrane component that is diffusible. The percent of the membrane component that is non-diffusible is:

$$100 - (Y/X \times 100)$$

Lateral mobility is directly related to the slope of the recovery process. A steeper slope indicates a faster recovery and higher diffusion rate. The diffusion rate D can be obtained from the following equation:

$$D = W^2/4t_{1/2}$$

where w is the width of the photobleaching beam and $t_{1/2}$ is the time required for the bleach spot to recover half of its initial intensity.

Single Particle Tracking (SPT)

While FRAP clearly has advantages over the original Frye-Edidin method, it still just gives an averaged diffusion rate for an ensemble of membrane residents. Single particle tracking (SPT), as the name implies, can follow diffusion of one molecule at a time [21]. Akiharo Kusumi (Figure 9.15) of the Kyoto University School of Medicine is recognized as the world leader in this emerging field [22]. In his laboratory, the temporal resolution of SPT has been pushed down to 25 μs and spatial resolution down to nanometers. As might be expected, this complex technology has resulted in a completely new paradigm in our understanding of membrane structure. STP development was driven by a basic observation that confused scientists for decades. If movement in a membrane is the result of free Brownian motion, why do membrane components diffuse 10–100 fold slower in a cell (plasma) membrane than they do in an artificial lipid bilayer membrane? In 1983 Michael Sheetz [23] proposed that trans-membrane proteins might be contained by the cytoskeleton ('fence' theory). However, membrane technology was not sufficiently advanced to confirm this hypothesis. Confirmation had to await the development of SPT.

FIGURE 9.15   Akiharo Kusumi. *Courtesy of Akihiro Kusumi*

SPT had its beginnings in a 1985 paper by De Brabander et al. [24], but was, and continues to be primarily advanced by Kusumi and colleagues. In what is now a classic 1993 biophysical Journal paper, Kusumi [25] reported on lateral membrane diffusion of the cell–cell recognition, adhesion receptor E-cadherin in cultured mouse karatinocytes. The paper compared SPT to the then more established technique of FRAP. Receptor movement was observed by SPT after E-cadherin was bound to an anti-cadherin monoclonal antibody coupled to a 40 nm gold particle. Motion was followed by video-enhanced differential interference contrast microscopy at a temporal resolution of 30 ms and at a nanometer special precision. After plotting the results as mean square displacement of the gold particle against time, four characteristic types of motion were observed (Figure 9.16):

A. Stationary mode where diffusion is less than 4.6×10^{-12} cm^2/sec (6%).
B. Simple Brownian diffusion mode (28%).
C. Directed diffusion mode indicated by unidirectional motion (2%).
D. Confined diffusion mode where Brownian diffusion is between 4.6×10^{-12} and 1×10^{-9} cm^2/s and diffusion is confined within a limited area by the cytoskeleton network (64%). Diffusion was confined into many small domains 300–600 nm in diameter.

Confined Diffusion represented almost two-thirds of the total diffusion patterns. Later Kusumi noticed that in rat kidney cells, diffusion appeared to be a connected series of confined diffusion mode domains. In other words the overall, low resolution diffusion measured by FRAP was composed of two parts, fast Brownian motion that is confined within corrals and slow 'hop diffusion' that connects the confined zones. Unfortunately, these observations could not be confirmed in erythrocytes or other cells. Kusumi realized that a faster methodology was needed. He obtained ultrafast filming technology from the field of explosives research [22]. This technology employed a camera that obtained 40,000 frames per second with a temporal resolution of 25 μs. This produced a 1,000 times faster resolution than before (30 ms for the E-cadherin experiment), allowing Kusumi to even measure hop diffusion for membrane lipids which hop at very fast rates (Figure 9.17).

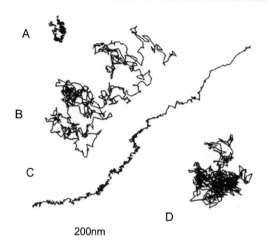

A

B

C

D

200nm

FIGURE 9.16   Four characteristic motions observed by SPT for E-cadherin in the plasma membrane of a cultured mouse karatinocyte. The four motions are: A: stationary mode; B: simple Brownian diffusion mode; C: directed diffusion mode; and D: confined diffusion mode. SPT trajectories of E-cadherins were recorded for 16.7 s (500 video frames) [25].

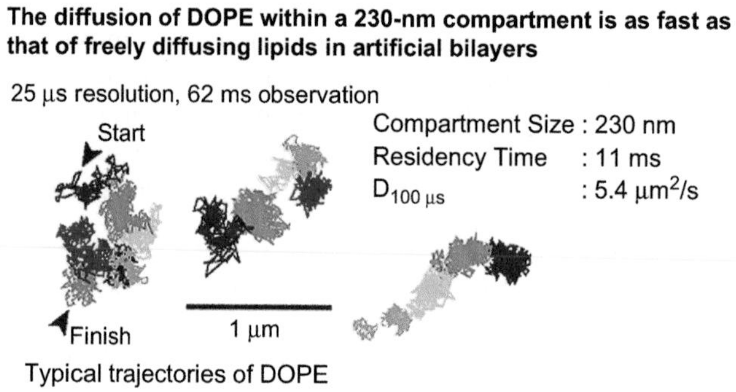

The diffusion of DOPE within a 230-nm compartment is as fast as that of freely diffusing lipids in artificial bilayers

25 μs resolution, 62 ms observation

Start

Compartment Size : 230 nm
Residency Time   : 11 ms
$D_{100\,\mu s}$: 5.4 μm²/s

Finish   1 μm

Typical trajectories of DOPE

FIGURE 9.17   Hop diffusion of the membrane phospholipid DOPE (18:1,18:1 PE). DOPE spends most of its time trapped in a compartment (~11 ms) before hopping out and diffusing to another compartment where it is again trapped. DOPE diffuses as fast within the confined domains as it does in a protein-free lipid bilayer [57].

The 'fence' (corral) model for membrane structure possesses some major questions that must be addressed. For example, what effect does the relatively enormous attached gold particle have on the measurements? Since it is likely that only the plasma membrane has a close association with the cytoskeleton, do the other intracellular membranes display corrals? How does 'raft' theory fit in with 'fence' theory? Does one negate the existence of the other or are they compatible?

Membrane Lateral Diffusion Rates: Conclusions

The Frye-Edidin experiment proved that membrane proteins were free to diffuse laterally in the cell plasma membrane at a rate sufficient to completely encircle a bacterial cell in

1 second and a eukaryotic liver cell in 1 minute. The diffusion rate was 5×10^{-11} cm^{2}s^{-1}. For a similar sized, but water-soluble protein (hemoglobin), the diffusion rate was much faster, at 7×10^{-7} cm^{2}s^{-1}. The effective membrane viscosity was therefore 7×10^{-7} cm^{2}s^{-1}/ 5×10^{-11} cm^{2}s^{-1} or ~10^3 to 10^4 times more viscous than water. Viscosity of the membrane bilayer interior therefore resembles that of light machine oil.

The diffusion rate for membrane proteins was shown to span a large range, from that approaching free phospholipids (~10^{-8} cm^{2}s^{-1}) through to essentially being immobile (bound to the cytoskeleton, ~10^{-12} cm^{2}s^{-1}). Interactions of many trans-membrane proteins with the cytoskeleton account for protein immobility and are now well established. A first indication of this occurred in the erythrocyte where removal of the cytoskeleton increased band 3 lateral diffusion 40 times. Phospholipid diffusion was shown to be independent of the head group (i.e. all phospholipids have about the same fast diffusion rate of ~10^{-8} cm^{2}s^{-1}), but diffusion is impacted by membrane phase (discussed below and in Chapter 10). For example, lipid diffusion in gel state bilayers is ~100 to 1,000 times slower than in fluid, liquid crystalline state bilayers. Lipid diffusion is also known to be impacted by the presence of cholesterol, generating the liquid ordered (l_o) state. This must be considered with regard to the structure, stability, and function of l_o-state lipid rafts. In general, membrane components diffused 10 to 100 times slower in biological membranes than in protein-free lipid bilayer membranes, implying hindered diffusion due to crowding (see Chapter 11).

It is generally accepted that biological membrane components, lipids, and proteins, are heterogeneously distributed into countless numbers of bewildering and short-lived domains. An early experiment by Michael Edidin employing lateral diffusion measurements supported the concept of membrane heterogeneity [26]. Edidin's experiment used FRAP to follow the diffusion of two carbocyanine dyes, one with two short lipid chains (C_{10}, C_{10}-DiI) and one with two long lipid chains (C_{22}, C_{22}-DiI). The esterified fatty acyl chains have very different T_ms. C-10 decanoic acid has its T_m at 31.6°C while the T_m of C-22 behenic acid is 79.9°C (see Table 4.3). At 37°C C-10 would be in the melted, liquid crystalline state while C-22 would be in the solid, gel state. Edidin predicted that the short chain dye (C_{10}, C_{10} DiI) would partition into a more fluid (disordered) phase while the longer chain dye (C_{22}, C_{22} DiI) would prefer a more ordered phase. If the membrane lipid bilayer was homogeneous, he reasoned, both dyes should have similar diffusion rates, but if the membrane bilayer was heterogeneous, the two fluorescent probes should exhibit different diffusion rates. When he tested this hypothesis on sea urchin and mouse eggs, he found large differences between the diffusion rates for each dye. His conclusion was that membrane bilayers have considerable patchiness that must exist for at least a few minutes. The nature of this patchiness is at the heart of membrane domain studies.

E. LIPID TRANS-MEMBRANE DIFFUSION (FLIP-FLOP)

Earlier in this chapter it was discussed that trans-membrane asymmetry is always absolute for proteins and carbohydrates. Therefore their rate of trans-membrane diffusion or flip-flop is zero. It never happens. However, lipids do exhibit partial lipid asymmetry and so one would assume that lipid flip-flop might be possible. However, if flip-flop was very fast, lipid

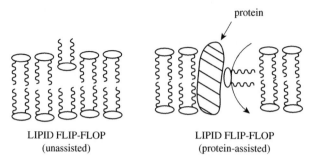

LIPID FLIP-FLOP   LIPID FLIP-FLOP
(unassisted)   (protein-assisted)

FIGURE 9.18   Lipid flip-flop, unassisted (left) and protein-assisted (right). A protein's surface can increase phospholipid flip-flop rate 10^3 to 10^5 fold.

asymmetry could not be maintained. If flip-flop was very slow, a lipid's asymmetry would reflect the membrane sidedness during its synthesis and it would be hard to explain how each membrane has a different asymmetry for each lipid. The well-established fact that all membranes exhibit partial lipid asymmetry indicates that the process of flip-flop might not be easy, but it should be possible. As a lipid undergoes flip-flop, two thermodynamically unfavorable events must occur simultaneously. The lipid hydrophobic tails must be exposed to water and the polar head group must be exposed to the hydrophobic bilayer interior. From this a rough estimate of the energy of flip-flop can be made. It requires ~2.6 kcal of free energy to transfer a mole of methane from a nonpolar medium to water at 25°C. Free energy required to transfer a mole of zwitterionic glycine from water to acetone is about 6.0 kcal at 25°C. Therefore it would very roughly require approximately 97 kcal to flip a phospholipid across a membrane (assuming an average chain length of 17.5 carbons per chain, and two acyl chains per phospholipid). This means that only 1 phospholipid in ~10^{15} would have enough energy to flip and so the inherent rate of flip-flop would be quite slow, but not impossible. Indeed this has been shown to be the case [27]. This slow rate can, however, be substantially increased (10^3 to 10^5 fold) by simply providing a trans-membrane protein surface (Figure 9.18) [28 for glycophorin and [29] for cytochrome b5]. Although lateral exchange of nearest lipid neighbors in the same membrane leaflet, and exchange across the membrane from one leaflet to another, traverse approximately the same distance (~4 nm), the lateral exchange rate is ~10^{10} x faster than is flip-flop. Nevertheless, lipid flip-flop is an essential component of membrane dynamics. Importance of flip-flop begins with biogenesis of the phospholipid and ends with cell death [30].

Chemical Method to Determine Flip-Flop Rates

Since flip-flop is essentially the kinetics of asymmetry, any method that can be used to measure lipid asymmetry should, in principle, also be applicable to measuring flip-flop. For example, the rate of flip-flop of the primary amine phospholipids PE and PS can be determined by combining the methods discussed above using trinitrobenzene sulfonate (TNBS) and isethionylacetimidate (IAI) for determining lipid asymmetry. The structure of these compounds is shown in Figure 9.8 and their application for determining flip-flop in Figure 9.19. If a sealed membrane vesicle (LUV) is exposed to TNBS, only the outer leaflet

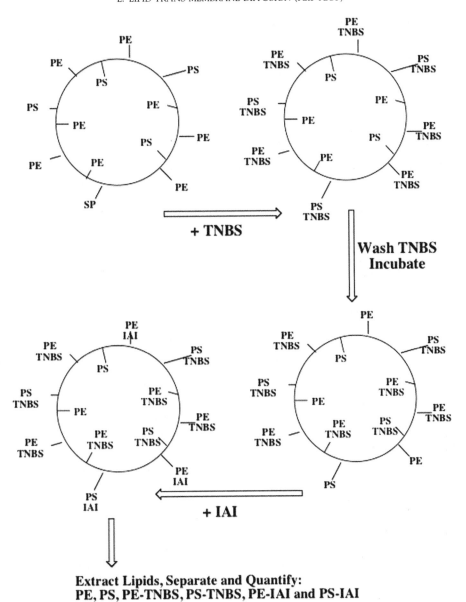

+ TNBS

**Wash TNBS
Incubate**

+ IAI

**Extract Lipids, Separate and Quantify:
PE, PS, PE-TNBS, PS-TNBS, PE-IAI and PS-IAI**

FIGURE 9.19   Determining flip-flop rates for the primary amine phospholipids PE and PS by use of the primary amine labels, TNBS and IAI.

PE and PS will be labeled since the reagent does not cross the membrane. The un-reacted TNBS is then washed away from the TNBS-modified vesicles and the vesicles are allowed to incubate for a time (t = X). During this time, some un-modified (inner leaflet) PE and PS can flip to the outer leaflet while roughly an equivalent amount of PE-TNBS and PS-TNBS can flip in. At t = X the vesicles are then exposed to another non-permeable primary amine reactive reagent, IAI. IAI then reacts with the PE and PS that flipped from the inner leaflet to the outer leaflet during the time interval X. After this, the non-reacted IAI is washed away from the vesicles and the lipids are extracted, separated, and quantified (see Chapter 13). From the measured amount of PE, PS, PE-TNBS, PS-TNBS, PE-IAI, and PS-IAI, the rate of flip-flop for PE and PS can be estimated.

Phospholipid Exchange Protein Method to Determine Flip-Flop Rates

Phospholipid exchange proteins (PLEPs, also called lipid transfer proteins, or TPs), offer a versatile approach to measure lipid asymmetry and flip-flop. PLEPs were first discovered by Wirtz and Zilversmit at Cornell University in 1968 [31]. These are a ubiquitous family of low-molecular weight proteins whose function is to shuttle various lipids from membrane to membrane [32]. Since many PLEPs exhibit absolute specificity for a particular phospholipid, they do not have to be totally purified but can be employed as a crude cellular mixture. The experiments can be constructed in almost limitless combinations to address lipid asymmetry and flip-flop. A simple example to determine the inherent flip-flop rate for PC in a protein-free lipid vesicle is depicted in Figure 9.20. Large unilamellar lipid vesicles are made from ^{32}P-PC (the method is discussed in more detail in Chapter 13). To the LUVs are added mito-chondria that provide a large excess of un-labeled (mitochondrial) PC. Mitochondria are chosen since they are easily obtained and are much bigger and heavier than the ^{32}P-PC LUVs. A simple low-speed centrifugation is sufficient to completely separate the LUVs and mitochondria. To start the experiment, a source of crude PLEP$_{(PC)}$ specific for PC is added. The mixture is incubated until all of the radio-labeled outer leaflet ^{32}P-PC (black head group in Figure 9.20) is exchanged for the non-labeled mitochondrial PC (white head group in Figure 9.20). The large excess of mitochondrial PC assures that all of the initial LUV exoleaflet ^{32}P-PC is replaced by non-radiolabeled mitochondrial PC. The now radiola-beled mitochondria and PLEP$_{(PC)}$ are then removed, leaving the LUVs with a radiolabeled ^{32}P-PC inner leaflet and a non-radiolabeled mitochondrial PC outer leaflet. The vesicles are allowed to incubate for a period of time t = X during which flip-flop can occur. At time = X, the vesicles are exposed to new mitochondria and PLEP$_{(PC)}$. Any ^{32}P-PC that has flipped from the inner leaflet to the outer leaflet during the time = X will now appear in the mito-chondria that can be centrifuged and counted. The method is sensitive enough to detect only a few flipped ^{32}P-PC lipids. This number (PCs flipped as followed by counts over time = X) is compared to the total counts in the vesicles where only the inner leaflet was radio-labeled. This experiment is then repeated for several longer incubations. From this an extrapolated half-life for PC flip-flop can be obtained.

Inherent phospholipid flip-flop rates, as determined in protein-free model lipid bilayer membranes, are very slow, often with half-lives in the days-to-weeks range [27]. In sharp contrast, James Hamilton [33] has shown that the flip-flop rates for free fatty acids are very fast, in the seconds-to-milliseconds range. Therefore it appears that moving the phospholipid polar head group into the hydrocarbon interior must be the major impediment to flip-flop. Supporting this is the very rapid flip-flop rate for diacylglycerol whose polar head group is only a simple uncharged alcohol. Indeed, Horman and Pownall [34] showed that at pH 7.4 flip-flop rates increased in the order PC < PG < PA < PE, where the rate for PE was at least

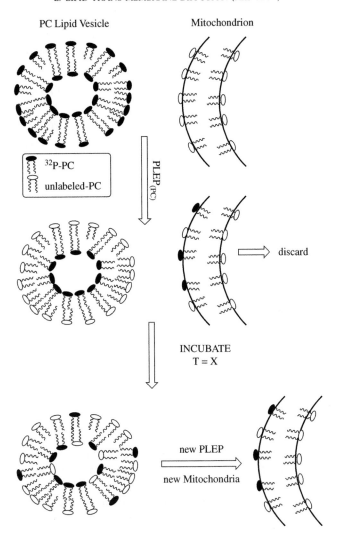

FIGURE 9.20   Use of a PLEP$_{(PC)}$ to determine the inherent rate of flip-flop for PC in a protein-free lipid bilayer (liposome). The PCs are either ^{32}P-PC (black head group) or non-labeled mitochondrial PC (white head group).

ten times greater than for the homologous PC derivative. They also reported that the flip-flop rate for PA increased 500 times when the solution pH was decreased from 7.4 to 4.0. At pH 4.0 the net negative charge on PA is much reduced compared to pH 7.4. Finally, Middelkoop et al. [35] reported the effect of acyl chain composition on the flip-flop rate of various PCs (16:0,16:0 PC < 18:1,18:1 PC < 16:0,18:2 PC < 16:0, 20:4 PC). They reported that for erythrocytes at 37°C flip-flop half-lives decreased with increasing number of double bonds.

Another way of looking at lipid flip-flop is to determine the effect of the bilayer hydrophobic environment (acyl chain components) on the flip-flop of a single lipid species. Armstrong et al. [36] measured flip-flop rates for the fluorescent phospholipid NBD-PE by

TABLE 9.1   Flip-Flop Rates (Expressed as Half-Lives) for NBD-PE as Followed by Quenching of NBD Fluorescence with Dithionite. Large Unilamellar Vesicles (LUVs) were Made from the Listed PCs with 1 % NBD PE Incorporated (25°C).

PC	Flip-Flop Half Life ($t_{1/2}$ (h))
16:0,16:0	>70
18:0,18:1	11.5 ± 1.6
18:1,18:1	8.1 ± 0.6
18:0,18:2	4.9 ± 0.9
18:2,18:2	1.9 ± 0.4
18:3,18:3	1.3 ± 0.3
18:0,22:6	0.29 ± 0.02
22:6,22:6	0.087 ± 0.006

quenching of NBD fluorescence with dithionite. As the number of double bonds comprising the bilayer increases, so does the flip-flop rate (Table 9.1). For example, flip-flop of NBD-PE in 22:6,22:6 PC with 12 double bonds is at least 10^3 times greater than in 16:0,16:0 PC with no double bonds. Flip-flop is related to the acyl chain double bond content in the lipid bilayer interior.

Flippase, Floppase, and Scramblase

From the very first determinations of lipid asymmetry in erythrocytes, it was evident that lipid asymmetry was complex. Particularly troublesome was the observation that the distribution of PS, unlike the other phospholipids, was almost totally asymmetric. Bretscher realized the importance of PS being found exclusively on the inner membrane leaflet. In 1972 [10] he proposed the existence of an energy-dependent membrane protein whose function was to flip PS from the outer membrane leaflet to the inner leaflet, thus preserving the total asymmetric distribution of PS. He even coined the term 'flippase' for this protein. An ATP-dependent flippase was later discovered by Seigneuret and Devaux in 1984 [12]. This protein proved to be only one of a family of related membrane proteins called 'flippases', 'floppases', and 'scramblases' whose functions are to correctly distribute the various phospholipids across membranes [15,16].

Newly synthesized phospholipids are initially incorporated into the inner membrane leaflet after which they are subsequently distributed by a variety of lipid translocators [15,16]. Although the term 'flippase' initially referred in general to any lipid translocator, it now refers to proteins that move lipids to the inner leaflet (cytofacial) membrane surface. They are ATP-dependent transporters. The flippase is highly selective for PS and functions to keep this lipid from appearing on the outer surface of the cell. Upon aging, PS starts to accumulate on the cell exterior, triggering apoptosis. Floppases are the opposite of flippases and are ATP-dependent transporters that move lipids to the outer leaflet (exofacial) membrane surface. They have been associated with the ABC class of transporters. The third family of lipid translocators is the scramblases, whose function is to move lipids bidirectionally across the membrane without use of ATP. Scramblases are inherently nonspecific and function to randomize the distribution of newly synthesized phospholipids and may be involved in membrane disruption events associated with cell death. In the case of the well-studied erythrocyte, net flip-flop rates are: PS > PE > PC > SM.

Some General Conclusions from Flip-Flop Studies

Flip-flop of proteins and carbohydrates (in the form of glycoproteins and glycolipids) is absolutely prohibited. Inherent rates of membrane phospholipid flip-flop are very slow, with half-lives varying from days to weeks. This rate can be substantially enhanced (10^3 to 10^5 fold) by the simple inclusion of a trans-membrane protein. Cholesterol, diacylglycerol, and free fatty acids have high inherent rates of flip-flop. Cell membranes have families of lipid translocators known as flippases, floppases and scramblases whose functions are to accurately maintain proper lipid asymmetry.

F. LIPID MELTING BEHAVIOR

The simple cartoon depiction of a lipid bilayer is deceptive. The lipid bilayer, existing in what is more correctly termed the 'lamellar phase', is highly dynamic and can actually exist in many related phases in addition to the lamellar state. Although the lamellar phase dominates biological membrane structure, many and perhaps even most membrane polar lipids, when isolated, would prefer to exist in exotic non-lamellar phases. It is the lipid mixtures that stabilize the lamellar phase. A variety of non-lamellar phases are discussed in Chapter 10. However, even the seemingly simple lamellar phase is complex and can exist in several variations. We will first discuss simple melting of the lamellar phase from the solid-like gel ($L_{\beta'}$ for PCs) to the fluid-like liquid crystalline (L_{α}) phase. Lipid bilayers should not be thought of as existing as either a hard solid or water-like fluid but more like what would be observed upon melting bacon grease. The gel state is a soft solid while the liquid crystalline state is a viscous fluid.

A major theme of Chapter 4 was the effect of fatty acyl chain length and degree of unsaturation on 'melting' temperature (expressed as T_ms). But why do fatty acyl chains melt at all? Figure 9.21 depicts the steps involved in melting of a saturated fatty acyl chain. A section of an un-melted, *trans* state, saturated fatty acid is shown in Figure 9.21a. Newman projections of a short segment of the chain are shown in Figure 9.21b. Upon heating the chain, rotation occurs around all $C - C$ bonds, for example the bond connecting carbons 2 and 3. The *trans* configuration (Figure 9.21b, left) has the lowest possible energy (Figure 9.21c) as the largest groups, carbon 1 and carbon 4, are as far apart as possible. Upon clockwise rotation, as the large C-4 passes a H, steric hindrance is encountered as depicted in the eclipsed state (Figure 9.21b, middle). The eclipsed state is an unfavorable, high-energy state (Figure 9.21c). Further rotation relieves some of the steric stress resulting in another low energy state referred to as the *gauche* state (Figure 9.21b, right). The *gauche* state is lower energy than the eclipsed state, but higher energy than the *trans* state, since

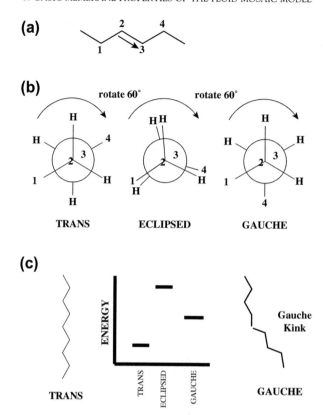

FIGURE 9.21 Melting of a saturated fatty acyl chain. Part (a) depicts a 4-carbon section somewhere along a saturated acyl chain existing in the un-melted *trans* state. Part (b) demonstrates Newman projections for the 4 sequential carbons depicted in (a). Upon heating the chain, rotation occurs around carbons 2 and 3. The *trans* configuration has the lowest possible energy (part c) as the largest groups, carbon 1 and carbon 4, are as far apart as possible. As the large C-4 passes a H, steric hindrance is encountered as depicted in the eclipsed state. The eclipsed state is an unfavorable, high-energy transient state (part c). Further rotation relieves some of the steric stress, resulting in another low energy state referred to as the *gauche* state. The *gauche* state is of lower energy than the eclipsed state, but higher energy than the *trans* state, since there is more steric interference between the closer, bulky C-1 and C-4 groups. This lower energy state creates a 'gauche kink' in the chain (part c). It is the accumulation of 'gauche kinks' in the chain that results in chain melting.

there is more steric interference between the closer, bulky C-1 and C-4 groups. This lower energy state creates a 'gauche kink' in the chain (Figure 9.21c, far right). It is the accumulation of 'gauche kinks' in the chain that results in melting. At 37°C there are ~2 kinks per acyl chain and the chain exists in the melted, liquid crystalline state. An acyl chain is essentially tethered to a bulky head group. The kinks then rapidly (~10^{-9} sec) work their way down the chain and are released at the bilayer interior. For this reason acyl chains exhibit minimal motion near the polar head and maximal motion at the methyl chain terminus (the omega end). Therefore, an acyl chain is more fluid at its omega end than at its alpha end.

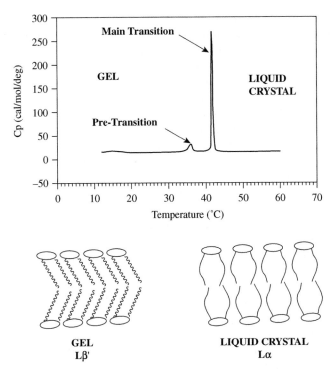

FIGURE 9.22 DSC scan of the gel ($L_{\beta'}$) to liquid crystal (L_α) transition for DPPC. For disaturated chain PCs in the gel state, the chains are tilted in relation to the polar head group. Upon melting to the liquid crystal state, the tilt disappears.

Due to the lack of '*gauche* kinks', saturated acyl chains below their melting point pack tightly while the same chains above their melting point pack loosely (Figure 9.22). Lipid packing as related to melting affects several important, basic membrane properties. Compared to gel-state membranes, liquid crystalline state membranes:

1. Are more poorly packed.
2. Occupy a larger area per lipid.
3. Are more fluid.
4. Are more permeable.
5. Are thinner.
6. Are less stable.
7. Are more dynamic.

Differential Scanning Calorimetry

The most direct method for monitoring melting in lipid bilayers is by differential scanning calorimetry (DSC). The technique was developed by E.S. Watson and M.J. O'Neill in 1960 and was rapidly moved into commercial production by 1963. DSC is a non-perturbing, analytical technique that can accurately monitor thermotropic phase behavior in lipid bilayers [37−42]. From DSC one can obtain T_ms, transition enthalpies, and information on phase. In general this technique involves heating a reference solution devoid of the membrane of interest and a sample containing a dilute aqueous subject membrane suspension. The technique measures the difference in the amount of heat required to increase the temperature of the sample and reference at a chosen rate. During the scan, the temperature of the sample and reference is very accurately kept the same.

The phospholipid most studied by DSC has been DPPC (16:0,16:0 PC), the workhorse of DSC membrane studies. This lipid is inexpensive and readily available in highly pure form from a number of commercial sources. The lipid is resistant to oxidation (it has two saturated acyl chains) and has a T_m that is not near the troublesome ice transition (0°C), nor is it so high as to make the lipid susceptible to massive heat-induced hydrolysis. A typical DSC scan for DPPC is shown in Figure 9.22. The main transition occurs at 41.3°C (Table 5.1). A much smaller transition, called the pre-transition, is observed at ~35.6°C. Since the pre-transition is not found in all phospholipid classes, it probably is related to the polar head group. For example DPPE (16:0,16:0 PE) has no pre-transition and has a T_m of 63°C. Profound differences in the DSC scans between DPPC and DPPE have been attributed to differences in head group size (Figure 9.23). PC has a much larger and more hydrated head group than does PE. As a result, at temperatures below T_m (in the gel state), the diameter of the PC head group (S) is wider than the combined diameters of the two all *trans* acyl chains (2Σ). Therefore $S > 2\Sigma$. The chains must then tilt relative to the head group until their combined diameters are the same size, $S = 2\Sigma$. Without tilting there would be a gap beneath the head that would have to be filled by water, a thermodynamic disaster. The larger the head group, the larger is the tilt angle. For example, in the gel state DMPC (14:0,14:0 PC) has a tilt angle of 12° while the much larger cerebroside (with a sugar head group) has a tilt angle of 41°. For PCs, upon melting the *gauche* kinks substantially increase the diameter of the acyl chains whereupon their sum becomes equal to, or greater than the head diameter ($S =/< 2\Sigma$), eliminating the tilt angle. For DPPE which has a small, poorly hydrated head group, in the gel state $S = 2\Sigma$ and the tilt angle is zero.

As depicted in Figure 9.23, gel state DPPC exhibits chain tilt, while DPPE does not. This may explain why PCs have a lower T_m than PEs. The chain tilt puts DPPC in a configuration more amenable to melting. The typical DSC scan shown in Figure 9.22 for DPPC shows both the main chain melting transition and the minor pre-transition. Both transitions are believed to be part of the same melting process. In the gel state ($L_{\beta'}$), the chains of DPPC are tilted. The tilt is lost upon transitioning into the melted liquid crystalline (L_α) state. Something unusual happens to DPPC in the temperature range spanning the pre-transition and main transition. Fluorescence and ESR methodologies (discussed below and in Chapter 10, respectively) indicate a co-existence of gel and fluid domains. More unusual, however, is the shape of the membrane over this temperature range. At the pre-transition, the flat membrane in the gel phase transforms into a periodically undulated bilayer, named the ripple or $P_{\beta'}$ phase (Figure 9.23). The ripple phase has been shown by electron density profiles obtained from small angle X-ray scattering to be an asymmetric undulation pattern resembling a saw-tooth with a wavelength of 120−160 Å, depending on acyl chain length. It is not at all clear if ripple structure has any biological importance as very small lipid 'contaminants,' including other membrane lipids, eliminate the pre-transition and hence the ripple phase. A biological

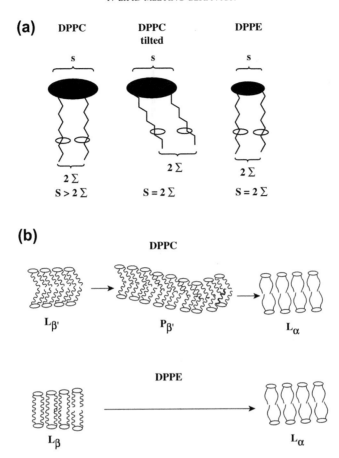

FIGURE 9.23   (a) Gel to liquid crystal phase transitions for DPPC and DPPE. Chain tilt noted for DPPC in the gel state ($L_{\beta'}$) is not seen with DPPE (L_{β}). Neither DPPC nor DPPE exhibit chain tilt in the liquid crystal ($L\alpha$) state. (b) Another unusual feature of the DPPC DSC scan that is not observed for DPPE membranes is the existence of 'ripple structure' ($P_{\beta'}$) between the pre-transition and the main transition. PEs exhibit neither a pre-transition nor ripple structure.

membrane contains hundreds to thousands of different lipids, greatly diminishing the likelihood of ripple phase. Ripple phase is also not observed for heteroacid molecular species that are predominant in membranes. Heteroacid phospholipids have different acyl chains attached to the sn-1 and sn-2 locations.

The DSC curve for DPPC in Figure 9.22 shows a narrow main transition. The narrower the transition, the more lipids melt as a single unit. Almost any lipid contaminant or a fast temperature (scan rate) will reduce and broaden the curve. In fact, an absolutely pure phospholipid scanned at an infinitely slow temperature ramp would theoretically result in a vertical line where all of the gel state lipid would melt at once. Lipid contaminants broaden and reduce the main transition. Figure 9.24 shows the effect of increasing amounts of cholesterol on the DPPC main transition. Even 1 mol % cholesterol noticeably affects the transition,

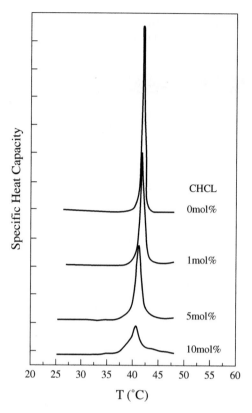

FIGURE 9.24   The effect of increasing levels of cholesterol on the main transition of DPPC as followed by DSC. The main transition shifts to lower temperature and exhibits a decrease in enthalpy (decrease in cooperativity) with increasing cholesterol. *Courtesy of Alfred C. Dumaual.*

and by 10 mol% cholesterol, most of the transition has disappeared. By 30 mol% cholesterol, the DPPC main transition is completely obliterated. From the shape of the curve, one can estimate the number of lipids melting as a unit. This empirical number, referred to as the 'cooperativity unit,' can be simply calculated from the DSC curve, as first described by Susan Mabrey and Julian Sturtevant in the mid 1970s [37] and nicely summarized in 1986 by Donald Small [39]. The cooperativity unit is defined as:

$$\text{Cooperativity Unit} = \Delta H^{\circ}_{vH}/\Delta H^{\circ}_{cal}$$

Where ΔH°_{vH} is the van't Hoff enthalpy and ΔH°_{cal} is the effective enthalpy which is the calculated area beneath the curve of the thermal transition expressed in cal/mol/gram. The van't Hoff enthalpy is defined by Julian Sturtevant as:

$$\Delta H^{\circ}_{vH} = 6.9T_m^2/\left[(T_2 - T_1)/2\right]$$

where T_m is the midpoint of the transition (°K) and the expression $(T_2 - T_1)/2$ is the peak width at one half the peak height. T_1 is the onset temperature and T_2 is the final (offset) temperature of the transition, both expressed in °K.

FIGURE 9.25   Julian M. Sturtevant (1908–2005). *Courtesy of Yale University*

Julian M. Sturtevant (Figure 9.25) was a pioneer in the field of biophysical chemistry with a world reputation on the application of thermochemistry to biological systems. He published into his 90s and died at the age of 97.

Since the DSC scan shape, and hence calculated cooperativity units, can be altered by changing the scan rate (a faster rate produces a broader curve and hence apparent lower cooperativity), the cooperativity unit calculations only have meaning compared to another lipid run under identical conditions. DSCs have the ability to scan over a wide range of temperature ramps. A 'normal' scan rate would be ~5°C/hr. This rate is sufficiently fast to allow for a complete scan in a reasonable amount of time, but not so fast as to obliterate curve structure. Also, lipids may become hydrolyzed or oxidized if the sample is left in the DSC crucible for too long. However, some very slow scan rates (~0.1°C/hr) may be employed to examine events occurring over a very narrow temperature range.

Fourier Transform Infrared (FT-IR) Spectroscopy

Although DSC is the best method to analyze lipid phase transitions, many other methodologies have successfully been employed. A very different method involves Fourier Transform Infrared (FT-IR) spectroscopy. FT-IR is a versatile, non-perturbing, powerful technique that can be used to obtain information on all regions of the phospholipid molecule simultaneously (for general discussion of FT-IR see [43,44]). Instead of (slowly) scanning through the entire spectrum as is done in most UV-visible applications, FT-IR irradiates the sample with all IR wavelengths at once and the entire signal is captured and subjected to Fourier analysis. This allows for multiple spectra to be rapidly obtained and the signals averaged resulting in vastly improved sensitivity. Nanogram samples can be accurately analyzed by FT-IR. Modern FT-IR spectrometers also have the advantage of being able to eliminate the large, troublesome water absorbance, allowing for the ability to measure membranes dispersed in aqueous solution [45,46].

FT-IR spectroscopy detects vibrations during which the electrical dipole moment changes. Of course this time is very short (~10^{11} s^{-1}). Most of the measured absorbance bands are associated with stretching and bending vibrations that occur at discrete positions in the mid-IR spectral range (between 2,500 and 25,000 nm). Traditionally, IR studies do not report wavelengths but rather wave numbers. The spectral range for FT-IR is therefore 4,000 to 400 cm^{-1}. Of particular importance to membrane studies are the absorbance bands at about 2,900 (C-H stretching), 1,740 (C=O stretching in esters and carboxylic acids), 1,400 (C-O stretching in carboxylates). 1,235 (P=O stretching in phosphate esters), and 1,000 (C-O stretching in carbohydrates) cm^{-1}.

For phospholipids, the acyl tails (including *trans/gauche* ratios) are followed by C-H vibrations (2,800–3,100 cm^{-1}), the glycerol-acyl chain interface by ester stretching (1,700–1,750 cm^{-1}) and the head group phosphate stretching (1,220–1,240 cm^{-1}) [45,46]. Therefore, FT-IR spectra provide a complete 'molecular snapshot' yielding information on phospholipid conformation, and dynamics from the tip of the polar head group to the omega end of the acyl tail. Figure 9.26 shows the gel to liquid crystal phase transition as determined by the absorbance maximum of the symmetric stretching band of CH$_2$ groups in liposomes composed of: a. 16:1,16:1 PC; b. 18:1,18:1 PC; c.14:0,14:0 PC; and d. 16:0,16:0 PC [46]. T$_m$s are in close agreement with values obtained by other methods including DSC.

Fluorescence Polarization

Another method that is commonly used to follow lipid phase transitions and fluidity in membranes is fluorescence polarization. Fluorescence was first reported, and the term coined, by George Gabriel Stokes (Figure 9.27) in 1852 [46,47]. Stokes noted that the mineral fluorite (calcium fluoride), emitted visible light when illuminated with 'invisible radiation' (later identified as ultra-violet light). It was from this simple, but seemingly unrelated

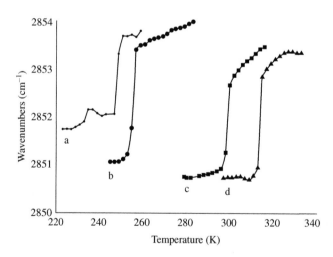

FIGURE 9.26   Gel to liquid crystal phase transition as determined by the absorbance maximum of the symmetric stretching band of CH$_2$ groups (by FT-IR) in liposomes composed of: a. 16:1,16:1 PC; b. 18:1,18:1 PC; c.14:0,14:0 PC; d. 16:0,16:0 PC. T$_m$s are in close agreement with values obtained by other methods including DSC [46].

FIGURE 9.27 George Gabriel Stokes (1819–1903) [58].

observation, that the most powerful and versatile technique in modern membrane studies, fluorescence, had its humble beginnings.

The technique of fluorescence is reviewed in two comprehensive books by Joseph Lakowitz [48,49]. Fluorescence occurs after a molecule (a fluorophore) absorbs light (is excited) at one wavelength and radiates (emits) light at a longer wavelength. Basic fluorescence is depicted in Figure 9.28. The specific frequency of excitation and emission are dependent on the characteristics of the fluorophore. The unexcited molecule has its electrons in the lowest energy or ground state, S_0. Upon absorbing light, a ground state electron is boosted to its first (electronically) excited state, S_1. A fluorophore in its excited S_1 state can relax by two basic pathways. It can undergo non-radiative relaxation (internal conversion) by losing heat to its surroundings, before returning to the ground state ($S_1 \rightarrow S_0$) in a process referred to as fluorescence. The electron spin direction does not change during fluorescence. However, if the electron spin changes during the excited state, a triplet state, T, results. Emitted radiation during return to the ground state ($T \rightarrow S_0$) is called phosphorescence. Since most membrane studies involve fluorescence, phosphorescence will not be discussed, and is not included in Figure 9.28.

The time domains involved in fluorescence are very fast. The process of light absorption and production of the excited state ($S_0 \rightarrow S_1$) takes $\sim 10^{-11}$ sec. It takes $\sim 10^{-8}$ sec to return from the excited to the ground state ($S_1 \rightarrow S_0$). The average time in which the fluorophore stays in its excited state before emitting a photon is referred to as the fluorescence lifetime. It is the time between these two events, light absorbance and fluorescence, that is important for membrane structure studies. In contrast, phosphorescence is relatively slow ($>10^{-3}$ sec).

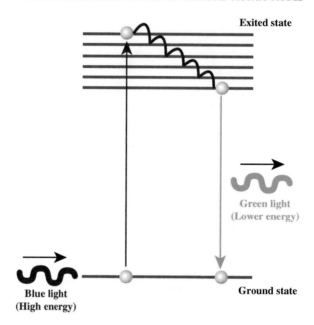

FIGURE 9.28   Basic fluorescence. Light is absorbed by the fluorophore and a ground state electron is boosted to an excited state (10^{-11} s). The electron then loses some of its energy by non-radiative relaxation (internal conversion) to a lower energy excited state before returning to the ground state in the process called fluorescence (~10^{-8} s). The emitted fluorescent light is of a longer wavelength (lower energy) than the light that was initially absorbed. *Courtesy of ScienceBlogs, LLC*

Fluorescence and ESR are 'fast' techniques (i.e. they can be used to monitor fast processes). Since NMR is greater than an order of magnitude slower, events that can be seen with fluorescence and ESR may not be detectable by NMR. Detectable frequencies are ~10^5 sec^{-1} for ^{2}H'NMR and ~10^8 sec^{-1} for ESR and fluorescence.

The fluorescence technique most often used to investigate membrane structure, fluorescence polarization (FP), was first described by the French physicist Jean Baptiste Perrin in 1926 and is depicted in Figure 9.29. Perrin was a diverse and prolific physicist in the first quarter of the 20th century, winning many awards including the 1926 Nobel Prize in Physics. Plane-polarized light, produced by passing light through a polarizer, is used to excite a fluorophore [48–50]. FP is based on the principle that if the fluorophore is immobile, fluoresced light is emitted in the same plane as the plane-polarized excited light. If the exciting light is polarized in the vertical direction, then the fluoresced light will also be polarized in the vertical direction. This, however, is dependent upon a lack of movement of the fluorophore during the excited state. If the fluorophore remains stationary from the time it absorbs light until the time it emits light, the fluorescent light will remain completely polarized. However, if the molecule rotates or tumbles during this time, some of the light will be polarized in different directions (it becomes depolarized). The extent of depolarization is monitored by comparing the intensity of vertically polarized fluorescent light (observed through a vertical polarizer) to depolarized light (observed

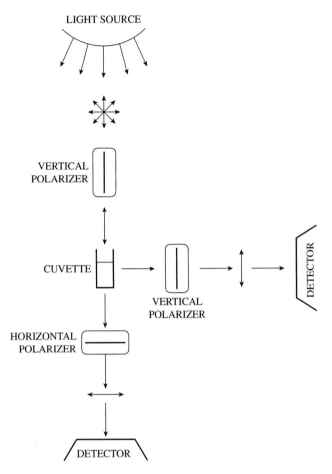

FIGURE 9.29 Method to measure fluorescence polarization. Light that is initially polarized in all planes is first passed through a polarizer selecting for light that will be polarized in only one plane. This light passes through a sample in a 4-sided cuvette. Upon emerging from the sample, the intensity of light polarized at planes 90° to one another (vertical and horizontal) is measured. The Anisotropy can then be calculated from these intensities as described in the text.

through a horizontal polarizer). If the fluorescing molecule is very large or immobile during the fluorescent excited state, all of the light will be seen through the vertical polarized light detector. The horizontal detector, attached behind the horizontal polarizer, will detect no light. If the fluorescing molecule is very small or rapidly tumbling, then equal amounts of emitted light will be observed through the vertical and horizontal polarizer detectors. The degree of depolarization can be expressed as either a Polarization or an Anisotropy value. Although the two values can be readily inter-converted, Anisotropy is more frequently encountered.

$$\text{Anisotropy} = I_{(V)} - I_{(H)}/I_{(V)} + 2I_{(H)}$$

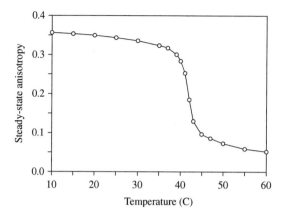

FIGURE 9.30   Gel-to-liquid crystal transition of DPPC as followed by the steady state anisotropy of DPH. *Courtesy of Charles University in Prague, Faculty of Science.*

where I is the emitted light intensity as measured through the vertical (V) or horizontal (H) polarizer detectors. If no molecular motion occurs between the time of absorbance and the time of emission, Anisotropy has a limiting value of 0.4. If rapid motion occurs, the limiting value for Anisotropy is zero. From the decay of the Anisotropy

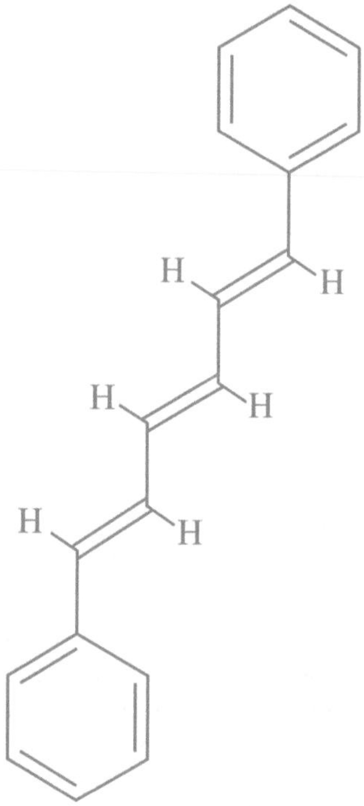

FIGURE 9.31   1,6-diphenylhexatriene (DPH).

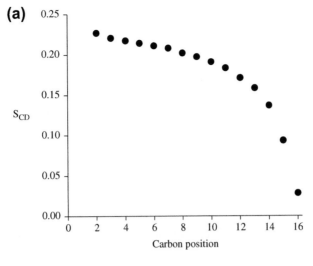

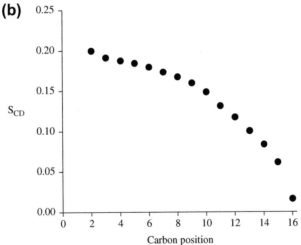

FIGURE 9.32   NMR order parameters (S_{CD}) for the palmitoyl group on the *sn*-1 chain of DPPC (16:0,16:0 PC) (a) and POPC (16:0,18:1 PC) (b). (Provided by Stephen R Wassall, Department of Physics, Indiana University-Purdue University Indianapolis.)

it is possible to obtain an order parameter that has the same definition as for ESR and NMR.

Fluorescence Polarization can also be used to monitor lipid phase transitions [51]. Figure 9.30 shows the gel-to-liquid crystal transition for DPPC as followed by the fluorescent probe 1,6-diphenylhexatriene (DPH, Figure 9.31). DPH has historically been the most used of all FP probes for membrane studies [52].

DPH possesses several important properties justifying its use in FP investigations. This fluorophore partitions strongly into membranes, is highly fluorescent, does not bind to

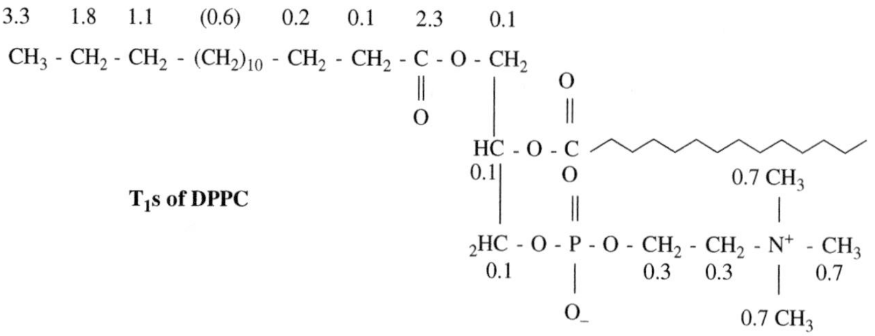

$$3.3 \quad 1.8 \quad 1.1 \quad (0.6) \quad 0.2 \quad 0.1 \quad 2.3 \quad 0.1$$

$$CH_3 - CH_2 - CH_2 - (CH_2)_{10} - CH_2 - CH_2 - C - O - CH_2$$

T₁s of DPPC

FIGURE 9.33 ^{13}C-NMR-derived T_1s for DPPC. A high T_1 means the carbon exhibits a lot of motion. A lower T_1 indicates less motion. 'Micro-viscosity' of the carbons of DPPC varies 70 fold from the most restricted carbons near the polar head group to the least restricted carbon at the omega terminus of the acyl chain.

proteins, and is sensitive to membrane physical state. The transition temperature, T_m, of DPPC as determined by FP (Figure 9.30) agrees with those reported by DSC (Figure 9.22) and FTIR (Figure 9.26).

G. MEMBRANE 'FLUIDITY'

The term 'fluidity' is one of the most used, yet one of the most poorly understood in the life sciences [53]. While everyone has a general feel for what 'fluidity' means, the devil is in the detail. A dictionary definition of a 'fluid' is something that is 'capable of flowing, not a solid'. In biology, fluidity refers to the viscosity of the lipid bilayer component of a cell membrane. For water, fluidity is defined as the inverse of viscosity. Historically, viscosity was easily estimated by simply measuring how fast a spherical marble falls through a solution. The concept of viscosity was therefore developed around isotropic motion with equal forces on all sides of a perfect sphere. But how much of this applies to a complex, highly anisotropic biological membrane?

In a biological membrane there are three complex types of motion to consider. First there is non-homogeneous lateral movement in the plane of the lipid bilayer. This is complicated by lipid domains of various compositions and properties, crowding by multiple proteins and protein complexes and interactions with the sub-membrane cytoskeleton. Next there is rotational movement around the longitudinal axis of the molecule. Finally there is trans-membrane lipid movement or flip-flop. In addition, regardless of how membrane 'fluidity' is defined, there will be a fluidity gradient from the aqueous interface through to the bilayer interior. The extreme heterogeneity of biological membranes makes analysis of 'fluidity' far more complicated than simply dropping a marble through a solution. Usually with membranes, molecular motion is monitored with either spin probes or fluorescent probes [54]. However, unlike marbles, the probes are not perfect spheres and so do not behave isotropically. Instead, they exhibit preferred orientations (they are anisotropic) in the bilayer and may preferentially partition into proteins or different membrane domains. As a result, membrane probes are sensitive to both rates of motion and to constraints on motion, and so report on both dynamics and order. Therefore, in biological membranes the term 'fluidity' is complex, technique-dependent, and quantifies a variety of parameters including:

1. Rotational correlation times
2. Order parameters
3. Steady state anisotropy
4. Partitioning of probes between the membrane and water

Although a number representing 'fluidity' can be obtained, its precise definition may be a matter of contention. Further complicating the problem is the enormous number of membrane probes that are commercially available. The major provider of membrane probes has for many years been Molecular Probes in Eugene, Oregon (now owned by Invitrogen, Carlsbad, CA). Chapter 13 of the Molecular Probes catalog, entitled 'Probes for Lipids and Membranes', provides an excellent review of the topic.

It is clear from several different types of measurements that 'fluidity' gradients exist from the aqueous interface through to the bilayer interior. This has often been demonstrated using fluorescence polarization as well as ESR. Both of these techniques require adding an anisotropic bulky, and hence perturbing, probe at different locations down the acyl chain. Both FP and ESR show a gradual, continuous increase in 'fluidity' from the carbons near the α or carboxylate end to the omega or terminal carbon [54,55]. Unfortunately, the bulkiness of the fluoro or spin probes disturbs the chains to such an extent that important detail of the 'fluidity' gradient can be lost.

NMR has been used to monitor the 'fluidity' gradient without perturbing the system. Without getting bogged down in experimental details, two types of NMR techniques (order parameter (Figure 9.32) and T_1 relaxation measurements) (Figure 9.33) have demonstrated that 'fluidity' increases at a steady but slow rate for about the first 8 to 10 carbons (called the plateau region), after which 'fluidity' progressively increases through the rest of the chain. This pattern is characteristic of all bilayers. However, details concerning the size and location of the plateau and the 'fluidity' of the most fluid terminal region vary depending on the lipid composition of the bilayer. In fact the shape of the 'fluidity' gradient can provide a crude fingerprint of the bilayer. The NMR-derived order parameter (S_{CD}) can be used to estimate how isotropic the motion is at a particular carbon. A low S_{CD} indicates a large range of motion and hence high 'fluidity'. The range of values can vary depending on what is being measured, but often range from ~0 (no motion) to ~1.0 (isotropic motion). Figure 9.32 presents S_{CD} values for the *sn*-1 palmitoyl chain of DPPC and POPC (16:0,18:1 PC). The typical trans-membrane 'fluidity' gradient pattern with a plateau region can be easily seen but it is also clear that the two profiles, while having a similar general shape, do exhibit differences. For example, the plateau region of POPC is more pronounced than for DPPC. The profile shown for DPPC was obtained at a relatively high temperature of 67°C. At a reduced temperature the plateau region would be longer and more pronounced.

Fast molecular motion can be followed by T_1 relaxation using NMR. A large T_1 indicates more rapid motion. Figure 9.33 shows T_1s for various carbons on DPPC. Again a substantial plateau region is evident. The viscosity varies about 70 fold from the highly fluid omega methyl terminus to carbons on the less 'fluid' polar head group.

The biophysical techniques used to study membrane 'fluidity' are best suited to compare *changes* in membrane physical properties brought about by environmental and membrane-soluble agents. It is predicted that anything that will increase membrane viscosity will decrease 'fluidity'. Examples include decreasing temperature, increasing pressure, increasing cholesterol content, and altering environmental pH and ionic strength. In addition, the nature of the phospholipid head groups and acyl chains, as well as favorable and unfavorable lipid–lipid and lipid–protein interactions, will also affect 'fluidity'. These concepts will be discussed in Chapter 10 under 'homeoviscous adaptation'.

SUMMARY

Every aspect of membrane structural studies involves parameters that are very small (micron to angstrom) and very fast (microseconds to picoseconds) requiring an array of esoteric biophysical methodologies. Membrane parameters discussed in this chapter include: membrane thickness, bilayer stability, membrane protein, carbohydrate and lipid asymmetry, lipid trans-membrane movement (flip-flop), membrane lateral diffusion, membrane-cytoskeleton interaction, lipid melting behavior, and membrane 'fluidity'. These basic membrane properties were discovered by countless investigators over ~7 decades and formed the basis of the Singer-Nicholson Fluid Mosaic model of membrane structure that is currently in vogue.

Chapter 10 will continue the discussion of membrane physical properties including: lipid affinities, lipid phases, lipid–protein interactions (hydrophobic match), and lipid interdigitation.

References

[1]   Fricke H. The electrical capacity of suspensions with special reference to blood. J Gen Physiol 1925;9:137–52.
[2]   Boys C Soap Bubbles: 1958. Their Colors and Forces which Mold Them. Dover Publications. (This is a reprint of the revised 1911 publication by C.V. Boys).
[3]   Mueller P, Rudin DO, Tien HT, Wescott WC. Reconstitution of cell membrane structure in vitro and its transformation into an excitable system. Nature 1962;194:979–80.
[4]   Tulenko TN, Chen M, Mason PE, Mason RP. Physical effects of cholesterol on arterial smooth muscle membranes: evidence of immiscible cholesterol domains and alterations in bilayer width during atherogenesis. J Lipid Res 1998;39:947–56.
[5]   Fernández-Morán H, Oda T, Blair PV, Green DE. A macromolecular repeating unit of mitochondrial structure and function. Correlated electron microscopic and biochemical studies of isolated mitochondria and submitochondrial particles of beef heart muscle. J Cell Biol 1964;22:63–100.
[6]   Sharon N, Lis H. Lectins. 2nd ed. Dordrecht, The Netherlands: Springer; 2007.
[7]   Edwards R. 2008. Poison-tip umbrella assassination of Georgi Markov reinvestigated. The Telegraph June 19, 2008.
[8]   Op den Kamp JAF. Lipid asymmetry in membranes. Annu Rev Biochem 1979;48:47–71.
[9]   Devaux PF. Static and dynamic lipid asymmetry in cell membranes. Biochemistry 1991;30:1163–73.
[10] Bretscher MSJ. Phosphatidyl-ethanolamine: Differential labeling in intact cells and cell ghosts of human erythrocytes by a membrane-impermeable reagent. Mol Biol 1972;71:523–8.
[11] Steck TL, Ye J, Lang Y. Probing red cell membrane cholesterol movement with cyclodextrin. Biophys J 2002;83:2118–25.
[12] Seigneuret M, Devaux PF. ATP-dependent asymmetric distribution of spin-labeled phospholipids in the erythrocyte membrane: relation to shape changes. Proc Natl Acad Sci USA 1984;81:3751–5.

[13] Devaux PF. Phospholipid flippases. FEBS Lett 1988;234:8−12.

[14] Daleke DL. Phospholipid flippases. J Biol Chem 2007;282:821−5.

[15] Daleke DL. Regulation of transbilayer plasma membrane phospholipid asymmetry. J Lipid Res 2003; 44:233−42.

[16] Boon JM, Smith BD. Chemical control of phospholipid distribution across bilayer membranes. Med Res Rev 2002;22:251−81.

[17] Pomorski T, Menon AK. Lipid flippases and their biological functions. Cell Mol Life Sci 2007;63:2908−21.

[18] Frye LD, Edidin M. The rapid intermixing of cell surface antigens after formation of mouse-human hetero-karyons. J Cell Sci 1970;7:319−35.

[19] Axelrod D, Koppel DE, Schlessinger J, Elson EL, Webb WW. Mobility measurements by analysis of fluores-cence photobleaching recovery kinetics. Biophys J 1976;16:1055−69.

[20] University of California Irvine, Developmental Biology Center. 2006.

[21] Saxton MJ, Jacobson K. Single Particle Tracking: Applications to membrane dynamics. Annu Rev Biophys Biomol Struct 1997;26:373−99.

[22] Abbott A. Cell biology: Hopping fences. Nature 2005;433:680−3.

[23] Sheetz MP. Membrane skeletal dynamics: role in modulation of red cell deformability, mobility of trans-membrane proteins, and shape. Semin Hematol 1983;20:175−88.

[24] De Brabander M, Geuens G, Nuydens R, Moeremans M, de Mey J. Probing microtubule-dependent intra-cellular motility with nanometre particle video ultramicroscopy (nanovid ultramicroscopy). Cytobios 1985;43:273−83.

[25] Kusumi A, Sako Y, Yamamoto M. Confined lateral diffusion of membrane receptors as studied by single particle tracking (nanovid microscopy) of calcium-induced differentiation in cultured epithelial cells. Biophys J 1993;65:2021−40.

[26] Wolf DE, Kinsey W, Lennarz W, Edidin M. Changes in the organization of the sea urchin egg plasma membrane upon fertilization: indications from the lateral diffusion rates of lipid-soluble fluorescent dyes. Dev Biol 1981;81:133−8.

[27] Thompson TE, Huang C. In: Andreoli TE, Hoffman JF, Fanestil DD, editors. Physiology of Membrane Disorders. New York: Plenum Press; 1978 [Chapter 2].

[28] Van der Steen AT, Taraschi TF, Voorhout WF, De Kruijff B. Barrier properties of glycophorin-phospholipid systems prepared by different methods. Biochim Biophys Acta 1983;733:51−64.

[29] Greenhut SF, Roseman MA. Cytochrome b5 induced flip-flop of phospholipids in sonicated vesicles. Biochemistry 1985;24:1252−60.

[30] Just W. Trans bilayer (flip-flop) lipid motion and lipid scrambling in membranes. FEBS Lett 2010;584:1779−86.

[31] Wirtz KWA, Zilversmit DB. Exchange of Phospholipids between Liver Mitochondria and Microsomes *in Vitro*. J Biol Chem 1968;243:3596−602.

[32] Wirtz KWA. Phospholipid transfer proteins. Annu Rev Biochem 1991;60:73−99.

[33] Hamilton JA. Transport of fatty acids across membranes by the diffusion mechanism. Prostag Leukot Essent Fatty Acids 1999;60:291−7.

[34] Homan R, Pownal HJ. Transbilayer diffusion of phospholipids: dependence on headgroup structure and acyl chain length. Biochim Biophys Acta 1988;938:155−66.

[35] Middelkoop E, Lubin BH, Op den Kamp JA, Roelofsen B. Flip-flop rates of individual molecular species of phosphatidylcholine in the human red cell membrane. Biochim Biophys Acta 1986;855:421−4.

[36] Armstrong VT, Brzustowicz MR, Wassall SR, Jenski LJ, Stillwell W. Docosahexaenoic acid and phospholipid flip-flop. Arch Biochem Biophys 2003;414:74−82.

[37] Mabrey S, Sturtevant JM. Investigation of phase transitions of lipids and lipid mixtures by high sensitivity differential scanning calorimetry. Proc Natl Acad Sci USA 1976;73:3862−6.

[38] Sturtevant JM. Some applications of calorimetry in biochemistry and biology. Ann Rev Biophys Bioeng 1974;3:35−51.

[39] Small DM. In: Handbook of Lipid Research. The Physical Chemistry of Lipids. From Alkanes to Phospho-lipids. New York: Plenum Press; 1986. p. 33−6.

[40] Lewis RN, Mannock DA, McElhaney RN. Differential scanning calorimetry in the study of lipid phase transitions in model and biological membranes: practical considerations. Methods Mol Biol 2007; 400:171−95.

[41] Sugar IP. Cooperativity and classification of phase transitions. Application to one- and two-component phospholipid membranes. J Phys Chem 1987;91:95–101.

[42] Heimburg T. Thermal biophysics of membranes tutorials in biophysics. Wiley-VCH; 2007.

[43] Griffiths PR, De Haseth JA. Fourier-Transform Infrared Spectroscopy. In: A Series of Monographs on Analytical Chemistry and its Applications. 2nd ed. Wiley-Interscience; 2007.

[44] Newport Corporation. Introduction to FT-IR Spectroscopy, www.newport.com; Irving, CA; 2011.

[45] Mantsch HH, McElhaney RN. Phospholipid phase transitions in model and biological membranes as studied by infrared spectroscopy. Chem Phys Lipids 1991;57:213–26.

[46] Pastiorius AMA. Biochemical applications of FT-IR spectroscopy. Spectrosc Eur 1995;7:9–15.

[47] Stokes GG. On the change of refrangibility of light. Philos Trans R Soc London 1852;142:463–562.

[48] Lakowicz JR. Principles of Fluorescence Spectroscopy. New York: Plenum Press; 1983.

[49] Lakowicz JR. Principles of Fluorescence Spectroscopy. 2nd ed. New York, NY: Kluwer Academic/Plenum; 1999.

[50] Albani JR. Fluorescence Polarization. In: Principles and Applications of Fluorescence Spectroscopy. Wiley Online Library; 2008 [Chapter 11].

[51] Papahadjopoulos D, Jacobson K, Nir S, Isac I. Phase transitions in phospholipid vesicles fluorescence polarization and permeability measurements concerning the effect of temperature and cholesterol. Biochim Biophys Acta 1973;311:330–48.

[52] Lentz B. Membrane 'fluidity' as detected by diphenylhexatriene probes. Chem Phys Lipids 1989;50:171–90.

[53] Kates M, Kuksis A, editors. Membrane Fluidity: Biophysical Techniques and Cellular Regulation. Clifton, New Jersey: The Humana Press; 1982.

[54] Anderson HC. Probes of membrane structure. Annu Rev Biochem 1978;47:359–83.

[55] Collins JM, Dominey RN, Grogan WM. Shape of the fluidity gradient in the plasma membrane of living HeLa cells. J Lipid Res 1990;31:261–70.

[56] IVIC (Instituto Venezolano de Investigaciones Científicas): http://en.wikipedia.org/wiki/File:Fernadez Moran.jpg

[57] Courtesy of the Kusumi Lab, Institute for Frontier Medical Sciences, Kyoto University, Kyoto University School of Medicine, Graduate School of Medical Sciences.

[58] Unknown: http://en.wikipedia.org/wiki/File:Ggstokes.jpg; public domain.

O U T L I N E

A. **Complex Lipid Interactions**   **176**
Lipid Affinities   176
A DSC Study   178
An X-Ray Diffraction Study   179
Lipid Raft Detergent Extractions   181

B. **Non-Lamellar Phases**   **182**
Inverted Hexagonal Phase (H_{II})   185
Cubic Phase   186
^{31}-P-NMR   187

C. **Lipid Phase Diagrams**   **190**
Single-Component Lipid Phase Diagrams   190
Two-Component Lipid Phase Diagrams   191
A Three-Component Lipid Phase Diagram   192

D. **Lipid–Protein Interactions**   **194**
1. Annular Lipids   194
2. Co-Factor (Non-Annular) Lipids   197
   Ca^{2+} ATPase   198
   Protein Kinase C   199
   Na^{+}/K^{+} ATPase   199
   Potassium Channel KcsA   199

3. Non-Anionic Co-Factor Lipids   200
   Rhodopsin   200
   D-β-Hydroxybutyrate
   Dehydrogenase   202
   Cytochrome c Oxidase   202
4. Hydrophobic Match   203
   The Hydrophobic Match Length:
   A DSC Study   203
   Biological Function of the
   Hydrophobic Match   205

E. **Lipid Interdigitation**   **206**
Interdigitated-Lipid Shape   206
Interdigitated Structure is Inherent
in Symmetrical Chain Phospholipids   208
Methods to Detect Interdigitation   209
Interdigitation Caused by
Perturbations   209
Biological Relevance   210

Summary   **210**

References   **211**

The previous chapter (Chapter 9) investigated some of the major principles of the Fluid Mosaic model. While these principles form the core of understanding basic membrane structure, there are many other, more poorly understood membrane properties and concepts that must be considered. Membranes are just very complex. Several of these properties will be discussed in Chapters 10 and 11.

An Introduction to Biological Membranes
http://dx.doi.org/10.1016/B978-0-444-52153-8.00010-6

A. COMPLEX LIPID INTERACTIONS

Lipid Affinities

A basic hallmark of all biological membranes is their extreme heterogeneity. This topic will come up over and over again in discussions of lipid—lipid and lipid—protein interactions. Since a single biological membrane is composed of hundreds or even thousands of different lipid molecular species, it is logical to assume that they will exhibit different affinities for one another. Each lipid may prefer to be in the company of one type of lipid while avoiding that of another type. One well studied example of this is the lipid raft. Lipid rafts (discussed in Chapter 8) are enriched in sphingolipids and cholesterol, the components responsible for the raft's basic physical properties. In fact, cholesterol is believed to be the molecular 'glue' that holds rafts together. Rafts are destroyed upon the addition of cyclodextrin, (Figure 10.1) a family of cyclic glucose oligosaccharides that are often used to extract cholesterol from membranes [1].

Through the years cholesterol has probably received more attention than any other membrane lipid. Cholesterol was first identified in 1769 as a solid in gallstones by François Poulletier de la Salle. However, the man most associated with early cholesterol studies was the French chemist Eugene Chevreul (Figure 10.2). Chevreul was an early pioneer on lipid

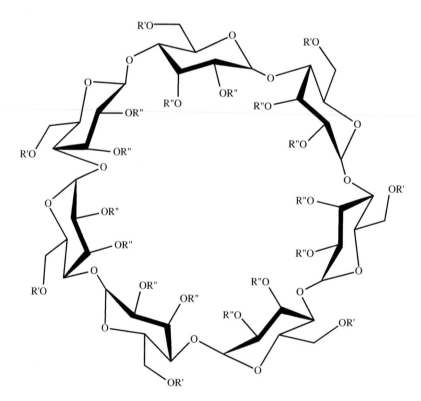

FIGURE 10.1   α-Cyclodextrin. *Courtesy of RDI Division of Fitzgerald Industries Intl*

FIGURE 10.2   Eugene Chevreul (1786–1889). *Engraved by C. Cook from a drawing by Maurir.*

components of animal fat. He was the first to isolate and name stearic and oleic acid, and also, in 1815 coined the term 'cholesterine' from the Greek word *chole* for bile. Chevreul is credited with creating margarine and, due to his longevity (102 years), he was also a pioneer in the field of gerontology. Therefore he was both an investigator and a subject of investigation!

Most of the attention given to cholesterol is due to its deleterious role in human atherosclerosis (heart disease) [2]. Just the word cholesterol has become synonymous with bad things. As if cholesterol has not been maligned enough, a recent report has even linked this unfortunate sterol to tail-chasing in dogs [3]! But cholesterol is not all evil. It does play an essential role in *some*, but not all, membranes. As discussed in Chapter 5, cholesterol is by far the major polar lipid in mammalian plasma membranes, comprising up to ~60 mol% of the polar lipids, yet it is almost totally absent in other mammalian intracellular membranes (e.g. the mitochondrial inner membrane). It is also hard to understand how cholesterol can play such an essential role in life processes in mammalian cells yet be essentially missing in fungi, plants, and bacteria. Simplistically, cholesterol is a 'membrane homogenizer'. It intercalates between membrane phospholipids and sphingolipids and helps to control membrane lipid packing. Lipid packing is an essential feature in membrane 'fluidity' (Chapter 9) and permeability (Chapter 14). In a tightly packed gel state, membrane insertion of cholesterol decreases lipid packing, thus increasing membrane 'fluidity' and permeability. In these membranes cholesterol behaves as a lipid-soluble contaminant. In contrast, in a liquid crystalline or fluid state, membrane cholesterol increases lipid packing, thus decreasing 'fluidity' and permeability. In these membranes cholesterol prevents the formation of *gauche* kinks that are responsible for acyl chain melting (see Chapter 9).

Mixing cholesterol and phospholipids often produces an unusual fluid membrane state called the liquid ordered (l_o) state [4–6]. The l_o state behaves as if it were half way between the gel (L_β) and liquid crystalline (L_α) states, having properties of both. In the l_o state, acyl chains are extended (they have fewer *gauche* kinks) and so in this sense behave like a gel state.

However, lateral diffusion in the l_o state is almost as great as in the liquid crystalline state. Since acyl chains are extended by cholesterol, membrane domains that are rich in cholesterol tend to be thicker than regions that are devoid of the sterol. Since lipid rafts are highly enriched in cholesterol, they are thicker than the surrounding non-raft membrane.

A DSC Study

In Figure 9.24 it was shown that cholesterol slightly lowers the T_m, substantially broadens the main transition, and diminishes the ΔH of DPPC in lipid bilayers. More than a decade before the advent of lipid rafts, van Dijck et al. [7] and Demel et al. [8] realized they could use the obliteration of T_ms to monitor relative affinities of cholesterol for various phospholipids. van Dijck's 1976 experiment, outlined in Figure 10.3, compared PC to PE. The experiments were based on having two phospholipids with T_ms sufficiently far apart that their

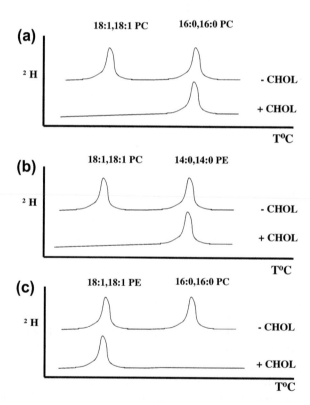

FIGURE 10.3   Relative affinity of cholesterol for PC and PE as detected by DSC. In Panel (a) cholesterol is shown to preferentially obliterate the lower melting transition (18:1,18:1 PC > 16:0,16:0 PC) when both transitions are for the same phospholipid type (PC in this example). In Panel (b) cholesterol is shown to preferentially obliterate the lower melting 18:1,18:1 PC over the higher melting 14:0,14:0 PE. In Panel (c), where the T_ms of PC and PE are reversed, cholesterol preferentially obliterates 16:0,16:0 PC even though its T_m is higher than that of 18:1,18:1 PE. This experiment was interpreted as evidence that cholesterol associates more strongly with PCs than with PEs. (*Experiment is redrawn from [7]*).

transitions are distinct and do not overlap (they are monotectic). If both the high melting and low melting phospholipids were of the same class (i.e. they were both PC or both PE), low levels of cholesterol first reduced the lower melting transition. This is depicted in Figure 10.3, panel (a) for 18:1,18:1 PC (low T_m) and 16:0,16:0 PC (high T_m). Higher levels of cholesterol eventually diminished the higher melting transition as well. Figure 10.3, panel (b) compares a low melting PC (18:1,18:1 PC) to a high PE (14:0,14:0 PE). Cholesterol first obliterated the low melting PC transition. This does not indicate whether cholesterol prefers PC over PE or whether cholesterol simply prefers the lower melting transition lipid. In Figure 10.3, panel (c), the transition temperatures of the PC and PE were reversed. In this experiment PE (18:1,18:1 PE) was the lower melting component while PC (16:0,16:0 PC) was the higher melting component. Here cholesterol first obliterated the higher melting, PC component. From these experiments it was concluded that cholesterol associates better with PC than PE. Later Demel et al. [8] extended these studies to compare several of the most common membrane lipids and reported that cholesterol has the following affinity:

$$SM > PS, \ PG > PC > PE$$

This DSC study was an early indication that cholesterol associates more strongly with SM than with other membrane lipids. A decade later Simons proposed the existence of lipid rafts, cell signaling membrane domains that are highly enriched in SM and cholesterol (Chapter 8).

An X-Ray Diffraction Study

Compatibility with cholesterol should depend not only on the type of phospholipid headgroup but also on the nature of the acyl chains. The structure of cholesterol with its 4 inflexible rings predicts that the sterol may not be compatible with highly flexible, poly-unsaturated chains. One method that has been used to test this premise involves measuring cholesterol solubility in membranes by X-ray diffraction (XRD) [9,10]. In Chapter 9 it was noted that a different application of this technique (an electron density profile, Figure 9.5) has been used to measure membrane thickness. Cholesterol's solubility in membranes can be determined by measuring radial intensity profiles plotted against reciprocal space (I-q plots) that show distinct peaks indicating how much cholesterol has been excluded from the membrane in the form of cholesterol monohydrate crystals. In the experiments outlined in Figures 10.4 and 10.5 and summarized in Table 10.1, Martin Caffrey and co-workers made lipid bilayer vesicles from different PC and PE molecular species with increasing mol fractions of cholesterol. At low levels, cholesterol was accommodated into the bilayer structure and no second order scattering peaks due to excluded cholesterol monohydrate crystals could be detected.

However, at some critical concentration, cholesterol exceeds the carrying capacity of the bilayer and is excluded. The excluded monohydrate crystals are observed in certain regions of the I-q plots (Figure 10.4). The integrated intensities of the scattering peaks 002 (0.3701 Å^{-1}), 020 (1.033 Å^{-1}), and 200 (1.044 Å^{-1}) were combined and plotted against the mol% cholesterol (Figure 10.5). Linear extrapolation of these plots to zero gives an accurate estimate (± 1.0 mol%) of cholesterol solubility limit in bilayers made from various phospholipids. The results are compiled in Table 10.1. This table confirms that PC bilayers can accommodate

TABLE 10.1   Limiting Solubility of Cholesterol in Various PC and PE
Lipid Membranes. Values were Obtained from X-ray
Diffraction (Figure 10.4) and I-q plots (Figure 10.5) [9].

PE	Cholesterol solubility (mol%)	PC	Cholesterol solubility (mol%)
16:0−18:1	51 ± 3^1	16:0−18:1	65 ± 3^4
16:0−18:2	49 ± 2^2		
16:0−20:4	41 ± 3^2	18:0−20:4	49 ± 1^5
16:0−22:6	31 ± 3^2	18:0−22:6	55 ± 3^6
22:6−22:6	8.5 ± 1^3	22:6−22:6	11 ± 3^6

more cholesterol than PE bilayers with identical acyl chain composition. Also, within a family
of either PCs or PEs, cholesterol is more excluded from bilayers as a function of increasing
acyl chain double bond content. Of particular interest is the extremely low solubility of
cholesterol in PC or PE membranes containing two chains of docosahexaenoic acid (DHA,
$22:6^{\Delta4,7,10,13,16,19}$), the most unsaturated fatty acid commonly found in membranes. This
experiment confirms the idea that the more flexible the acyl chain, the less accommodating
it is for cholesterol.

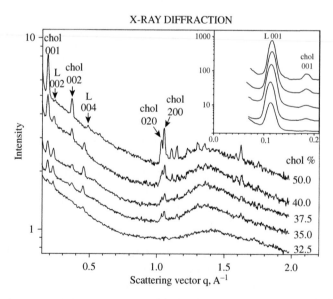

FIGURE 10.4   I-q plots for 16:0, 22:6 PE membranes showing the intensities of cholesterol monohydrate peaks
with added cholesterol. The integrated intensities of the scattering peaks 002 (0.3701 Å^{-1}), 020 (1.033 Å^{-1}), and 200
(1.044 Å^{-1}) were combined and plotted against the mol% cholesterol in Figure 10.5 [9,10].

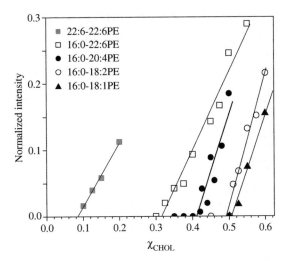

FIGURE 10.5   Plot of the normalized scattering peaks for cholesterol monohydrate crystals excluded from the membrane *vs* the mol fraction of cholesterol added to the membrane. The sum of the normalized cholesterol monohydrate peaks are extrapolated to 0, representing the membrane carrying capacity for cholesterol. The values are reported in Table 10.1 [9].

Lipid Raft Detergent Extractions

In its initial form, the Fluid Mosaic model (Singer and Nicolson, 1972 [11]) did not appreciate lipid heterogeneities. However, it soon became obvious that lipid patchiness must exist. By 1974 a series of biophysical studies started to appear supporting the basic concept of lipid microdomains in membranes [12]. Of particular importance were cholesterol- and sphingolipid-enriched domains studied in model bilayer membranes in the late 1970s by Biltonen and Thompson [13]. The lipid microdomain concept was formalized in a classic 1982 paper by Karnovsky et al. [12]. Therefore cholesterol/sphingolipid microdomains predated the concept of lipid rafts by more than a decade.

The pre-raft experiments strongly implied that there was a preferential affinity of appreciate for sphingolipids. This observation on model membranes was given a biological link with the discovery of lipid rafts in biological membranes (see Chapter 8) [14,15]. Rafts can be extracted as an insoluble fraction at 4°C with non-ionic detergents such as Triton X-100 or Brij-98. Isolated rafts contain about twice the amount of cholesterol and also are enriched in sphingolipids by about 50% when compared to the surrounding plasma membrane bilayer. Cholesterol is purported to be the 'glue' that holds rafts together, since rafts fall apart when cholesterol is extracted by cyclodextrin. Also, cholesterol is responsible for rafts existing in the tightly packed, liquid ordered (l_o) state and being thicker than the surrounding non-raft membrane. A characteristic of most, but apparently not all raft lipids, is their having long, mostly saturated acyl chains. Figure 10.6 shows molecular models of SM and cholesterol. Note that in this representation the inverted cone shape of SM fits nicely next to the cone-shaped cholesterol, enhancing affinity between these two lipids. Fitting lipids of different shapes together like pieces of a jigsaw puzzle is known as 'complementarity'. However, it is not certain whether raft-insolubility in cold non-ionic detergent results from extraction of intact rafts or is the result

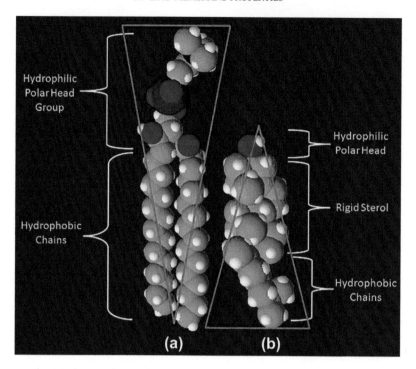

FIGURE 10.6   Space filling models of two major components of lipid rafts, SM (left) and cholesterol (right). Note, in this depiction SM is drawn as a slight inverted cone, while cholesterol is cone-shaped. The two molecules fit together, thus displaying 'complementarity' [79].

of similar, coincidental insolubility of sphingolipids and cholesterol in the detergent solutions. Non-detergent methods for raft extractions from membranes have now been developed [16] and are discussed in Chapter 13. In general these newer methods support the basic conclusions derived from the older detergent methods. Sphingolipids and cholesterol do appear to have an albeit non-perfect affinity for one another.

An interesting medical problem related to raft structure has recently come to light [17]. Heavy, long term use of cholesterol-lowering statin drugs, primarily in the elderly, has produced a result similar to that of cyclodextrin on lipid raft stability. Statin-induced hypocholesterolemia has resulted in the functional failure of cholesterol-rich lipid rafts in processes involving exocytosis and endocytosis. For example, a five year trial showed a 30% increase in the incidence of diabetes associated with a cholesterol reduction therapy. Long-term statin use has also been linked through lipid raft failure to bone loss and fractures. These results surprisingly linking heavy statin use, cholesterol reduction by cyclodextrin, and lipid raft function in some important human maladies may represent a new paradigm for medical interventions.

B. NON-LAMELLAR PHASES

When picturing a biological membrane, the first thing that comes to mind is the simple lipid bilayer, but we have seen how deceptive this can be. Even the lipid bilayer (actually the lamellar phase) can exist in a variety of very different states. In prior sections we have used the terms; bilayer, lamellar, gel, solid, S_o, L_β, $L_{\beta'}$, ripple, $P_{\beta'}$, fluid, liquid crystalline, L_α, liquid disordered, L_d, liquid ordered and l_o to describe the 'simple' lipid bilayer. But there is even more, far more, to structures of polar lipids dispersed in water. Depending on the polar lipid and conditions of the aqueous solvent (e.g. temperature, water content, ionic strength, polyvalent cations, pH, pressure etc.), lipid dispersions can exist in a wide variety of additional forms. Multiple long-range structures that amphipathic (polar) lipids can take when dispersed in water are referred to as 'lipid polymorphism' [18–21]. Two of the major questions about membrane lipids concern 'lipid diversity' and 'lipid polymorphism'. Lipid diversity addresses the question of why there are so many different lipids in membranes. Lipid polymorphism addresses the question of why amphipathic lipids form so many different long-range structures when dispersed in water.

While there are dozens of highly unusual long-range structures that amphipathic lipids can assume, only four are believed to be stable at biologically high water levels and only one of these, the lamellar state, predominantly, perhaps exclusively, exists in a living membrane. The four lipid phases that are stable at high water concentrations [22] are:

1. Lamellar
2. Inverted Hexagonal (H_{II})
3. Cubic Q^{224}
4. Cubic Q^{227}

What phase an amphipathic lipid prefers is dependent on the lipid's molecular shape. Membrane lipids can be roughly divided into the three basic shapes shown in Figure 10.7: truncated cone (panel a); cylinder (panel b); and inverted cone (panel c) [23]. Cylindrical lipids prefer the lamellar phase, while lipids with very truncated or inverted cone shapes in pure form prefer being in other, non-lamellar structures. All biological membranes are composed of hundreds or thousands of different lipids of all three basic shapes, yet membranes are found almost entirely in the lamellar phase. This is particularly perplexing since most of the world's membrane lipids are not cylindrical. For example, the most prevalent membrane lipid found on planet Earth is the photosynthesis-associated lipid monogalactosyldiacylglycerol (MGDG, Figure 5.25), which prefers the H_{II} phase. This suggests that cylindrical lipids have a stronger influence on overall membrane structure than do the other

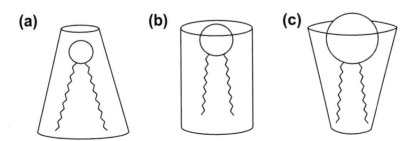

FIGURE 10.7 Basic membrane lipid shapes: (a) truncated cone; (b) cylinder and; (c) inverted cone. Circles represent the polar head group of different sizes [23].

non-cylindrical lipids. Indeed, only 20–50% of the total lipids need be cylindrical to maintain the normal lamellar phase.

A general description of lipid shape is not entirely satisfying as lipids actually exist at all locations on the continuum, from truncated cone, through cylinder, to inverted cone [24]. Therefore attempts have been made to quantify lipid shape. Two of these approaches are discussed below.

The first approach, proposed by Pieter Cullis in 1979 [19] defines lipid shape by a dimensionless parameter S:

$$S = v/a_o l_c$$

where a_o is the optimum cross-sectional area per molecule at the lipid–water interface, v is the volume per molecule and l_c is the length of the fully extended acyl chain. Since these parameters are not easy to obtain, a simplified version was suggested, where a_o is the cross-sectional area of the head group at the aqueous interface and a_h is the cross-sectional area at the bottom of the acyl chains. Therefore if:

$$a_o/a_h > 1 \text{ then } S > 1 \rightarrow \text{inverted cone}$$

$$a_o/a_h = 1 \text{ then } S = 1 \rightarrow \text{cylindrical}$$

$$a_o/a_h < 1 \text{ then } S < 1 \rightarrow \text{cone}$$

Unfortunately, S is not a fixed parameter of the lipid's shape, as it may vary with the membrane's environment (i.e. bathing solution pH, ionic strength, atmospheric pressure, lateral pressure, temperature etc.).

A second approach, the equilibrium curvature (R_o) concept, was later proposed by Sol Gruner [25]. R_o is a quantitative measure of the propensity of a lipid to form a non-lamellar structure. It is an estimate of the tendency of a monolayer made from a particular lipid to curl, thus resulting in hydrocarbon packing strain. R_o is normally derived experimentally

TABLE 10.2   The Spontaneous Radius of Curvature (R_o) for Several Important Membrane Lipids.

Lipid	Spontaneous radius of curvature R_o (Å)
DOPS	+150
Lyso PC	+38 to +68
DOPE	−28.5
DOPC	−80 to −200
Cholesterol	−22.8
α-Tocopherol	−13.7
DOG	−11.5

The table was adapted from [23].

from X-ray diffraction of lipid films. A large negative or positive value ($> \pm 100$ Å) suggests that there is little inherent curvature in the lipid (i.e. the lipid prefers a cylindrical shape). A lower absolute value ($< \pm 100$ Å) suggests that the lipid will add a curvature stress (negative or positive) and thus has a cone or inverted cone shape. Table 10.2 lists the R_os of a few common membrane lipids [23]. The cylindrical lipids, DOPS and DOPC, have R_os >100 and so form a lamellar structure. Lyso PC has a small positive R_o indicating it is an inverted cone and so prefers the micellular phase. DOPE, cholesterol, α-tocopherol, and DOG (dioleoylglycerol) all have low negative R_os indicating they are cone-shaped and so prefer a non-lamellar phase like H_{II}.

Inverted Hexagonal Phase (H_{II})

The most studied of the non-lamellar phases is the inverted hexagonal phase, referred to as H_{II} [18,20,26]. H_{II} phase (Figure 10.8) consists of 6 long, parallel cylindrical tubes of indefinite length that are distributed into a hexagonal pattern, thus giving rise to the designation inverted hexagonal phase. Water fills the cylinders and hydrocarbon chains fill the voids between the hexagonally packed cylinders, holding the entire H_{II} structure together. The molecular shape of H_{II}-preferring lipids (wide base, narrow head) gives negative curvature strain to adjacent lipids or proteins.

A full H_{II} phase would require a substantial number of cone-shaped lipids to be recruited to one location in a biological membrane. This seems highly unlikely and many unsuccessful attempts have been made to clearly identify H_{II} structure in complex biological membranes. But this does not necessarily mean that the H_{II} phase is totally absent. Perhaps characteristic signals from the limited H_{II} phase are simply lost due to the 'population weighted average problem'. Even though it is unlikely that any significant level of H_{II} actually exists in biological membranes, this non-lamellar phase has received, and continues to receive, considerable attention in the world of membrane biophysics. Primarily through model membrane studies, H_{II} phase-preferring lipids have been strongly linked to fusion and other membrane contact phenomena [27]. It was proposed that during fusion, the lipids of two stable bilayers must intermingle through some kind of unstable intermediate. This process is known as

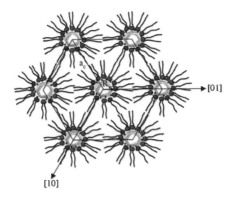

FIGURE 10.8   The reverse hexagonal H_{II} phase consists of an array of six cylinders. In each cylinder, water fills the channel interior that is lined by the lipid polar head groups. The outward facing acyl chains interact with each other, stabilizing the entire structure via the hydrophobic effect [18].

hemi-fusion and is well-documented in model membranes. In 1986, David Siegel [28] suggested that an inverted micelle structure, resembling one cylinder of H_{II} structure, represents the primary point defect in lamellar to H_{II} transition (see Figure 10.9) [20,27]. Importantly, factors known to promote H_{II} phase also promote membrane fusion, and lipid-soluble fusogens that induce cell—cell fusion *in vitro* induce H_{II} formation in model membranes.

Assigning a non-lamellar phase role in membrane fusion was compelling since it could start to explain some aspects of the 'lipid polymorphism' conundrum. Unfortunately, sometimes even beautiful stories disintegrate when confronted with harsh reality. Many vital processes in cells require *rapid* and *precise* control of fusion. Biological fusion occurs on the order of milliseconds and must be turned on and off in an orderly fashion. Model lipid vesicle fusion is by orders of magnitude too slow to start, and once begun has no mechanism to end. In addition it is hard to envision how H_{II}-favoring lipids can overcome the large negative charge repulsion on fusing-cell surfaces. Speed and control are two problems the H_{II}-phase lipid fusion theory could not answer. In 1993 James Rothman (Figure 10.10) and colleagues proposed a protein-based theory of membrane fusion called the SNARE hypothesis [29,30]. In the years since its proposal, this hypothesis has successfully addressed many of the complex problems associated with fusion. In a 2009 paper in *Nature* [31], Rothman reviewed the central aspects of how the proteins SNARE and SM (Sec1/Munc18-like) in concert with Ca^{2+} organize lipid microdomains in fusing membranes. H_{II} phase was not even mentioned!

Cubic Phase

Another non-lamellar phase that is stable in excess (biological) levels of water includes the cubic phases, Q^{224} and Q^{227} [21,25]. Like the H_{II} phase, the cubic phase is at most a very rare component of biological membranes and remains a phase in search of a function. Q^{227} has been proposed to repair the damage of a lipolytic agent at a local spot by forming a water-tight

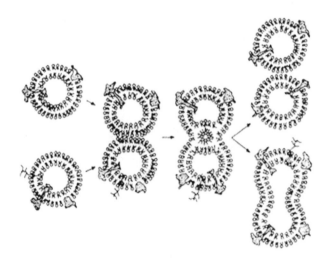

FIGURE 10.9 Diagram of the mechanism of lipid-driven membrane fusion including one central H_{II}-like cylinder [20,27].

FIGURE 10.10   James Rothman. *Courtesy of Yale University.*

patch. Like H_{II}, the cubic phase has also been proposed to be involved in fusion. However, the role of the cubic phase in fusion is at best uncertain, as this phase has been far less extensively investigated than H_{II}. The cubic phase is a lipid bilayer that is convoluted into a complex bicontinuous surface. Although the role of the cubic phase in biological membranes is unknown, it has found a home in a totally different arena. Q^{224} (normally composed of mono-olein) has been successfully employed to nucleate and grow membrane protein crystals for X-ray crystallography [32]. The protein partitions into the cubic membrane phase while the detergent used in the protein's isolation (discussed in Chapter 13) partitions into the aqueous phase. There is great hope that this method may provide a new paradigm for crystallizing difficult membrane proteins.

^{31}-P-NMR

Perhaps the most employed and easiest to interpret technique that can distinguish membrane phase is ^{31}P-NMR. Although X-ray diffraction is the classic technique used to define the parameters of model membranes, data interpretation is difficult and its application to biological membranes limited. ^{31}P-NMR is sensitive to phospholipid head group orientation and its spectra can clearly distinguish lamellar, H_{II}, and micellar phases (Figure 10.11) [19]. Lamellar phase spectra are distinguished by a broad (40–50 PPM), low-field shoulder and a high-field peak. In total contrast, H_{II} phase spectra are narrow (~20 PPM, half the width of the lamellar phase spectra) and feature a low-field peak and high-field shoulder. A third type of spectra is a narrow featureless peak that may indicate the presence of micellar structure, small rapidly tumbling vesicles or even cubic or rhombic phase. Unlike X-ray diffraction, ^{31}P-NMR can be used to follow lamellar → H_{II} transitions. In one example, various mixtures of egg PC and soy PE were made into large unilamellar vesicles (methods discussed in Chapter 13). The vesicles were then subjected to ^{31}P-NMR analysis. The pure egg PC vesicles produced a spectra indicating lamellar phase only (i.e. it had a low-field shoulder and a high-field peak).

Phospholipid phases   Corresponding ^{31}P-NMR spectra   Corresponding Fracture-Faces

Bilayer

Hexagonal (H$_{II}$)

Phases where
isotropic motion occurs
1. Vesicles
2. Inverted micellar
3. Micellar
4. Cubic
5. Rhombic

├──── 40 ppm ────┤   H→

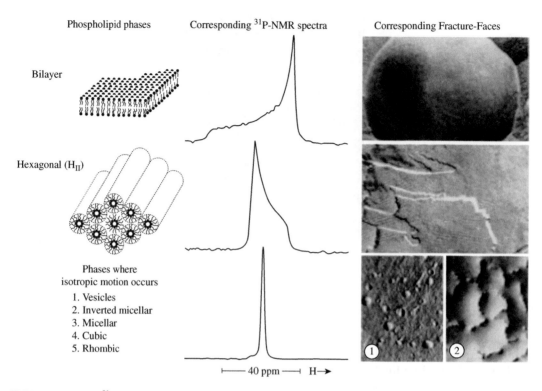

FIGURE 10.11 ^{31}P-NMR spectra of lamellar (bilayer) phase, inverted hexagonal (H$_{II}$) phase and other (isotropic) phase. Left Column, membrane structure; Middle Column, ^{31}P-NMR spectra; Right Column, freeze fracture electron micrograph [19].

The pure soy PE vesicles produced a spectra indicating H$_{II}$ phase only (i.e. it had a low-field peak and a high-field shoulder). As PC was added to the PE vesicles the H$_{II}$ phase spectra started to diminish. By 15 mol% PC the major spectral component was a shoulderless peak, indicating a transition state. Increasing levels of PC served to amplify the lamellar spectra. By 50 mol% PC the spectra indicated only the presence of lamellar phase. Therefore it was concluded that PC has a larger influence on lipid phase than does PE.

So we are still left with the same perplexing question. What is the biological function of 'lipid polymorphism'? After almost 4 billion years of chemical and biological evolution, Nature would not produce something so profound without a very good reason. We are just at the beginning of addressing this problem. Table 10.3 lists some basic membrane lipids and their preferred phase. It appears that cylindrical lipids prefer the lamellar phase, while cone-shaped lipids prefer the H$_{II}$ (or non-lamellar) phase, and inverted cone-shaped lipids prefer the micellar phase. Even this is complicated by the ability of a particular lipid to have its preferred phase altered upon changes in its environment. For example, at pH 7 PS and PA have a cylindrical shape and prefer the lamellar phase. However, if the pH is reduced to less than 4, the head group negative charge is reduced, thus decreasing charge repulsion. This reduces the apparent head group diameter (a$_o$) without affecting the tail diameter (a$_h$). When reduced to pH 4, PS and PA switch from a cylindrical (lamellar) shape to a cone (H$_{II}$)

TABLE 10.3   Common Membrane Lipids and their Preferred Phase.

Lipid	Comments	Phase
PC:	Cylindrical under most conditions	Lamellar
	Extreme dehydration, high temperature	H_{II}
PE:	High unsaturation	H_{II}
	Saturated chains, low temp, gel state	Lamellar
	Saturated chains, high temp, fluid state	H_{II}
	Unsaturated chains, pH 9	Lamellar
SM:	Cylindrical under most conditions	Lamellar
PS:	pH 7, cylindrical	Lamellar
	pH < 4 (reduce negative charge repulsion)	H_{II}
PA:	pH 7, cylindrical	Lamellar
	pH < 4 (reduce negative charge repulsion)	H_{II}
CL:	pH 7, cylindrical	Lamellar
	pH 7, + Ca^{2+}	H_{II}
MGDG	Cone-shaped	H_{II}
DGDG	Cylindrical	Lamellar
Cerebroside	Cylindrical	Lamellar
Ganglioside	Inverted cone	Micelle
Lyso PC	Inverted cone	Micelle
Mixtures:		
Unsaturated chains → + Ca^{2+}		
	PS-PE	Lamellar → H_{II}
	PG-PE	Lamellar → H_{II}
	PA-PE	Lamellar → H_{II}
	PI-PE	Lamellar → H_{II}
	PE-PS-Cholesterol (high salt)	H_{II}

shape. And this is for a single pure molecular species. Biological membranes are comprised of hundreds to thousands of molecular species that exhibit different and largely unknown affinities for one another. To date, only a very few lipid mixtures have been tested for lipid phase preference and this has been shown to be complicated. For example, mixtures of PS-PE, PG-PE, PA-PE or PI-PE all undergo a lamellar → H_{II} transition upon addition of Ca^{2+}. What all of this means is unknown. Perhaps the preference of each molecular species for a particular phase is just a reflection of lipid molecular shape that manifests itself by imposing molecular strain on adjacent proteins. This in turn would affect the protein's conformation and thus activity. This is further discussed below. Perhaps non-lamellar phases are just a curiosity observed only in well-defined lipid bilayers.

C. LIPID PHASE DIAGRAMS

Membrane lipids come in an overwhelming variety of chemical structures ('lipid diversity') and each molecular species, when isolated in pure form, can aggregate into a large number of long-range structures or phases ('lipid polymorphism'). Environmental conditions (e.g. temperature, lateral pressure, extent of hydration, aqueous solution ionic strength, presence of polyvalent metals, pH etc.) can further affect the lipid phase. The situation becomes far more complicated with multi-component lipid mixtures, as the lipids interact with one another resulting in phase separation into discrete lipid microdomains. And all of this is before proteins are even incorporated into the membrane!

For years lipid biophysicists have attempted to pictorially describe various lipid interactions and phases through what are known as 'phase diagrams' [33]. Phase diagrams can take a wide variety of forms that report lipid behavior from different perspectives. As the number of lipid components increases, the phase diagrams become ever more complex. Anyone who is not confused by a three component phase diagram just does not understand membrane structure. There are now a countless number of phase diagrams that have been compiled [34]. Below are five examples of phase diagrams, starting with simple one-component DPPC phase diagrams and progressing through to a complex, three-lipid phase diagram of a lipid raft model.

Single-Component Lipid Phase Diagrams

We have already encountered one single-lipid phase diagram in the form of a DSC scan of DPPC. This is depicted in Figure 9.22. The DSC scan follows the phase transitions of DPPC from low temperature gel phase ($L_{\beta'}$) to high temperature liquid crystalline phase (L_α). With pure DPPC, there are two transitions, a lower temperature pre-transition (at ~35.6°C) and a higher temperature main transition, T_m (at ~41.3°C). In the region between the pre-transition and main transition the bilayer exists in a non-flat ripple structure ($P_{\beta'}$). Therefore the DSC-derived phase diagram shows the following transitions:

$$L_{\beta'} \to P_{\beta'} \to L_\alpha$$

A second single-component phase diagram looks at the DPPC transitions from a different perspective. Figure 10.12 shows the effect of hydration on DPPC lipid phase [35]. The lipid solution becomes more dilute from left to right. The dotted line indicates maximum adsorption of water at the membrane surface. At any point to the right of the dotted line DPPC is saturated with water. Normal biological water levels are > 99.9%. Note that at almost all hydration levels DPPC undergoes the same transitions reported in the DSC scan shown in Figure 9.22 ($L_{\beta'} \to P_{\beta'} \to L_\alpha$). However at very low water content and high temperature, additional non-lamellar phases, including cubic and H_{II} appear. H_{II} phase is not normally

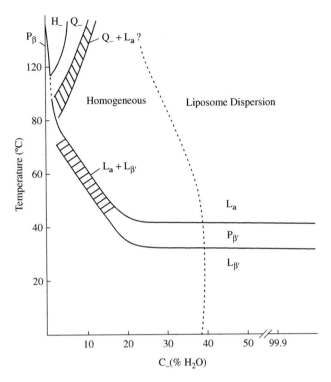

FIGURE 10.12   Phase diagram of DPPC showing the effect of water content (hydration) on lipid phase. At high (biological) levels of water, DPPC bilayers go from $L_{\beta'} \rightarrow P_{\beta'} \rightarrow L_a$. Other unusual phases exist under conditions of extreme dehydration [35].

reported for any PC, but can form under extreme dehydration. Importantly, dehydration may occur at localized places on a membrane surface in the presence of chaotropic agents or divalent metals including Ca^{2+}. These agents are known to replace waters bound to a phospholipid head group, thus creating a localized area of extreme dehydration.

Two-Component Lipid Phase Diagrams

A phase diagram for bilayers composed of mixtures of DMPC (14:0,14:0 PC) and DPPC (16:0,16:0 PC) is shown in Figure 10.13, top panel (a). The T_m of DMPC is 23.6°C and DPPC 41.3°C, as depicted on the Y-axis. Since the T_ms are so close, mixtures produce a broad transition between the two single phospholipid T_m extremes. The dotted line represents the theoretical T_m for various DMPC/DPPC mixtures. The solid lines on either side of the dotted line are the experimentally derived melting values. The thick arrow follows a 1:1 (mol:mol) mixture of DMPC/DPPC as it is heated from a temperature where the lipid mixture is entirely in the gel ($L_{\beta'}$) state through the melting transition T_m until the mixture is in the liquid crystalline (L_α) state. In the region between temperatures T_1 and T_2 the membrane consists of co-existing gel and liquid crystalline states. The corresponding DSC scan is shown in the bottom

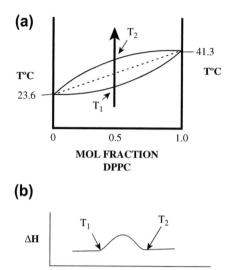

FIGURE 10.13   (a) Phase diagram for mixtures of DMPC (14:0,14:0 PC) and DPPC (16:0,16:0 PC). The T_m of DMPC is 23.6°C and DPPC 41.3°C. LUVs made from a 1:1 mixture of DMPC/DPPC are subjected to increasing temperature from low, where both lipids are in the gel state, to high temperature, where both lipids are in the liquid crystalline state (thick arrow). (b) Corresponding DSC scan for the DMPC/DPPC mixed membrane.

panel (b). As the gel state mixture is heated, the membrane begins to melt at temperature T_1. The first component to melt is enriched in the lower melting lipid, DMPC. In the DSC scan, T_1 is the temperature where the membrane starts to melt and the scan leaves the baseline. As the melting continues, the mixture becomes more enriched in the higher melting DPPC. At T_m the mixture is half gel and half liquid crystal. At temperature T_2 the entire mixture melts as the membrane enters the liquid crystalline (L_a) phase where the DSC scan returns to the base-line. The last component to melt is highly enriched in DPPC, and is depleted of DMPC.

A second two-component phase diagram is shown in Figure 10.14 for mixtures of DPPC and cholesterol [36]. Clearly this phase diagram is far more complex than that of DMPC/DPPC. At 0 mol% cholesterol the lipid melts at 41.3°C, confirming it is DPPC. At low choles-terol concentrations (< 5 mol%), DPPC can simultaneously exist in either a liquid crystalline (l_d) or gel (S_o) state, depending on the temperature. At high cholesterol levels (>30 mol%), DPPC only exists in the liquid ordered (l_o) state. Liquid ordered (l_o) is a state mid-way between liquid crystalline (l_o or L_α) and gel (S_o or $P_{\beta'}$) states, having properties of both. Liquid ordered state is characterized by having tightly packed lipids (like in gel) that nevertheless support rapid lateral diffusion (like in liquid crystalline). Between 5–30 mol% cholesterol, DPPC exists in mixed states, $l_d - l_o$ or $S_o - l_o$.

A Three-Component Lipid Phase Diagram

Three-component phase diagrams are normally quite complex and not easy to inter-pret [37]. Figure 10.15 shows an example of a lipid raft model phase diagram by de Almeida et al. (2003) [38]. Membranes were composed of cholesterol/POPC/PSM where

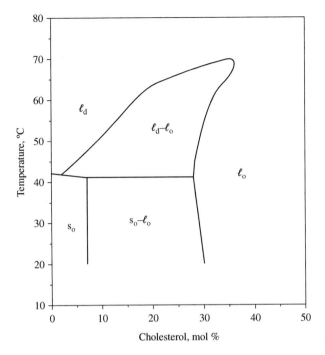

FIGURE 10.14 Phase diagram for DPPC/cholesterol mixtures. l_d is the liquid disordered (liquid crystalline) state. S_o is the solid or gel state. l_o is the liquid ordered state [36].

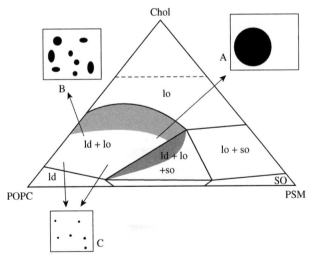

FIGURE 10.15 Phase diagram (23°C) of a three-component lipid raft model composed of cholesterol/POPC/PSM where POPC is 16:0,18:1 PC and PSM is palmitoyl-sphingomyelin. Abbreviations are the same as in Figure 10.12. l_d is the liquid disordered (liquid crystalline) state. S_o is the solid or gel state and l_o is the liquid ordered state. Lipid rafts are present in the light gray, light area inside the curve labeled 1d + 1o, and dark gray-shaded areas. In the light gray area l_o predominates over l_d. In the light area l_d predominates over l_o, while in the dark gray area l_o, l_d, and S_o coexist. Raft size varies. Insert A has large rafts (> 75–100 nm); B has intermediate size rafts (~20 nm to ~75–100 nm), and C has small rafts (< 20 nm) [38].

POPC is 16:0,18:1 PC and PSM is sphingomyelin where the variable chain is palmitic acid. Assembling this complex phase diagram required a combination of various techniques from several sources. The diagram gives the composition and boundaries of the lipid rafts. Rafts are present in several phases (see Figure 10.15 legend for details). Raft size varies: insert A has large rafts (>75–100 nm); B has intermediate size rafts (~20 nm to ~75–100 nm), and C has small rafts (<20 nm).

Since three-component phase diagrams are very hard to construct and interpret, it is hard to imagine what a biologically relevant phase diagram with hundreds of lipids would look like! Phase diagrams do have significant practical limitations.

D. LIPID–PROTEIN INTERACTIONS

1. Annular Lipids

One of the first, and most basic questions asked about membrane structure concerned the possible existence of a single layer of special lipids tightly associated with membrane proteins [39]. It is obvious that an integral protein must be solvated by lipids in the surrounding bilayer. This layer came to be known as 'annular' or 'boundary' lipids. But does this layer have the same lipid composition as the surrounding bilayer, or is it somehow different? The name 'annular' or 'boundary' implies a static lipid layer with some permanence surrounding the protein. This would require a significant attractive interaction between the annular lipid and the integral protein. But does such a unique layer actually exist and is it of different composition than the surrounding bulk membrane? An early attempt to investigate the annular lipid ring compared the rate of nearest neighbor exchange between an annular lipid and an adjacent bulk bilayer lipid to the lateral exchange rate between two bulk bilayer lipids (Figure 10.16).

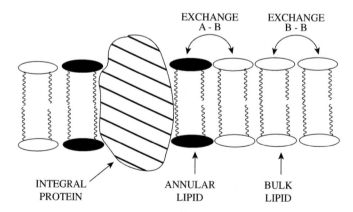

FIGURE 10.16   Nearest neighbor exchange between annular (A) and bulk (B) bilayer lipids (A − B exchange) and between two bulk bilayer lipids (B − B exchange).

Initially, electron spin resonance (ESR) of spin-labeled phospholipids was used to distinguish annular from bulk bilayer lipids [39]. The approach is based on motional restriction of any spin probe-lipid that is adjacent to an integral protein (annular lipids). Restricted motion results in a much broader ESR spectrum than is measured for freely rotating spin-labeled phospholipids (bulk bilayer lipids, Figure 10.17) [40]. Therefore, addition of a spin-labeled phospholipid to a membrane containing protein(s) could produce a two component spectrum if there is a motional difference between annular and bulk bilayer lipids that exists for at least $10^{-8} - 10^{-7}$ sec, the shortest time scale detectable by ESR. A two component spectrum was indeed detected by ESR, suggesting the possible existence of annular lipids. This was corroborated by use of another rapid time technique, fluorescence quenching. Brominated phospholipids [41] were used to quench the inherent fluorescence of tryptophans in the integral protein. Fluorescence quenching is very sensitive to distances on the nm range. Brominated annular lipids, being closer to the tryptophans,

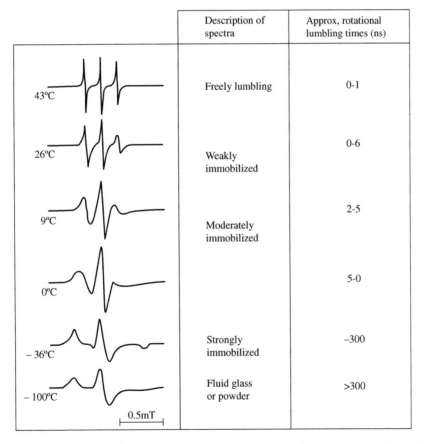

	Description of spectra	Approx, rotational lumbling times (ns)
43°C	Freely lumbling	0-1
26°C	Weakly immobilized	0-6
9°C	Moderately immobilized	2-5
0°C		5-0
– 36°C	Strongly immobilized	–300
– 100°C	Fluid glass or powder	>300

0.5mT

FIGURE 10.17   ESR spectra of a nitroxide spin probe from freely tumbling (top) to totally restricted (glass state, bottom). Note the freely tumbling spectrum is narrow and sharp while motional restrictions broaden the spectra.

quench better than do brominated lipids in the more distant bulk lipid bilayer. However, when similar experiments were repeated using ^{2}H-NMR instead of a spin- or brominated probe, no difference between annular and bulk bilayer lipids could be detected. This suggests that a distinct ring of annular lipids does not exist. Discrepancies between the experiments are attributed to the inherently slower times that are detectable by NMR ($\sim 10^{-6}$ sec) compared to ESR and fluorescence. Therefore, it was concluded that annular lipids, distinct from bulk bilayer lipids, may exist for short periods of time ($<10^{-6}$ sec). If annular lipids do exist, their association with integral proteins must be transient and hence very weak. A transient existence suggests that lipid composition of annular lipid closely track composition of the bulk bilayer.

For decades searches have been conducted to find binding sites for specific lipids on the surface of membrane integral proteins. Of particular interest would be relatively long-lived lipid protein complexes that are stable on the order of the time required for the turnover of a typical enzyme (10^{-3} sec). The large numbers of different lipid species that comprise a typical membrane (see 'lipid diversity' discussed in Chapters 4 and 5) strongly argue for the existence of these complexes. A large number of lipid molecular species would be required to fit the many different binding sites that might exist on the surfaces of membrane proteins. Such binding sites would help to explain the perplexing problem of 'lipid diversity'. However, the existence of specific, tight lipid-binding sites and weak binding of annular lipids seem contradictory.

Another related question that had to be addressed concerned the number of phospholipids required to completely solvate a specific integral protein and the lipid specificity, if any, of the solvation. In his 1988 book *Biomembranes: Molecular Structure and Function* [42], Robert Gennis presented a crude calculation to estimate the number of phospholipids that might surround a 'typical' membrane integral protein. Some assumptions and estimates had to be employed. The 'typical' protein was assigned a molecular weight of 50,000 and comprised 50% of the membrane by weight. This is well within the range of 20–75% protein for membranes (Chapter 1). Assuming an average molecular weight of ~830 for a phospholipid, the phospholipid-protein ratio was 60:1 by number, again a realistic assumption. The protein was envisioned to be a cylinder that extended 10 Å above and below the bilayer surface, and the radius of the cylinder was 18 Å. Since phospholipids have a radius of ~4.4 Å, it would require ~16 phospholipids on each leaflet of the bilayer to completely solvate the protein. Therefore a total of ~32 lipids would be required to form the annulus. That would leave only 28 phospholipids left to form the next lipid layer around the protein. Since this second layer must be larger than the annular layer, there are not enough lipids to complete a second lipid shell.

Actual experiments agree with Gennis's rough estimate. For example, the integral membrane enzyme, Ca^{2+} ATPase was isolated free of phospholipids. Phospholipids were then slowly added back and the protein's activity monitored [43]. Activity did not return until the lipid-protein ratio reached ~30:1 (Figure 10.18). The assumption was that, at this ratio, activity returned because the protein was completely solvated by lipid. Additional lipid added to the growing bilayer did not further increase activity. Confirming experiments were done adding a spin-labeled phospholipid to a de-lipidated protein [44]. The process was followed using ESR. Initially the ESR spectra were broad, indicating close association of the lipid and protein. At some lipid-protein ratio all of the annular lipid sites became filled. Subsequent lipid

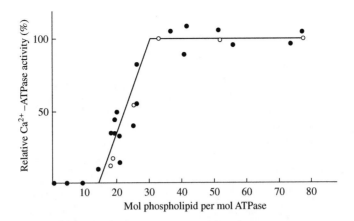

FIGURE 10.18   Activity of the Ca^{2+} ATPase isolated from the sarcoplasmic reticulum as a function of phospholipid content. It required ~30 phospholipids to support full activity (open circles). This provided an estimate of the number of lipids required to completely surround the Ca^{2+} ATPase once (the size of the annular lipid boundary). Since in mixed PC/cholesterol membranes Ca^{2+} ATPase activity exactly followed PC content and was independent of cholesterol (filled circles), it was proposed that cholesterol was excluded from the annulus [43].

then produced a second, narrow component to the spectra, indicating formation of the bulk bilayer.

These and numerous other related reports call into question the classic definition of a biological membrane being 'proteins floating like icebergs in a sea of phospholipids'. Clearly the lipid bilayer 'sea' must be limited and very, very crowded. Figure 10.19 shows a typical example of a membrane cartoon drawing. Similar drawings can be found in most biology, biochemistry, physiology, and cell biology textbooks. In this depiction a protein, contrary to what is actually found, is surrounded by ~11 lipid bilayer rings! In the mitochondrial Cristae, a very busy membrane having ~75% protein by weight, a simple calculation indicates there is not enough lipid to totally surround the resident proteins even once. Many of the mitochondrial proteins must actually be touching. On average, a plasma membrane protein would have at best enough lipid to form ~2 concentric layers around a protein. And one of these rings must form the annulus. An additional plasma membrane problem is related to its very high cholesterol content, ~50 mol% (or more) of the polar lipids. A report by Warren et al. [43] indicates that cholesterol is excluded from the annulus. If this is in fact true (see discussion below), the second lipid ring, that surrounding the annulus, must be composed almost entirely of cholesterol. But cholesterol doesn't form bilayers! In the membrane with the lowest protein content, the myelin sheath, the proteins would be surrounded by several lipid rings that could keep the proteins ~11 Å apart. So how large and crowded is the bilayer 'sea', and where does all that cholesterol reside?

2. Co-Factor (Non-Annular) Lipids

So far we have seen that most of the annular lipids, if distinct annular lipids even exist at all, are transient and weakly bound to integral proteins and the bilayer 'sea' is very small and

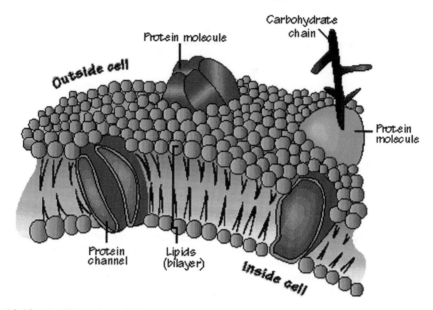

FIGURE 10.19 An illustration of the components of a 'typical' plasma membrane. Similar drawings are commonly found in many textbooks. Note the unrealistically high number of phospholipid shells surrounding each protein (~11 lipid rings). *Courtesy of ThinkQuest.*

crowded. Despite this, the activity of many membrane proteins is known to be significantly affected by specific lipids [45–47]. Annular lipid selectivity is weak and may reside with either the lipid head group or hydrophobic tails. However, it does appear that at least some integral proteins, in addition to the weakly associated annular lipids, also have a few binding sites for specific lipids. These are called 'non-annular' or 'co-factor' lipids. The best example of strong preferential affinity for membrane proteins has been demonstrated for anionic lipids binding to the Ca^{2+} ATPase, Protein Kinase C, Na^+/K^+ ATPase, and the Potassium Channel KcsA.

Ca^{2+} ATPase

One integral membrane protein that has contributed substantially to understanding lipid–protein interactions is the Ca^{2+} ATPase, a protein that is easily isolated from the sarcoplasmic reticulum of skeletal muscle. Discovery of the enzyme and its function, to pump Ca^{2+} out of the sarcoplasmic reticulum, dates from the early 1960s. One advantage of this enzyme is that it has two measurable activities, trans-membrane Ca^{2+} transport and ATP hydrolysis. Importantly, the enzyme can be reconstituted using a deoxycholate dialysis procedure (discussed in Chapter 13).

In a 1975 paper in *Nature*, Warren et al. reconstituted the Ca^{2+} ATPase into liposomes made from a variety of lipid mixtures [43]. When these investigators plotted PC content versus enzyme activity they estimated that activity returned at a PC-protein ratio of ~30 (Figure 10.18). They interpreted this number as an estimate of the number of lipids required to completely surround the Ca^{2+} ATPase once (hence the size of the annular lipid boundary).

Since in mixed PC/cholesterol membranes Ca^{2+} ATPase activity exactly followed PC content and was independent of cholesterol, they proposed that cholesterol was excluded from the annulus (Figure 10.18). This very interesting concept was later shown via binding assays to not be true. Later work from A.G. Lee and co-workers [44,45,48,49] confirmed the size of the annular lipid ring, reporting a minimum number of annular phospholipid sites of 32 and 22 at 0 °C and 37°C, respectively. Lee also extended the lipid/Ca^{2+} ATPase interaction studies and found that anionic phospholipids PA, PS, and PI activate the protein while binding at the general annular lipid sites and are involved in binding of Mg^{2+}ATP. In addition, Ca^{2+} ATPase activity is modified by bilayer acyl chain length, reaching a maximum at C-18 and falling off at C-14 and C-22.

Protein Kinase C

Perhaps the most thoroughly studied membrane protein is protein kinase C (PKC), a key enzyme in signal transduction [50]. The enzyme was only discovered by Yasutomi Nishizuka in the late 1970s [51]. Phosphatidylserine (PS) is an essential cofactor that specifically binds to and activates PKC. In fact all of the numerous isoforms of PKC are strictly dependent on PS for activity. PKC specifically recognizes 1,2-sn-phosphatidyl-L-serine (the D isomer is ineffective) [52]. This and the fact that PS is not believed to be involved in cell signaling through the formation of active metabolites, as is the case with PI, implies that specific, non-annular cofactor binding sites for PS must exist on PKC. Also, PKC's specificity for PS requires diacylglycerol. In the absence of diacylglycerol PKC binds anionic membranes with no discrimination between phospholipid headgroups beyond requiring a negative charge. PKC activity is further modulated by bulk bilayer lipids related to curvature strain (discussed below). The absolute specificity for PS is remarkable for known lipid–protein interactions.

Na$^+$/K$^+$ ATPase

The plasma membrane-bound Na^+/K^+ ATPase is responsible for maintaining essential trans-membrane Na^+ and K^+ ion gradients. Its mechanism of action is discussed in Chapter 14. The Na^+/K^+ ATPase was first described by Jens Skou (Figure 10.20) in 1957 [53], for which he was awarded a 1997 Nobel Prize in Chemistry. The enzyme, isolated from *Squalus acanthus,* was shown to bind 60 phospholipid molecules, comprising the annulus, with a preference for negatively charged lipids (K = 3.8 for CL/PC, 1.7 for PS/PC, and 1.5 for PA/PC). The effect of various lipids on stabilization of the Na^+,K^+-ATPase activity has been reported by Haim et al. [54]. These authors reported that acidic phospholipids are required and PS is somewhat better than PI. Also, optimal stabilization is achieved with heteroacid PSs having a saturated fatty acid ($18:0 \geq 16:0$) in the sn-1 and unsaturated fatty acid ($18:1 > 18:2$) in the sn-2 chain. In addition they also hypothesized that PS and cholesterol interact specifically with each other near the α_1/β_1 subunit interface, thus stabilizing the protein. A general role for anionic lipids at protein–protein interfaces is becoming more evident.

Potassium Channel KcsA

Recently Marius et al. [55] investigated the absolute requirement of anionic lipids for potassium channel KcsA activity. The protein was isolated from *Streptomyces lividans*. High-resolution X-ray crystallography showed an anionic lipid molecule bound to each of the four protein–protein interfaces comprising the channel. In addition to the annular or

FIGURE 10.20   Jens Skou (1918–). *Courtesy of Find the Best.com, Inc*

boundary lipids that surround the channel, there are four non-annular (co-factor) sites that bind the anionic phospholipid PG. The non-annular sites are only ~60–70% occupied in bilayers composed entirely of PG. Increasing the anionic lipid content of the membrane leads to a large increase in open channel probability, from 2.5% in the presence of 25 mol% PG to 62% in 100 mol% PG. A model was proposed where three or four of the four non-annular lipid sites in the KcsA homo-tetramer have to be occupied by an anionic lipid for the channel to open.

3. Non-Anionic Co-Factor Lipids

Although the major effort to identify specific, non-annulus lipid-binding to membrane proteins has focused on the anionic head group lipids, a few examples involving other types of lipid head groups or acyl chains have been reported. The most specific of these involves rhodopsin and the PUFA docosahexaenoic acid (DHA).

Rhodopsin

Rhodopsin is the membrane integral protein that functions as the visual receptor of the eye rod outer segment (ROS). The protein is solvated by ~24 phospholipids and shows very little preference of PC over CL. The ROS is highly enriched in DHA, the longest and most unsaturated fatty acid commonly found in membranes (see Chapter 4) [56]. In the ROS, DHA levels can approach 50% of the total acyl chains and so would be expected to have a dominant effect on this membrane's structure and function. In fact there is so much DHA in the ROS that homo-acid di-DHA phospholipid species have been identified. Many years ago it was predicted that DHA may possess an unusual structure, perhaps accounting for rhodopsin's activity [57,58]. Easy rotation about each methylene that separates the 6 double bonds results in a molecule that exists in a wide variety of highly contorted conformations. Two extreme

Docosahexaenoic Acid Structure

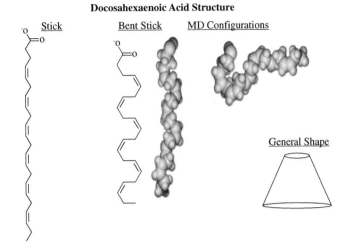

FIGURE 10.21   Several views of the structure of docosahexaenoic acid (DHA) [59].

conformers of the many possibilities, as determined by molecular dynamics (MD) simulations, are shown in Figure 10.21 [59]. The average of all conformations makes DHA a cone-shaped molecule (Figure 10.21) that, when incorporated into phospholipids, would prefer being in a non-lamellar phase. DHA-phospholipids are loosely packed and can induce negative curvature strain to its immediate membrane neighborhood. These concepts will be discussed in detail below. Combined MD simulations of rhodopsin and DHA by Scott Feller of Wabash College have resulted in a startling observation. DHA finds and fits perfectly into a groove on rhodopsin's surface (Figure 10.22) [60]! This, of course, suggests the existence of

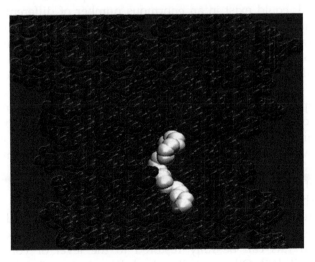

FIGURE 10.22   Molecular dynamics simulation of rhodopsin (blue) and docosahexaenoic acid (DHA, white). DHA fits into a groove on the surface of rhodopsin [60].

a DHA co-factor receptor site on rhodopsin that may explain the need for DHA in vision. It is known that upon complete removal of DHA and its metabolic precursors from the diet, mammals will ferociously retain ROS DHA at the expense of all other lipids [56].

D-β-Hydroxybutyrate Dehydrogenase

D-β-hydroxybutyrate dehydrogenase is a lipid requiring enzyme that is found on the inner surface of the inner mitochondrial membrane where it catalyzes the reaction:

$$(R)\text{-}3\text{-hydroxybutanoate} + NAD^+ \leftrightarrow acetoacetate + NADH + H^+$$

What makes this enzyme so unusual is its absolute specificity for phosphatidylcholine (PC) [61,62]. When purified free of lipid the enzyme is inactive. Activity is regenerated by adding PC or lipid mixtures containing PC. The active form of the enzyme is the enzyme-PC complex. Many membrane enzymes can be functional in bilayers made from PC, but D-β-hydroxybutyrate dehydrogenase is the only one with an absolute requirement for PC.

Cytochrome c Oxidase

The most logical place to look for an absolute link between a specific lipid and a mammalian function would be cardiolipin (CL) and mitochondrial electron transport and oxidative phosphorylation. CL was first isolated from beef heart in 1942 [63] and its presence was soon shown to be characteristic of bioenergetic membranes (bacterial plasma membrane and the mitochondrial inner membrane). Cytochrome oxidase is the terminal component in electron transport where its function is to reduce O_2 to H_2O. CL comprises about 20% of the total lipid composition of the mitochondrial inner membrane. The structure of CL (see Chapter 5) is clearly the most unusual of all phospholipids. It is two phosphatidic acids held together by a glycerol. At physiological pH, its large head group is a di-anion and it has four acyl chains. Due to its high negative charge density, CL can be induced by Ca^{2+} to undergo a lamellar-to-hexagonal (L_a-H_{II}) phase transition. An almost limitless number of possible acyl chain combinations would imply that CL could exist in an enormous number of different molecular species. However, the vast majority of acyl chains in CL are linoleic acid ($18{:}2^{\Delta 9,12}$) with lesser amounts of oleic acid ($18{:}1^{\Delta 9}$) and linolenic acid ($18{:}3^{\Delta 9,12,15}$). This severely limits the number of possible molecular species to a manageable number.

So is CL, with all of its unusual properties, absolutely required for cytochrome oxidase activity? Are some of the reputed ~40−55 annular lipids required to solvate cytochrome oxidase bound to specific co-factor sites? A number of reports have linked CL levels and full electron transport activity, particularly of cytochrome oxidase. In one report, Paradies exposed rat liver mitochondria to a free radical generating system that peroxidized the CL [64]. Concomitant with the loss of CL was a decrease in cytochrome oxidase activity. Lipid-soluble antioxidants prevented the loss of CL and preserved the enzyme activity. External addition of CL, but not any other phospholipid, prevented the loss of cytochrome oxidase activity. This experiment demonstrated a close correlation between oxidative damage to CL and reduced cytochrome oxidase activity. In a related study these same investigators also demonstrated that lower cytochrome c oxidase activity observed in heart mitochondria from aged rats can be fully restored to the level of young control rats by exogenously-added CL [65]. Other mitochondrial diseases (e.g. Tangier disease and Barth syndrome) have also been linked to CL and can be treated by controlling CL levels.

Several of the mitochondrial electron transport/oxidative phosphorylation complexes have been shown to require CL in order to maintain full enzymatic function. Examples include Complex IV (cytochrome c oxidase) that requires 2 CLs, Complex III (cytochrome bc1) and Complex V (F1 ATPase) that requires 4 CLs. While all of these various examples clearly indicate an important role for CL in mitochondrial bioenergetics, no absolute specificity has yet been demonstrated. Mitochondrial function can be maintained by PC without CL, albeit at reduced levels and the activity ratio for CL-PC is only about 5.

4. Hydrophobic Match

One partial explanation for the large number of membrane lipid species ('lipid diversity') is the requirement for lipids of different lengths to properly solvate the varying hydrophobic surface areas associated with integral membrane proteins. Relationship between length of the trans-membrane protein hydrophobic surface and the hydrophobic thickness of the neighboring lipid bilayer is referred to as the 'hydrophobic match' [66,67]. A basic assumption is that these two lengths should be similar and if they are not, the mismatch must be compensated for by alterations in lipid and/or protein structure.

It has been shown that size of the hydrophobic segment on the membrane protein varies from protein to protein even within the same membrane and the length of the lipid bilayer hydrophobic interior varies with the lipid shape, acyl chain length, and number and location of any double bonds. Temperature, lateral pressure, cholesterol content, and presence of divalent metal ions can also affect thickness of the bilayer. A major membrane bilayer 'thickening' agent is cholesterol. Cholesterol-rich domains, including lipid rafts, are partially characterized by being significantly thicker than the neighboring non-raft bilayer. In proteins, the hydrophobic trans-membrane segments are related to the types and sequence of the amino acids comprising the segment. Of particular importance is the number of trans-membrane α-helices present. As a result, the hydrophobic match length may vary considerably. For example, in *E. coli*, the hydrophobic length of the inner membrane leader peptidase is 15 amino acids long while the lactose permease is 24 ± 4 amino acids long. Lipids can affect protein activity, stability, orientation, state of aggregation, localization, and conformation. Proteins in turn can affect lipid chain order, phase transition, phase behavior, and microdomain formation. Therefore, simultaneously lipids can affect membrane integral proteins and integral proteins can affect nearby membrane lipids. The effect of hydrophobic mismatch must be concurrently viewed from the perspective of both the lipid and the protein.

The Hydrophobic Match Length: A DSC Study

Determining the hydrophobic match length is not a straightforward measurement. Although X-ray crystallography can determine the precise location of every atom in a protein, it cannot identify with any certainty the location of the hydrophobic span that would accommodate the lipid bilayer. One unusual and indirect method that has been used to determine the length of the hydrophobic match involves differential scanning calorimetry (DSC) (see Chapter 9). In one report, Toconne and collaboraters [68] reconstituted bacteriorhodopsin isolated from *Halobacterium halobium* into lipid bilayers composed of phosphatidylcholines (PCs) containing acyl chains of different lengths. The PCs were dilauroyl PC (DLPC, 12:0,12:0 PC), dimyristoyl PC (DMPC, 14:0,14:0 PC), dipalmitoyl PC (16:0,16:0 PC), and distearoyl PC (DSPC 18:0,18:0 PC). Bacteriorhodopsin-induced shifts in the expected T_m of each lipid were determined by DSC. Shifts in T_ms were consistent with Mouritsen and Bloom's 'Mattress Model' (see Figure 10.23, right panel) [69]. According to this model, a protein with a larger hydrophobic match length than the solvating lipid can stretch the lipid to fit its hydrophobic segment. This results in an increase in lipid order that is associated with an increase in the lipid's T_m. In contrast, a protein with a shorter hydrophobic match length than the solvating lipid can shrink the lipid to fit its hydrophobic segment. This results in a decrease in lipid order that is associated with a decrease in the lipid's T_m. Tocanne found that bacteriorhodopsin reconstituted into DLPC bilayers increased the lipid's T_m +40°C (from 0°C to ~40°C) (Figure 10.23, left panel). Therefore, the length of DLPC (2.4 nm) is shorter than the hydrophobic match length of bacteriorhodopsin. The T_m of DMPC was also increased by bacteriorhodopsin, but only by +23°C (from 23.6°C to ~46.6°C). Therefore, the length of DMPC (2.8 nm) is also shorter than the hydrophobic match length of bacterio-rhodopsin. Upon incorporation of bacteriorhodopsin into the longer lipid DSPC (3.7 nm), T_m of the lipid *decreased* from 58°C to 45°C (−13°C). This indicates that the hydrophobic match length must be shorter than 3.7 nm (DSPC), but longer than 2.8 nm (DMPC). When bacterio-rhodopsin was incorporated into DPPC bilayers (3.2 nm), no shift in T_m (41.3°C) was observed. This indicates that the length of the hydrophobic match for bacteriorhodopsin is ~3.2 nm. Other methodologies agree, placing the hydrophobic match length for

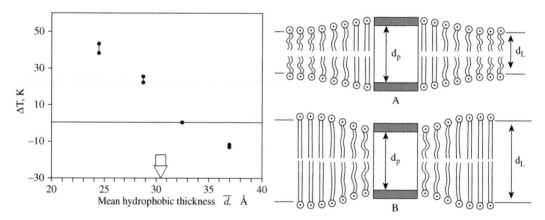

FIGURE 10.23 The Mouritsen and Bloom 'Mattress' model [69] (right panel). The hydrophobic match length of an integral protein is depicted by d_P, and the length of the lipid bilayer hydrophobic interior by d_L. If $d_P > d_L$, the lipid must stretch to match the protein hydrophobic length, thus increasing its T_m (right panel, part A). If $d_P < d_L$, the lipid must shrink to match the protein hydrophobic length, thus decreasing its T_m (right panel, part B). This principle was tested by Toconne and co-workers [68] for bacteriorhodopsin reconstituted into bilayers made from either DLPC (12:0,12:0 PC), DMPC (14:0,14:0 PC), DPPC (16:0,16:0 PC) or DSPC (18:0,18:0 PC) (left panel). The bacteriorhodopsin-induced change in expected T_ms for each lipid was determined by DSC and plotted against the mean hydrophobic thickness. Bacteriorhodopsin reconstituted into DLPC and DMPC bilayers increased the expected T_m and so both lipids are shorter than the protein's hydrophobic match. Bacteriorhodopsin reconstituted into DSPC bilayers decreased the expected T_m and so is longer than the protein's hydrophobic match. Bacterio-rhodopsin had no effect on the T_m of bilayers made from DPPC. Therefore bacteriorhodopsin's hydrophobic match length is about the same as the hydrophobic length of DPPC, ~30.5 Å [68].

bacteriorhodopsin at 3.1–3.2 nm. This experiment supports some important basic conclusions. First, proteins of the same hydrophobic match length can be accommodated into bilayers of different thickness and integral proteins can severely affect neighboring lipid structure.

Single span α-helices are particularly sensitive to hydrophobic mismatch [66–70]. If length of the trans-membrane α-helix exceeds the bilayer hydrophobic thickness, the helix may tilt to shorten its effective length, or it may bend or even slightly compress or may adjust its conformation into a different type of helix altogether. An incorrect helix length may cause the protein to aggregate. If the helix is much too short, the protein may even be excluded from the membrane interior, resulting in a new surface location. Multiple span proteins are resistant to tilt and it is hard to assess the size and importance of mismatch. For example, it has been reported that the multi-span protein rhodopsin needs a substantial lipid mismatch of 4 Å thicker or 10 Å thinner than the protein's hydrophobic length in order to aggregate.

It is likely that all integral proteins may select lipids of appropriate match length from the myriad of lipids that comprise the surrounding bilayer. Many, perhaps most, of the bilayer lipids have a cone-shape and so prefer non-lamellar structure. The presence of can-shaped lipids stabilizes the lamellar structure, but the presence of so many cone-shaped lipids puts the lamellar bilayer under stress. This is referred to as a 'frustrated bilayer' that is susceptible to factors that may drive the bilayer into non-lamellar structure. Among these factors are proteins exhibiting a hydrophobic mismatch.

Biological Function of the Hydrophobic Match

Many studies have linked protein activity to hydrophobic mismatch [66,67]. A good example comes from the $(Ca^{2+} + Mg^{2+})$-ATPase isolated from the sarcoplasmic reticulum [71]. The enzyme was reconstituted into bilayers made from PCs with monounsaturated (cis Δ9) chains of length n = 12 to n = 23. Enzyme activity increased from n = 12 to n = 20, then decreases for n = 23. Mixtures of any of these two lipids supported an activity that was the average of the two lipids. Also, the addition of decane to n = 12 increased activity. This experiment indicates that biological activity of the $(Ca^{2+} + Mg^{2+})$-ATPase was maximal at a hydrophobic match length of 18–20 carbons. These same investigators later reported similar findings for the reconstituted Na^+, K^+ ATPase [72].

Biological membrane thickness, and hence hydrophobic match length, is now believed to play a crucial role in membrane trafficking. In biological membranes, cholesterol content and membrane thickness increase from the endoplasmic reticulum to the Golgi, and finally to the plasma membrane. The average plasma membrane protein's hydrophobic match length is a full 5 amino acids longer than those of the Golgi. It is believed that proteins with a shorter hydrophobic match remain in the Golgi, while proteins with a longer hydrophobic match are transported to the plasma membrane. Therefore the plasma membrane is much thicker than the endoplasmic reticulum due to its high cholesterol and SM content, central components of lipid rafts.

A different type of hydrophobic match occurs between membrane lipids, the best documented being between cholesterol and phospholipid acyl chains. Decades ago it was noted that the rigid ring structure of cholesterol fitted nicely next to a stretch of nine saturated carbons from acyl chain C-1 to C-9. Interruption of this stretch by a double bond decreases cholesterol affinity. Therefore cholesterol interacts well with long saturated fatty acyl chains (e.g. palmitic and stearic acid) or with fatty acyl chains that have double bonds at position 9 (e.g. oleic acid, $18:1^{\Delta 9}$) or farther (e.g. linoleic acid, $18:2^{\Delta 9,12}$) down the chain. Therefore cholesterol was said to fit into a '$\Delta 9$ pocket'. Cholesterol avoids fatty acyl chains with double bonds before position $\Delta 9$ (e.g. arachidonic acid, $20:4^{\Delta 5,8,11,14}$; eicosapentaenoic acid, $20:5^{\Delta 5,8,11,14,17}$; and docosahexaenoic acid, $22:6^{\Delta 4,7,10,13,16,19}$). Hydrophobic match between cholesterol and a saturated fatty acid is required for the formation of lamellar liquid ordered (l_o) phase, the phase of lipid rafts.

Hydrophobic match is far more complex than it would initially appear. There are many possible ways in which membranes can deal with a mismatch. Hydrophobic match is a tool to regulate local bilayer thickness and hence membrane enzyme activity and membrane trafficking.

E. LIPID INTERDIGITATION

Another unusual example of lipid polymorphism involves lipid interdigitation, where one or both acyl chains extend beyond the bilayer mid-plane [73–75]. When this occurs, the penetrating chain takes up residence in both membrane leaflets, profoundly affecting basic membrane structure and hence function. Therefore interdigitation is yet another way to vary structural properties of the bilayer. Interdigitation normally occurs when one chain is significantly longer than the other (Figure 10.24), but can also be induced in symmetrical chain phospholipids. The most common membrane lipids that normally exhibit substantial differences in their acyl chain lengths are the large family of sphingolipids. Sphingolipids have a relatively short, permanent sphingosine chain and a very long (often 24-carbons) variable chain (see Chapter 5). Therefore it has been proposed that membrane domains that are enriched in sphingolipids (i.e. lipid rafts) may exhibit considerable interdigitation into the opposite leaflet.

There are three basic interdigitated states termed partially interdigitated, mixed interdigitated, and fully interdigitated (Figure 10.24). Each interdigitated state is characterized, in part, by the number of acyl chains subtended directly under the polar head group; 2 for the partially interdigitated, 3 for the mixed interdigitated and 4 for the fully interdigitated state. It is clear from the illustration in Figure 10.24 that each state is characterized by a different bilayer thickness. The thickest membrane would be associated with the non-interdigitated state, while the thinnest bilayer would characterize the fully interdigitated state. Membrane thickness is most accurately assessed by X-ray and neutron diffraction (Chapter 9), although other methodologies including DSC, Raman and FTIR, NMR, electron microscopy, ESR, and fluorescence have been employed. These other methodologies, however, must be 'calibrated' with respect to X-ray diffraction.

Interdigitated-Lipid Shape

Most membrane lipids that have a tendency to interdigitate do so in the gel state and have a large discrepancy in the length of the two acyl chains [75]. This discrepancy can be quantified as the 'chain inequivalence' [74]:

$$\text{Chain inequivalence} = \Delta C/CL$$

16:0,10:0 PC

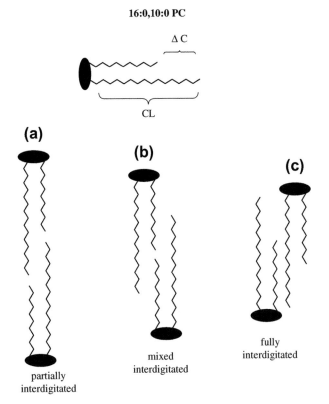

FIGURE 10.24   Lipid interdigitation. Top: An asymmetric phospholipid, 16:0,10:0 PC. Bottom: Three lipid interdigitated states. (a) partially interdigitated; (b) mixed interdigitated and; (c) fully interdigitated.

Where CL is the length of the longer chain and ΔC is the absolute difference in the chain lengths (see Figure 10.24) and is given by:

$$\Delta C = n_1 - n_2 + 1.5$$

where n_1 and n_2 are the number of carbons in the sn-1 and sn-2 acyl chains, respectively. The factor $+1.5$ is added to account for the sn-1 chain extending into the bilayer interior ~1.5 carbons deeper than the sn-2 chain (discussed below and shown in Figure 10.25). The chain inequivalence parameter therefore represents the magnitude of the chain length inequivalence normalized by the length of the hydrocarbon region of the bilayer.

For a family of phosphatidylcholines, C18,CX PC where the sn-1 chain is 18-carbons and the sn-2 chain, X, varies from 18- to 0-carbons. C18,C18 PC is fully non-interdigitated while C18,C0 PC (a lyso-PC) is fully interdigitated. Other interdigitated states are found for X between 18- and 0-carbons. The chain inequivalence parameter for C18,C10 PC is close to 0.5, characteristic of the mixed interdigitated state (Figure 10.24, B). Mixed interdigitation requires one chain to be about half the length of the other. Upon undergoing a thermotropic phase transition, the mixed interdigitated gel phase bilayers transform into partially

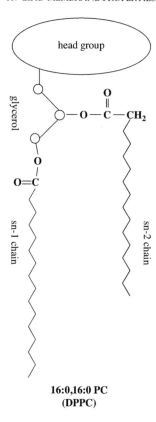

16:0,16:0 PC
(DPPC)

FIGURE 10.25   Symmetric acyl chain 16:0,16:0 PC (DPPC) demonstrating that the *sn*-1 chain protrudes directly into the bilayer interior while the first two carbons of the *sn*-2 chain run parallel with the membrane surface before also bending into the bilayer interior.

interdigitated bilayers with 2 acyl chains subtended under the head group (Figure 10.24, A). So, even in the melted, liquid crystalline state, lipids with large asymmetries in acyl chain length may still retain significant interdigitation.

The maximum possible chain asymmetry exists for lyso lipids or lipids where the *sn*-2 chain consists of either a proton or a very short acetyl ester. An example of a biologically important interdigitated lipid is C18,C2 PC, the platelet-activating factor. These lipids are fully interdigitated in the gel state where the short *sn*-2 chain lies parallel to the plane of the membrane, but upon undergoing a thermotropic phase transition develops membrane disrupting micellar structure.

Interdigitated Structure is Inherent in Symmetrical Chain Phospholipids

Surprisingly, even phospholipids with identical *sn*-1 and *sn*-2 chains exhibit chain inequivalence. While the *sn*-1 chain drops straight down into the bilayer interior, the first two carbons of the *sn*-2 chain run parallel to the membrane plane, before bending into the bilayer interior, where it then runs parallel to the *sn*-1 chain (Figure 10.25) [76]. As a result the *sn*-2 chain behaves as if it is shorter than the *sn*-1 chain. In the gel state the two chain terminal methyls are out of register by about 1.8Å or 1.5 carbon bond lengths. In biological membranes this is compensated for with the *sn*-2 chain being about 2 carbons *longer* than the *sn*-1 chain, thus preventing interdigitation.

Methods to Detect Interdigitation

Since diffraction methods can directly measure bilayer thickness, they are the gold standard for monitoring interdigitation. As discussed in Chapter 9, the electron density profiles of all membranes look very similar by X-ray diffraction. They consist of a low density bilayer interior flanked by two high density regions characteristic of the polar head groups (see Figure 9.5). With interdigitation there is a loss of the deep depression in electron density that is characteristic of free motion associated with terminal methyls in noninterdigitated bilayers. Therefore interdigitated electron density profiles are more shallow than those associated with noninterdigitated bilayers. Also, interdigitated bilayers are thinner than noninterdigitated bilayers.

Many other membrane biophysical methodologies have been used to indirectly monitor interdigitation. For example, freeze fracture EM images of noninterdigitated bilayers show a smooth surface resulting from fractionation running down the uninterrupted bilayer midplane. In sharp contrast are images obtained from mixed interdigitated gel phase bilayers indicating discontinuous fracture patterns where the fracture planes are interrupted by up and down steps. Another approach measures order parameters using ESR (electron spin resonance) [77] where spin probes, particularly DOXYL-stearic acids, are employed. This homologous series of probes has an unpaired electron (free radical) stabilized in the DOXYL structure and attached at different locations down the stearic acyl chain. Figure 10.26 shows the structure of 5-DOXYL stearic acid.

The measured order parameters decrease from near the head group down the chain to near the highly disordered chain terminus. In one experiment order parameters at the ordered 5-position were compared to order parameters at the highly disordered 16-position. In noninterdigitated bilayers the order parameters for the two positions were distinct, with the 5-position being considerably higher than the 16-position. However, upon interdigitation, the 16- and 5-positions reside near the same location and so have similar order parameters. Above T_m both probes exhibit fluid phase isotropic motion.

Interdigitation Caused by Perturbations

There are many types of perturbations that can drive noninterdigitated bilayers into a fully interdigitated gel state. The most studied of these perturbants are short-chain alcohols

FIGURE 10.26   5-DOXYL Stearic acid.

including methanol, ethanol [78], and glycerol and some anesthetics including chlorproma-zine, tetracaine, and benzyl alcohol. The perturbants must be amphipathic and localize at the aqueous interface where they replace bound waters and increase the effective size of the lipid head groups. One of the earliest interdigitation experiments involved DSC measurements of glycerol on DPPC bilayers. In the presence of glycerol (and later other alcohols), the T_m of DPPC was initially observed to decrease until a temperature was reached whereupon addi-tional glycerol resulted in an *increase* in T_m. This change in T_m was referred to as the 'biphasic effect' and is characteristic of a shift from a noninterdigitated to a fully interdigitated state. The 'biphasic effect' works better for longer chain PCs but is not observed for PEs. Also observed with DSC of fully interdigitated state bilayers are hysteresis of heating and cooling scans and a shift to lower temperature and disappearance of the pre-transition. Unfortunately, the requirement of pure phospholipids and unrealistically high alcohol levels (often greater than 1M!), seriously question any biological role for alcohol-induced interdigitation.

Other perturbants include chaotropic agents (e.g. thiocyanate at 1M), tris buffer, myelin basic protein, and polymyxin with anionic lipids and increased pressure. Also it must be noted that ether-linked phospholipids often prefer interdigitation when compared to analo-gous ester-linked phospholipids. For example, in the gel state, DPPC is noninterdigitated while DHPC (hexadecyl ether linked-PC) spontaneously interdigitates.

Biological Relevance

While it is clear that a vast array of conditions can drive the normal noninterdigitated state to several types of interdigitated states, it is not clear whether these transformations are merely esoteric curiosities or whether they can have some biological relevance. Suggested biological functions for interdigitation include:

1. Creation of novel membrane micro-domains.
2. Trans-membrane coupling between opposing membrane leaflets.
3. Reduction in charge density per unit area of a membrane surface.
4. Varying membrane thickness, thus altering the hydrophobic match.

SUMMARY

All membrane lipids exhibit different affinities and aversions for one another. An example of this is the strong affinity of cholesterol for sphingolipids that stabilizes lipid rafts. Cholesterol-rich lipid mixtures produce a liquid ordered phase whose properties are midway between gel and liquid crystalline. Cholesterol—lipid affinity follows the sequence: SM > PS, PG > PC > PE and is stronger with saturated than polyunsaturated acyl chains. In addition to the conventional lipid bilayer (more correctly the lamellar phase), dozens of other phases, whose functions are unknown, exist. Various lipid mixtures prefer different phases and complex phase diagrams have been developed that pictorially describe various lipid phases and can take a wide variety of forms. For an integral protein to achieve maximal activity, the length of its trans-membrane hydrophobic surface must match the hydrophobic length of the surrounding bilayer. The Hydrophobic Match is involved in membrane protein activity and trafficking. Lipids that normally exhibit substantial differences in their acyl chain lengths (e.g. sphingolipids) can be involved in trans-membrane coupling (interdigitation) between the inner and outer membrane leaflets.

Chapter 11 will conclude the trilogy of chapters on membrane physical properties by discussing 'packing free volume', lipid area/molecule, collapse point, 'lipid condensation', surface elasticity, molecular 'squeeze out', long-range order, crowding, membrane domains,' and homeoviscous adaptation.

References

[1] Ilangumaran S, Hoessli DC. Effects of cholesterol depletion by cyclodextrin on the sphingolipid micro-domains of the plasma membrane. Biochem J 1998;335:433–40.

[2] Manolio TA, Pearson TA, Wenger NK, Barrett-Connor E, Payne GH, Harlan WR. Cholesterol and heart disease in older persons and women. Review of an NHLBI workshop. Annu Epidemiol 1992;2:161–76.

[3] Viegas J. Dog tail-chasing linked to high cholesterol. Discovery News 2009. March 24, 2009.

[4] Quinn PJ, Wolf C. The liquid-ordered phase in membranes. Biochim Biophys Acta 2009;1788:33–46.

[5] Van der Goot FG, Harder T. Raft membrane domains: from a liquid-ordered membrane phase to a site of pathogen attack. Semin Immunol 2001;13:89–97.

[6] Brown DA, London E. Structure and origin of ordered lipid domains in biological membranes. J Membr Biol 1998;164:103–14.

[7] van Dijck PWM, De Kruijff B, van Deenen LLM, de Gier J, Demel RA. The preference of cholesterol for phosphatidylcholine in mixed phosphatidylcholine-phosphatidylethanolamine. Biochim Biophys Acta 1976;455:576–87.

[8] Demel RA, Jansen JW, van Dijck PW, van Deenen LL. The preferential interaction of cholesterol with different classes of phospholipids. Biochim Biophys Acta 1977;465:1–10.

[9] Shaikh SR, Cherezov V, Caffrey M, Soni S, Stillwell W, Wassall SR. Molecular organization of cholesterol in unsaturated phosphatidylethanolamines: X-ray diffraction and solid state ^{2}H NMR studies. J Am Chem Soc 2006;128:5375–83.

[10] Wassall SR, Shaikh SR, Brzustowicz MR, Cherezov V, Siddiqui RA, Caffrey M, et al. Interaction of poly-unsaturated fatty acids with cholesterol: A role in lipid raft phase separation. In: Danino D, Harries D, Wrenn SP, editors. Talking About Colloids. Macromol Symp, Vol. 219. Wiley-VCH; 2005. p. 73–84.

[11] Singer SJ, Nicolson GL. The fluid mosaic model of the structure of cell membranes. Science 1972;175:720–31.

[12] Karnovsky MJ, Kleinfeld AM, Hoover RL, Klausner RD. The concept of lipid domains in membranes. J Cell Biol 1982;94:1–6.

[13] Estep TN, Mountcastle DB, Barenholz Y, Biltonen RL, Thompson TE. Thermal behavior of synthetic sphin-gomyelin-cholesterol dispersions. Biochemistry 1979;18:2112–7.

[14] Simons K, Ikonen E. Functional rafts in cell membranes. Nature 1997;387: 569–372.

[15] Brown DA, London E. Functions of lipid rafts in biological membranes. Annu Rev Cell Develop Biol 1998;14:111–36.

[16] Smart EJ, Ying Y-S, Mineo C, Anderson RGW. A detergent-free method for purifying caveolae membrane from tissue culture cells. Proc Natl Acad Sci USA 1995;92:10104–8.

[17] Wainwright G, Mascitelli L, Goldstein MR. Cholesterol-lowering therapies and cell membranes. Arch Med Sci 2009;5:289–95.

[18] Tilcock CPS. Lipid polymorphism. Chem Phys Lipids 1986;40:109–25.

[19] Cullis PR, de Kruijff B. Lipid polymorphism and the functional roles of lipids in biological membranes. Biochim Biophys Acta 1979;559:399–420.

[20] Cullis PR, Hope MJ, Tilcock CPS. Lipid polymorphism and the roles of lipids in membranes. Chem Phys Lipids 1986;40:127–44.

[21] Gruner SM. Lipid Polymorphism. The molecular basis of non-bilayer phases. Ann Rev Biophys Biophys Chem 1985;1985(14):211–38.

[22] Luzzati V. Biological significance of lipid polymorphism: the cubic phases. Curr Opin Struct Biol 1997;7:661–8.

[23] Atkinson J, Epand RF, Epand RM. Tocopherols and tocotrienols in membranes: a critical review. Free Radical Biol Med 2008;44:739–64.

[24] Kumar VV. Lipid molecular shapes and membrane architecture. Ind J Biochem Biophys 1993;30:135–8.

[25] Gruner SM. Intrinsic curvature hypothesis for biomembrane lipid composition: a role for nonbilayer lipids. Proc Natl Acad Sci USA 1985;82:3665–9.

[26] Cullis PR, De Kruijff B. The polymorphic phase behaviour of phosphatidylethanolamines of natural and synthetic origins. A ^{31}P NMR study. Biochim Biophys Acta 1978;513:31–42.

[27] Cullis PR, de Kruijff, Verkleij AJ, Hope MJ. Lipid polymorphism and membrane fusion. Biochem Soc Trans 1986;14:242–5.

[28] Siegel DP. Inverted micellar intermediates and the transitions between lamellar, cubic, and inverted hexagonal lipid phases. II. Implications for membrane–membrane interactions and membrane fusion. Biophys J 1986;49:1171–83.

[29] Sollner T, Whiteheart SW, Brunner M, Erdjument-Bromage H, Geromanos S, Tempst P, et al. SNAP receptors implicated in vesicle targeting and fusion. Nature 1993;362:318–24.

[30] Rothman JE, Warren G. Implications of the SNARE hypothesis for intracellular membrane topology and dynamics. Curr Biol 1994;4:220–33.

[31] Sudhof TC, Rothman JE. Membrane fusion: Grappling with SNARE and SM proteins. Science 2009;323: 474–7.

[32] Landau EM, Rosenbusch JP. Lipidic cubic phases: A novel concept for the crystallization of membrane proteins. Proc Natl Acad Sci USA 1996;93:14532–5.

[33] Feigenson GW. Phase diagrams and lipid domains in multicomponent lipid bilayer mixtures. Biochim Biophys Acta 2009;1788:47–52.

[34] Koynova R, Caffrey M. An index of lipid phase diagrams. Chem Phys Lipids 2002;115:107–219.

[35] Sackman E. In: Hoppe Walter, Lohmann Wolfgang, Markl Hubert, Ziegler Hubert, editors. Physical Foundations of the Molecular Organization and Dynamics of Membranes. New York: Springer-Verlag; 1983. p. 425–57.

[36] Sankaram HB, Thompson TE. Cholesterol-induced fluid-phase immiscibility in membranes. Proc Natl Acad Sci USA 1991;88:8686–90.

[37] Goni FM, Alonso A, Bagatolli LA, Brown RE, Marsh D. Phase diagrams of lipid mixtures relevant to the study of membrane rafts. Biochim Biophys Acta 2008;1781:665–84.

[38] De Almeida RFM, Fedorov A, Prieto M. Sphingomyelin / phosphatidylcholine / cholesterol phase diagram: boundaries and composition of raft structures. Biophys J 2003;85:2406–16.

[39] Davoust J, Schoot BM, Devaux PF. Physical modifications of rhodopsin boundary lipids in lecithin- rhodopsin complexes: A spin-label study. Proc Natl Acad Sci USA 1979;76:2755–9.

[40] Campbell LD, Dwek RA. Biological Spectroscopy. Menlo Park, CA: Benjamin Cummings Publishing; 1984.

[41] Jones OT, McNamee MG. Annular and nonannular binding sites for cholesterol associated with the nicotinic acetylcholine receptor. Biochemistry 1988;5:2364–74.

[42] Gennis RB. Biomembranes. Molecular Structure and Function. New York, NY: Springer-Verlag; 1988.

[43] Warren GB, Houslay MD, Metcalfe JC, Birdsall NJM. Cholesterol is excluded from the phospholipid annulus surrounding an active calcium transport protein. Nature 1975;255:684–7.

[44] East JM, Melville D, Lee AG. Exchange rates and numbers of annular lipids for the calcium and magnesium ion dependent adenosinetriphosphatase. Biochemistry 1985;24:2615–23.

[45] Lee AG. How lipids and proteins interact in a membrane: a molecular approach. Mol Biosyst 2005;1: 203–12.

[46] Slater SJ, Kelly MB, Yeager MD, Larkin J, Ho C, Stubbs CD. Polyunsaturation in cell membranes and lipid bilayers and its effects on membrane proteins. Lipids 1996;31:S-189–92.

[47] Spector AA, Yorek MA. Membrane lipid composition and cellular function. J Lipid Res 1985;26:1015–35.

[48] Lee AG. How lipids interact with an intrinsic membrane protein: the case of the calcium pump. Biochim Biophys Acta 1998;1376:381–90.

[49] Dalton KA, East JM, Mall S, Oliver S, Starling AP, Lee AG. Interaction of phosphatidic acid and phosphatidylserine with the Ca^{2+}-ATPase of sarcoplasmic reticulum and the mechanism of inhibition. Biochem J 1998;329:637–46.

[50] Newton AC. Protein kinase c: poised to signal. Am J Physiol Endocrinol Metab 2010;298:E395–402.

[51] Takai Y, Kishimoto A, Inoue M, Nishizuka Y. Nucleotide-independent protein kinase and its proenzyme in mammalian tissues. I. Purification and characterization of an active enzyme from bovine cerebellum. J Biol Chem 1977;252:7603−9.

[52] Johnson JF, Zimmerman ML, Daleke DL, Newton AC. Lipid structure and not membrane structure is the major determinant in the regulation of protein kinase C by phosphatidylserine. Biochemistry 1998;37:12020−5.

[53] Skou J. The influence of some cations on an adenosine triphosphatase from peripheral nerves. Biochim Biophys Acta 1957;23:394−401.

[54] Haim H, Cohen E, Lifshitz Y, Tal DM, Goldshleger R, Karlish SJD. Stabilization of Na^+, K^+-ATPase purified from *Pichia pastoris* membranes by specific interactions with lipids. Biochemistry 2007;46:12855−67.

[55] Marius P, Zagnoni M, Sandison ME, East M, Morgan H, Lee AG. Binding of anionic lipids to at least three nonannular sites on the potassium channel KcsA is required for channel opening. Biophys J 2008;94:1689−98.

[56] Salem Jr N, Kim HY, Yergey JA. Docosahexaenoic acid: membrane function and metabolism. In: Simopoulos AP, Kifer RR, Martin R, editors. The Health Effects of Polyunsaturated Fatty Acids in Seafoods. New York: Academic Press; 1986. 363−317.

[57] Dratz EA, Deese AJ. The role of docosahexaenoic acid in biological membranes: examples from photoreceptors and model membrane bilayers. In: Simopoulos AP, Kifer RR, Martin R, editors. The Health Effects of Polyunsaturated Fatty Acids in Seafoods. New York: Academic Press; 1986. p. 319−51.

[58] Mitchell DC, Straume M, Litman BJ. Role of sn-1-saturated, sn-2-polyunsaturated phospholipids in control of membrane receptor conformational equilibrium: Effects of cholesterol and acyl chain unsaturation on the metarhodopsin I - metarhodopsin II equilibrium. Biochemistry 1992;31:662−70.

[59] Feller SE, Gawrisch K, MacKerrell Jr AD. Polyunsaturated fatty acids in lipid bilayers: intrinsic and environmental contributions to their unique physical properties. J Am Chem Soc 2002;124:318−26.

[60] Feller SE, Gawrisch K, Wolfe TB. Rhodopsin exhibits a preference for solvation by polyunsaturated docosahexaenoic acid. J Am Chem Soc 2003;125:4434−5.

[61] Maurer A, McIntyre JO, Churchill S, Fleischer S. Phospholipid protection against proteolysis of D-β-hydroxybutyrate dehydrogenase, a lecithin-requiring enzyme. J Biol Chem 1985;260:1661−9.

[62] Loeb-Hennard C, McIntyre JO. (R)-3-Hydroxybutyrate dehydrogenase: selective phosphatidylcholine binding by the C-terminal domain. Biochemistry 2000;39:11928−38.

[63] Pangborn M. Isolation and purification of a serologically active phospholipid from beef heart. J Biol Chem 1942;143:247−56.

[64] Paradies G, Ruggiero FM, Petrosillo G, Quagliariello E. Peroxidative damage to cardiac mitochondria: cytochrome oxidase and cardiolipin alterations. FEBS Lett 1998;424:155−8.

[65] Paradies G, Ruggiero FM, Petrosillo G, Quagliariello E. Age-dependent decline in the cytochrome c oxidase activity in rat heart mitochondria: role of cardiolipin. FEBS Lett 1997;406:136−8.

[66] Killian JA. Hydrophobic mismatch between proteins and lipids in membranes. Biochim Biophys Acta 1998;1376:401−16.

[67] Killian JA, de Planque MRR, van der Wel PCA, Salemink I, de Kruijff B, Greathonse DV, et al. Modulation of membrane structure and function by hydrophobic mismatch between proteins and lipids. Pure &App Chem 1998;70:75−82.

[68] Dumas F, Lebrun MC, Tocanne J-F. Is the protein/lipid hydrophobic matching principle relevant to membrane organization and function? FEBS Lett 1999;458:271−7.

[69] Mouritsen OG, Bloom M. Mattress model of lipid−protein interactions in membranes. Biophys J 1984;46(1984):141−53.

[70] Jensen MO, Mouritsen OG. Lipids do influence protein function − the hydrophobic matching hypothesis revisited. Biochim Biophys Acta 2004;1666:205−26.

[71] Johansson A, Keightley CA, Smith GA, Richards CD, Hesketh TR, Metcalfe JC. The effect of bilayer thickness and n-alkanes on the activity of the $(Ca^{2+} + Mg^{2+})$-dependent ATPase of sarcoplasmic reticulum. J Biol Chem 1981;256:1643−50.

[72] Johannsson A, Smith GA, Metcalfe JC. The effect of bilayer thickness on the activity of $(Na^+ + K^+)$-ATPase. Biochim Biophys Acta 1981;641:416−21.

[73] Slater JL, Huang C-H. Interdigitated bilayer membranes. Prog Lipid Res 1988;27:325−59.

[74] Slater JL, Huang C-H. Lipid bilayer interdigitation. In: Yeagle PL, editor. The Structure of Biological Membranes. Boca Raton, FL: CRC Press; 1991. p. 175—210.

[75] Mason JT. Properties of mixed-chain length phospholipids and their relationship to bilayer structure [Chapter 8]. In: Lasic DD, Barenholz Y, editors. Handbook of Nonmedical Applications of Liposomes: Theory and Basic Sciences, Vol 1. Boca Raton, FL: CRC Press; 1996. p. 195—221.

[76] Pascher I, Sundell S. Molecular arrangements in sphingolipids. The crystal structure of cerebroside. Chem Phys Lipids 1977;20:175—99.

[77] Boggs JM, Mason JT. Calorimetric and fatty acid spin label study of subgel and interdigitated gel phases formed by asymmetric phosphatidylcholines. Biochim Biophys Acta 1986:863,231—242.

[78] Zeng J, Smith KE, Chong PL-G. Effects of alcohol-induced lipid interdigitation on proton permeability in L-α-dipalmitoylphosphatidylcholine Vesicles. Biophys J 1993;65:1404—14.

[79] Wlstutts: http://en.wikipedia.org/wiki/File:Space-Filling_Model_Sphingomyelin_and_Cholesterol.jpg; public domain.

Long-Range Membrane Properties

O U T L I N E

A. **Membrane Bilayer Lipid Packing** 215
  Lipid Packing Free Volume 215
  Lipid Monolayers (Langmuir Film Balance) 217
  Area/Molecule 219
  Lipid 'Condensation' 220
  Surface Elasticity Moduli, C_s^{-1} 221
  Lipid 'Squeeze Out' 221

B. **Membrane Protein Distribution** 222
  Freeze Fracture Electron Microscopy 222

C. **Imaging of Membrane Domains** 223
  1. Macrodomains 223
    Membrane Macrodomains 224
    Gap Junctions 224

    Clathrin-Coated Pits 226
    Caveolae 226
  2. Lipid Microdomains 227

D. **Homeoviscous Adaptation** 232
  Acyl Chain Length 232
  Acyl Chain Double Bonds 233
  Phospholipid Head Groups 233
  Sterols 233
  Isoprene Lipids 233
  Anti-freeze Proteins 233
  Divalent Metals 233
  Don Juan Pond 234

Summary 235

References 235

Chapter 11 will discuss a few large-scale membrane properties including lipid bilayer packing, membrane protein distribution, membrane lipid micro-domains, and finally a process that keeps all of the various membrane properties in synch, homeoviscous adaptation.

A. MEMBRANE BILAYER LIPID PACKING

Lipid Packing Free Volume

It has been well-documented that the large-scale state of the lipid bilayer componemt of membranes has a profound impact on the activity of resident membrane proteins [1,2].

One measure of this, a parameter known as the 'packing free volume' (f_v), was first introduced in a series of papers in the late 1980s by Straume and Litman to quantify bulk phospholipid acyl chain packing [3]. This parameter was derived from time-resolved anisotropy decay of the fluorescent lipid bilayer probe DPH (1,6-diphenyl-1,3,5-hexatriene, see Chapter 9) which characterizes the volume available for probe re-orientational motion in the anisotropic bilayer relative to that available in an unhindered, isotropic environment. The DPH-containing membrane (where DPH displays partially hindered, anisotropic motion) is placed in the fluorescence sample cuvette, while the reference cuvette contains POPOP (1,4-bis(5-phenyloxazol-2-yl)benzene) in absolute ethanol. POPOP in ethanol displays totally unhindered, isotropic motion. Totally hindered motion has an f_v of 0.0, while the f_v of totally isotropic motion is 1.0. The measured f_v for a membrane will fall between 0 and 1. The f_v is particularly adept at measuring small changes in lipid packing that can profoundly affect integral membrane protein conformation.

Mitchell and Litman [4–6] have extensively used the packing free volume parameter to explore the effect of membrane lipid unsaturation on the visual receptor metarhodopsin I ↔ metarhodopsin II equilibrium. Metarhodopsin II is the active form of the receptor and occupies a larger volume than does metarhodopsin I. Table 11.1 presents f_vs for several PCs commonly found in the rod outer segment membrane [7,8]. At near physiological temperature (40°C), the values increase from liquid crystalline 14:0,14:0 PC with no double bonds ($f_v = 0.101$) to 22:6,22:6 PC with 12 double bonds ($f_v = 0.201$). This demonstrates an increase in packing free volume with increasing phospholipid unsaturation. Also, f_v decreases for every phospholipid at lower temperature (10°C) and increasing cholesterol content. Lower temperature decreases the number of *gauche* kinks in the acyl chains, thus decreasing the packing free volume. And, as discussed below, cholesterol 'condenses' fluid state phospholipids, also decreasing the packing free volume. In addition, Mitchell and Litman measured the

TABLE 11.1   Values of the Packing Free Volume Parameter, f_v, in Phosphatidylcholine Bilayers.

Acyl Chain Composition	f_v		
	40°C[1]	10°C[1]	40°C[2], 30 mol% cholesterol
di 22:6n3	0.201 ± 0.013	0.133 ± 0.01	0.122 ± 0.009
di 20:4n6	0.278 ± 0.016	0.205 ± 0.01	0.11 ± 0.01
di 18:1n9	0.147 ± 0.005	0.119 ± 0.005	0.062 ± 0.002
16:0, 22:6n3	0.154 ± 0.005	0.096 ± 0.005	0.074 ± 0.003
16:0,20:4n6	0.155 ± 0.003	0.124 ± 0.004	0.067 ± 0.002
16:0,18:1n9	0.130 ± 0.005	0.073 ± 0.003	0.056 ± 0.002
di 14:0	0.101 ± 0.006	-	0.011 ± 0.002

[1]Values from Mitchell and Litman. 1998. Biophysical Journal 74, 879–891. [7]
[2]Values from Mitchell and Litman. 1998. Biophysical Journal 75, 896–908. [8]
Values kindly provided by Drake Mitchell, Department of Physics, Portland State University.

TABLE 11.2 Packing Free Volume (f_v) and Metarhodopsin I $\leftrightarrow$ Metarhodopsin II Equilibrium (K_{eq}) for Various PCs Commonly Found in the Rod Outer Segment.

Phosphatidylcholine	meta I $\leftrightarrow$ meta II (K_{eq})	Packing free volume (f_v)
22:6,22:6 PC	6	0.201
20:4,20:4 PC	4.7	0.278
16:0,22:6 PC	3.3	0.154
16:0,20:4 PC	2.7	0.155
14:0,14:0 PC	1.7	0.101

metarhodopsin I $\leftrightarrow$ metarhodopsin II equilibrium, expressed as K_{eq}, in bilayer membranes made from five of the same lipids reported in Table 11.1 [7,8]. Table 11.2 shows a general relationship between formation of the more voluminous, active metarhodopsin II (expressed as K_{eq}) and the packing free volume (expressed as f_v). A larger f_v drives the metarhodopsin I $\leftrightarrow$ metarhodopsin II equilibrium to the right. These investigators also reported a reduction in K_{eq} with both lower temperature and cholesterol. From their observations, Mitchell and Litman proposed a molecular reason for the naturally high levels of docosahexaenoic acid in the rod outer segment and the requirement of this fatty acid for vision [9].

Lipid Monolayers (Langmuir Film Balance)

The first reported membrane physical properties were determined on lipid monolayers using a Langmuir film balance or trough. This device was named after Irving Langmuir, an early pioneer in the field (see Chapter 2). A schematic of a Langmuir film balance is depicted in Figure 11.1. The trough is normally made of an inert, non-wettable material such as Teflon or polytetrafluoroethylene (PTFE). The monolayer is formed by adding a small volume of a volatile organic solution containing the dissolved lipid to the surface of an

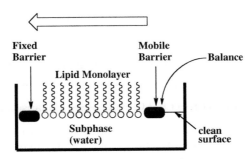

FIGURE 11.1 Diagram of a Langmuir Film Balance. A lipid monolayer is deposited on a clean aqueous surface between two barriers. Movement of the mobile barrier from right to left compresses the monolayer as the lateral pressure is continuously monitored via a sensitive balance attached to a Wilhelmy plate inserted through the monolayer.

aqueous sub-phase between two Teflon barriers, one or both of which can move parallel to the side walls of the trough. The barriers are in constant contact with the top of the subphase. After the organic solvent has evaporated, the barrier(s) are moved and the monolayer is compressed. Barrier movement, and hence the size of the monolayer. is accurately controlled by computer. The temperature of the water subphase is also carefully controlled. During the compression, the surface tension of the monolayer is continuously monitored by use of a Wilhelmy plate attached to an electronic linear-displacement sensor, or electrobalance (discussed in Chapter 3). Monolayer surface pressure (π) is calculated by subtracting the surface tension of the subphase with the floating monolayer (γ) from the surface tension of the pure subphase with no monolayer (γ_0). γ_0 for water, the normal subphase, is a constant at a single temperature (e.g. $\approx$ 73 mN/m at 20°C) [10].

$$\pi = \gamma_0 - \gamma$$

Surface pressure varies with the molecular area of the compressed monolayer. Surface tension measurements are exquisitely sensitive to any contamination, as contaminants will often accumulate at the water/air interface where they compete with the monolayer lipids. Since even 1 ppm contaminant can radically change monolayer behavior, maximal cleanliness and purity of components must be observed. Experiments are often run in a clean room to prevent airborne contaminants and on a vibration-free table. Dilute solutions of the membrane lipid to be tested (~1mg/ml) are made in an organic carrying solvent that must dissolve the amphipathic lipids while also being volatile. Examples of these solvents currently in use include hexane/2-propanol (3:2) or ethanol/hexane (5:95). The lipids rapidly spread over the clean aqueous interface with their polar head groups in the water and their hydrophobic tails extended into the air. After 4–5 minutes to allow for the carrying solvent to dissipate, the compression is begun.

Figure 11.2 shows pressure-area (Π-A) isotherms for a simple, saturated fatty acid (e.g. palmitic or stearic acid) on the left and a heterochain phospholipid (e.g. 18:0, 18:1 PC) on the right. Note that the phospholipid isotherm is far more complicated than that observed for the simple fatty acid [11]. At the lowest possible lipid densities, the lipids are not touching and the monolayer exhibits a quasi two-dimensional gas state (G). At ~78 $Å^2$/molecule, depending on the phospholipid, the lipids come into contact with one another and enter the liquid-expanded (LE) phase. Note the LE phase is missing for un-esterified, saturated chain fatty acids. In the LE phase, lipids are in contact but without molecular order. At ~5 mN/m a transition occurs from the LE phase to the liquid-condensed (LC) phase that exhibits order (LE to LC). Further compression will result in the solid (S) phase and eventually lead to collapse of the monolayer (Collapse Point), where the lipids exhibit maximal possible density. The sequence of states that a phospholipid monolayer goes through upon compression is therefore:

$$G \rightarrow G\text{-}LE \rightarrow LE \rightarrow LE/LC \rightarrow LC \rightarrow S \rightarrow \text{Collapse Point}$$

It is estimated that the lateral pressure of a biological membrane is ~30–35 mN/m, considerably less than the Collapse Point, but substantially greater than the LE phase.

Figure 11.3 shows Π-A curves for lipids of different compressibilities [12]. Curve 'a' is the highly incompressible 16:0,16:0 PC which remains in the solid-like phase throughout the compression; curve 'b' is compressible 22:6,22:6 PC which is similar to the curve for 14:0,14:0 PE discussed in Figure 11.2; and curve 'c' is mixed chain egg PC which remains

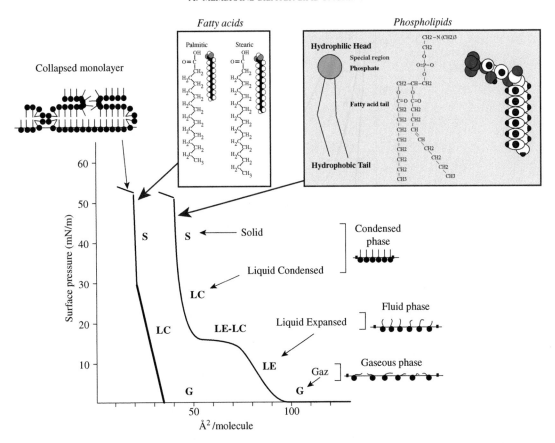

FIGURE 11.2   Pressure-area (Π-A) isotherms for a simple, saturated fatty acid on the left and a heterochain phospholipid on the right. Note that the phospholipid isotherm is far more complicated than that observed for the simple fatty acid. Phases are: **G**, quasi, two-dimensional gas state; LE, liquid-expanded phase; LC, liquid-condensed phase; S, solid phase; and finally the Collapse Point. The sequence of states that a phospholipid monolayer goes through upon compression is: G → G-LE → LE → LE-LC → LC → S → Collapse Point. *Reprinted with Permission* [56].

in a liquid-like phase throughout the transition. Very steep Π-A curves indicate poor compressibility (e.g. long chain di-saturated phospholipids or cholesterol) while shallow curves indicate the lipid is compressible (e.g. lipids containing polyunsaturated chains).

A variety of important physical parameters can be derived from Π-A isotherms. These include the area/molecule as a function of lateral pressure, Collapse Point (maximum lipid packing density and therefore the minimal area/molecule), effect of polar head groups on the area/molecule, effect of acyl chains on the area/molecule, 'condensation' between two or more lipids, surface elasticity, and molecular 'squeeze out' of a component from a membrane. A few of these properties are briefly discussed below.

Area/Molecule

The area/molecule for membrane lipids at different lateral pressures or at different temperatures can be directly read off Π-A isotherms. One interesting application of these

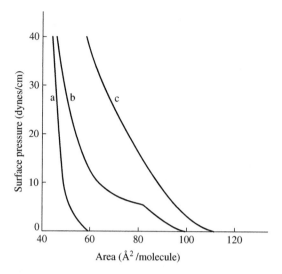

FIGURE 11.3   Π-A curves for lipids of different compressibilities. Curve a is the highly incompressible 16:0,16:0 PC which remains in the solid-like phase throughout the compression; curve b is compressible 22:6,22:6 PC; and curve c is mixed chain egg PC which remains in a liquid-like phase throughout the transition [12].

measurements involves the melting (gel $\rightarrow$ liquid crystal) behavior of a di-saturated PC. In the gel state DSPC (18:0,18:0 PC) has a cross-sectional area of ~48 Å^2. Upon melting to the liquid crystalline state the area increases to ~70 Å^2. Interestingly, DOPC (18:1,18:1 PC) occupies almost the same cross-sectional area (~72 Å^2) as liquid crystalline DSPC. Therefore it can be concluded that the ~30° kink imparted by a *cis* double bond in an acyl chain increases the molecular area of a phospholipid by approximately the same amount as that which arises upon increasing the *gauche* kinks upon chain melting.

Lipid 'Condensation'

'Condensation' is an expression of how closely two lipids behave from ideal with respect to their areas per molecule. If two molecules behave as non-compressible (ideal) spheres, when mixed their combined measured areas should equal the sum of the area/molecule determined independently for each component [13].

$$A_{ideal} = \chi_1 (A_1)_\pi + (1 - \chi_1)(A_2)_\pi$$

Where χ_1 is the mol fraction of component 1 and $(A_1)_\pi$ and $(A_2)_\pi$ are the mean molecular areas of components 1 and 2 at surface pressure π. Negative deviations of the experimentally measured value (A_{exp}) from A_{ideal} represent attraction ('condensation') while positive deviations represent repulsion. If A_{ideal} = the experimentally measured value, the lipids behave ideally. The extent of non-ideal behavior is expressed as '% condensation':

$$\% \text{ condensation} = [(A_{ideal} - A_{exp})/A_{ideal}] \times 100$$

where A_{ideal} is the mean molecular area calculated assuming ideal additivity and A_{exp} is that observed experimentally.

The most studied example of membrane condensation is with phospholipids and cholesterol [14]. For decades cholesterol has been known to condense fluid state membranes, thus decreasing their permeability, fluidity, packing free volume, and increasing membrane thickness [15–17]. However, when cholesterol is added to gel state bilayers, membrane packing is decreased while permeability, fluidity, and packing free volume are increased. Bilayer thickness is also decreased. It has been proposed that cholesterol-induced condensation, which is extremely high for sphingolipids and di-saturated phosphatidylcholines, may even be responsible for the formation and stability of lipid rafts (see Chapter 8). Cholesterol associates best with acyl chains that have no double bonds before position $\triangle 9$. In contrast, cholesterol avoids close association with chains that have double bonds before position $\triangle 9$ (e.g. γ-linolenic, arachidonic, eicosapentaenoic, and docosahexaenoic acids). In the normal biological motif where the *sn*-1 chain is saturated and the *sn*-2 chain unsaturated, cholesterol renders area condensation measurements relatively insensitive to structural changes in the *sn*-2 chain. Therefore it has been proposed that cholesterol likely orients adjacent to the saturated, *sn*-1 side of a phospholipid.

Surface Elasticity Moduli, C_s^{-1}

The surface elasticity moduli, C_s^{-1}, has been proposed to provide a more accurate assessment of cholesterol–phospholipid interactions than condensation [18].

$$C_S^{-1} = (1 - A)(d\pi/dA)\pi$$

Note that C_s^{-1} is a *change* in surface pressure with area and so, unlike condensation, is dynamic.

Smaby et al. (1997) [18] used the in-plane elasticity moduli (inverse of the lateral compressibility moduli) to address the question of cholesterol interaction with mixed acyl chain PCs. The PCs had a saturated *sn*-1 chain and an *sn*-2 chain composed of either 14:0, 18:1, 18:2, 20:4 or 22:6. At biological lateral pressure ($\geq 30 \text{ mN/m}$) they reported that cholesterol caused the in-plane elasticity of all of the mixed monolayers to decrease. PCs with more double bonds, however, were less affected by cholesterol and so these PC/sterol mixtures maintain relatively high in-plane elasticity. For di-saturated PCs the decrease in interfacial elasticity is ~6 to 7 fold with equimolar cholesterol, while with di-polyunsaturated PCs the decrease is only ~1.7 fold. The unsaturated *sn*-2 acyl chain strongly modulates the elasticity of the mixed-chain PC films due to the poor association of the rigid α-surface of the sterol ring with the unsaturated *sn*-2 chains. These surface pressure measurements on lipid monolayers indicate that cholesterol will preferentially associate with saturated versus unsaturated acyl chains. This conclusion is in agreement with results derived from a variety of other methodologies (see Chapter 10, section A. Complex Lipid Interactions). Also, it can be concluded that cholesterol will associate more strongly with the saturated *sn*-1 chain of a mixed acyl chain phospholipid, particularly when the *sn*-2 chain is polyunsaturated.

Lipid 'Squeeze Out'

In some lipid mixtures increasing lateral pressure may force a membrane component to be excluded or 'squeezed out' of the monolayer [19]. An example of this behavior is shown in

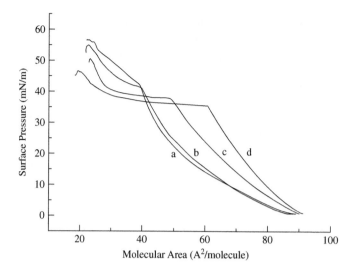

FIGURE 11.4   Π-A curves demonstrating 'squeeze out' in SM/18:0,22:6 PE/cholesterol monolayers. Monolayers are composed of various mixtures of SM and 18:0,22:6 PC to which 0.05 mol fraction of cholesterol was added. The mol fraction of 18:0, 22:6 PE in SM are: Curve a, 0.1; Curve b, 0.2; Curve c, 0.3; Curve d, 0.4. The 'squeeze out' of 18:0,22:6 PE is reflected by the plateau component that is most evident in curves b, c, and d. [13].

Figure 11.4 [13]. Monolayers were made of mixtures of SM and 18:0,22:6 PE containing an additional 0.05 mol fraction cholesterol. This lipid mixture represents a model plasma membrane/lipid raft. Upon increasing the lateral pressure, an 18:0,22:6 PE-dependent plateau region emerged. The interpretation was that 18:0,22:6 PE was 'squeezed out' from the SM/cholesterol (lipid raft) domain. The percent 'squeeze out' in the plateau portion of π/A curves is defined as:

$$L = 1 - (A_e/A_b) \times 100$$

Where L is the percentage of molecules lost or 'squeezed out', and A_b and A_e are the beginning and end of the surface area of the near-horizontal ('plateau') region of the curve.

B. MEMBRANE PROTEIN DISTRIBUTION

Freeze Fracture Electron Microscopy

Freeze fracture electron microscopy (EM) was developed in the 1960s. It is an unusual technique that has been used to investigate membrane structure from the perspective of both the membrane surface and the hydrophobic interior [20]. From this technique the distribution of membrane integral proteins in the lipid milieu can be directly estimated. Freeze fracture EM is based on rapidly freezing a membrane at very cold temperature ($< -100°C$) followed by fracturing using a knife. The fraction plane follows the weak point in the frozen sample, the center of the membrane bilayer interior. This exposes the inner surface of both leaflets

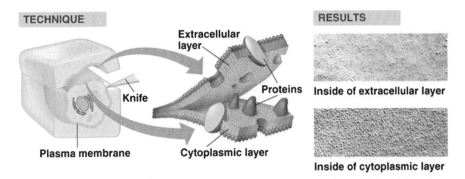

FIGURE 11.5   The technique of freeze fracture electron microscopy. The process is briefly discussed in the text and is discussed in much more detail in a 2007 review paper by Severs [20]. Note there are many more proteins projecting from the cytoplasmic leaflet than from the extracellular leaflet [21].

of the membrane (Figure 11.5) [21]. The integral membrane proteins appear as particles of about 80 to 100 Å in diameter. Usually they are randomly dispersed, but may be arranged in groups. Some particles are preferentially associated with one membrane leaflet face or the other (as demonstrated in Figure 11.5). Freeze fracture EM supplies direct evidence that proteins reside inside membranes and are not just stuck on the surface. The particle distribution also indicates that there is no long-range order, instead observed order is only a few tenths of a micron and the membrane bilayer sea is indeed very crowded [22].

Examples of long-range order due to cytoskeletal attachments have been observed. In an early report from 1976, Yu and Branton [23] showed that egg lecithin liposomes have very smooth membrane leaflet surfaces that were devoid of protein particles. When the erythrocyte protein band 3 was reconstituted into the liposome membranes, dispersed particles appeared in freeze fracture that closely resembled what was imaged for natural erythrocyte membranes. When the pH was dropped to 5.5, the particles observed in the erythrocyte clumped together due to action of the cytoskeleton. Interestingly, when spectrin and actin, components of the cytoskeleton, were added to the reconstituted band 3-egg lecithin liposomes and the pH was decreased to 5.5, particle clumping was also observed.

Unfortunately, freeze fracture EM can be fraught with potential artifacts. Just the process of rapid freezing in liquid nitrogen produces destructive ice crystals. Therefore pretreatment with a cryoprotectant (glycerol) and glutereraldehyde is often employed. The biological membrane itself is not directly imaged. Instead a replica is made by shadowing the sample with platinum at a 45° angle to highlight the topography. The replica is further strengthened with a thin layer of carbon. The replica is then carefully washed off the sample and imaged by EM. Every additional step in a complex process is a potential source of artifacts.

C. IMAGING OF MEMBRANE DOMAINS

1. Macrodomains

The term 'domain' has come to be used extensively in the life sciences literature. Unfortunately, 'domain' means quite different things to different investigators. Funk and Wagnalls'

Standard College Dictionary defines the term domain as 'a territory over which dominance is exerted'. In proteins, domains are a component of the total protein structure and they exist and function essentially independently of the rest of the protein. Many proteins consist of several functional domains that can be very large, varying in length between 25 to 500 amino acids. Among the various types of protein domains are those that bind to specific portions of a nucleic acid that are also referred to as a domain. So the term domain can refer to functional patches of proteins, nucleic acids or, as we will see, membranes!

Quite different from protein and nucleic acid domains are domains that comprise membrane structure. It is universally accepted that biological membranes are not homogeneous mixtures of lipid and protein but instead consist of patches of widely differing and often rapidly changing composition called domains. Domains exist in a bewildering array of sizes, stabilities, lipid and protein compositions, and functionalities. Membrane domains can be roughly divided into large, stable macrodomains and small unstable lipid microdomains. Most macrodomains are stable for extended time periods and so are isolatable and fairly well defined. The major part of most biological membranes, however, is likely composed of an enormous number of poorly understood and less stable lipid microdomains that are difficult to study, and so far impossible to isolate in pure form. Through the years the reputed size and stability of lipid microdomains has continuously decreased from microns down to tens of molecules or so with associated lifetimes into the un-biological nanosecond range [24]. While large macrodomains are easily imaged, their analogous lipid microdomain counterparts in biological membranes are more elusive, far smaller, and thus harder to image. In the following sections, examples of macroscopic and lipid microdomain imaging are presented.

Membrane Macrodomains

Most macrodomains owe their discovery to the development of electron microscopy in the 1940s and 1950s. Some examples including basolateral and apical halves of epithelial cells, thylakoid grana and stroma, sperm head and tail, tight junctions and bacteriorhodopsin 2-D patches in the purple membrane of *Halobacterium halobium* have been known for decades and will not be considered here. Images of the macroscopic domains; gap junctions, clathrin-coated pits, and caveolae will be briefly discussed.

Gap Junctions

Gap junctions were probably first observed by the pioneering electron microscopist J.D. Robertson in 1953 [25]. Robertson's 'Unit Membrane' model was based on his many EM images. Gap junctions were eventually characterized in the late 1950s and early 1960s by several investigators including George Palade and his wife Marilyn Farquahr in 1963 [26]. Gap junctions are tube-like structures that connect the cytoplasms of adjacent cells [27,28]. Unlike tight junctions that press adjacent cells together, gap junctions maintain a 2 to 3 nm spatial gap between cells. Their function is to allow for rapid communication between cells through exchange of various small molecules (~1,000 molecular weight limit) and ions. Important examples of gap junction-transported solutes include the second messengers cAMP, IP_3, and Ca^{2+}. Cells with gap junctions are therefore in direct electrical and chemical contact with each other [29]. Gap junction intercellular connections are through hollow cylinders called connexons. Each connexon is a circular arrangement of six subunits of a protein

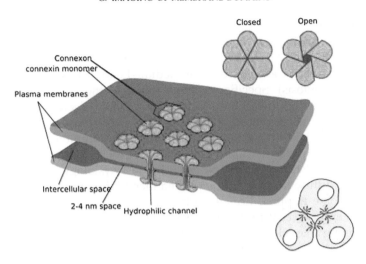

FIGURE 11.6   Schematic drawing of a field of gap junctions [57].

called a connexin (Figure 11.6). Since connexins come in about 15 different types, small variations in connexon structure are known. Each connexon spans the plasma membrane and protrudes into the gap between the cells where it connects to the protruded connexon of an adjacent cell, thus forming a completed channel. The channel is only about 3 nm in diameter at its narrowest, thus limiting the size of transported solutes. Gap junctions can be a single connexon or may be found in a field of hundreds or thousands of adjacent connexons called a plaque. Gap junctions are especially abundant in muscle and nerve that require rapid intracellular communication. Gap junction defects have been linked to certain congenital heart defects and deafness.

George Palade (Figure 11.7) is perhaps the most accomplished cell biologist of all time. As an electron microscopist, he was involved in identifying and describing many cell structures including gap junctions and Golgi apparatus. In 1974 he shared the Nobel Prize in Physiology and Medicine with Albert Claude and Christian de Duve 'for discoveries concerning the functional organization of the cell that were seminal events in the development of modern cell biology.'

FIGURE 11.7   George Palade (1912–2008). *Courtesy of UC San Diego School of Medicine Cellular and Molecular Medicine Department.*

Clathrin-Coated Pits

Plasma membranes are in part characterized by having structures that facilitate uptake of required solutes into the cell. An example of such a structure is the clathrin-coated pit, a structure that is involved in the process of receptor-mediated endocytosis (also referred to as clathrin-dependent endocytosis). Specific receptors found in the outer leaflet of the plasma membrane are involved in the internalization of macromolecules (ligands) including hormones, growth factors, enzymes, serum proteins, cholesterol-containing lipoproteins, antibodies, and ferritin-iron complexes. The best known example of this is the uptake of cholesterol-containing low density lipoprotein (LDL) particles [29], work that won Michael Brown and Joseph Goldstein the 1986 Nobel Prize in Physiology and Medicine. Ligand-receptor complexes diffuse laterally in the plasma membrane until they encounter a coated pit where they become trapped. Proteins attached to the plasma membrane inner leaflet including clathrin and two coat proteins, COPI and COPII, cause the pit to invaginate until it pinches off, forming a vesicle free in the cytoplasm (Figure 11.8, [30]). The vesicle has three possible fates: 1. It can travel to the lysosome for degradation; 2. It can go to the TGN (Trans-Golgi Network) for transport through the endomembrane system; or 3. It can return back to the plasma membrane at the opposite side of the cell where it is secreted as part of a process called transcytosis [31]. Coated pits are quite abundant in the plasma membrane [32]. Since they occupy ~20% of the total plasma membrane surface area and the entire process of internalization only takes about 1 minute, there may be 2,500 coated pits invaginating into vesicles per minute in a cell. Another protein found associated with coated pits is caveolin, a major component of caveolae, a type of lipid raft. Coated pits are twice the size of caveolae, but share some common properties with caveolae and lipid rafts.

Caveolae

Caveolae are complex plasma membrane structures whose properties appear to place them between coated pits and lipid rafts (discussed in Chapter 8). They are small

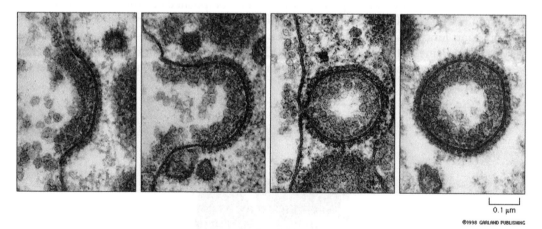

0.1 µm

©1998 GARLAND PUBLISHING

FIGURE 11.8   Sequence of formation of coated vesicles (right) from clathrin-coated pit (left) during receptor-mediated endocytosis on the surface of the plasma membrane. Pictures are electron micrographs of yolk protein on a chicken oocyte [30].

(50–100 nm) invaginated membrane structures that superficially resemble coated pits. In fact, in 1955 Yamada [33] proposed the descriptive name 'caveolae' which is Latin for *little caves*. Caveolae were first described by the electron microscopist George Palade in 1953 and are abundant in many vertebrate cell types, especially endothelial cells and adipocytes where they may account for between 30 to 70% of the total plasma membrane surface area. Caveolae, however, are not a universal feature of all cells as they are totally absent in neurons. Like lipid rafts, caveolae are partially characterized by being enriched in sphingo-lipids and cholesterol and also participate in signal transduction processes [34,35]. In fact caveolae are often described as being 'invaginated lipid rafts' that differ primarily by the presence of a family of marker proteins called caveolins. But, similar to coated pits, caveolae may even play a role in endocytosis.

2. Lipid Microdomains

Although macrodomains are large, stable, and relatively easy to image, most of a biological membrane likely consists of many unstable heterogeneous patches of lipids and proteins known as lipid microdomains. These structures are well known in model membrane systems where they are driven by lipid lateral phase separations. However, if lipid microdomains exist at all in biological membranes, they are fleeting and very small. Their possible existence is based primarily on homologies with many experiments performed for decades on model lipid monolayers and bilayers. The documented phase separations have been primarily induced by changes in temperature, pressure, ionic strength, divalent cations, and cationic peripheral proteins. In contrast to macrodomains, lipid microdomains have to date been impossible to isolate and directly study as pure entities. Instead their existence is often inferred from biophysical techniques where the experimental measurements demonstrate co-existing multiple membrane populations, i.e. lipid microdomains.

In early experiments with sea urchin and mouse eggs (discussed in Chapter 9), Michael Edidin employed lateral diffusion measurements using fluorescence recovery after photo-bleaching (FRAP) to support the concept of membrane heterogeneity in membranes [36]. Edidin's experiment followed the diffusion of two carbocyanine dyes, one with two short, saturated lipid chains (C_{10}, C_{10} DiI) and one with two long saturated lipid chains (C_{22}, C_{22} DiI). The measured difference in diffusion rates between the two dyes was consistent with the existence of lipid microdomains.

Membrane microdomains have also been inferred from the activity of reconstituted enzymes. In one interesting model [37,38], the sarcoplasmic reticulum Ca^{2+} ATPase was reconstituted into liposomes made from either zwitterionic DOPC (18:1, 18:1 PC), anionic DOPA (18:1, 18:1 PA) or a DOPC/DOPA (1:1) mixture, and the enzyme activity measured. Activity was high in DOPC and low in DOPA. For the DOPC/DOPA (1:1) liposome, an inter-mediate activity was observed. Upon the addition of Mg^{2+} to the Ca^{2+} ATPase reconstituted in DOPC, the activity did not change and remained high. In contrast, Mg^{2+} reduced activity of the Ca^{2+} ATPase reconstituted into DOPA to zero. The precipitous decrease in activity was attributed to Mg^{2+} binding to the anionic DOPA and inducing an isothermal phase transition resulting in the DOPA liposome being driven into the totally inactive gel state. When Mg^{2+} was added to the DOPC/DOPA mixed liposome, the Ca^{2+} ATPase activity was observed to *increase* as DOPA was removed from the fluid mixture as a gel. The induced isothermal phase transition resulted in the enzyme accumulating in the favorable DOPC fluid state domain. Thus the Ca^{2+} ATPase activity increased.

Direct observations (images) of lipid microdomains have routinely been reported for lipid monolayers and bilayers using fluorescent probes. It should be pointed out that lipid microdomains in model protein-free membranes are orders of magnitude larger than microdomains observed in biological membranes and so are much more amenable to imaging. A few examples follow:

Figure 11.9 shows epifluorescence images demonstrating simple liquid crystalline/gel phase separations as a function of lateral pressure on Langmuir Trough lipid monolayers

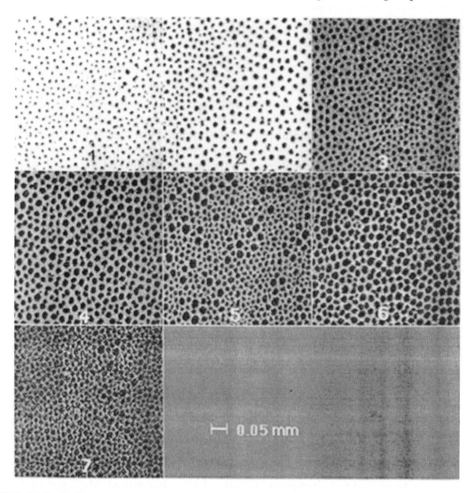

FIGURE 11.9 Epifluorescence images demonstrating simple liquid crystalline/gel phase separations as a function of lateral pressure on Langmuir Trough lipid monolayers. Monolayers were made from 70 mol% DPPC (16:0, 16:0 PC) and 29 mol% SDPC (18:0, 22:6 PC) to which 1 mol% N-rhodamine-PE was added as the membrane fluorescent imaging agent. Compressions were run and images taken at every 5 mN/m (23°C). Plate 1, 5 mN/m; Plate 2, 10 mN/m; Plate 3, 15 mN/m; Plate 4, 20 mN/m; Plate 5, 25 mN/m; Plate 6, 30 mN/m. Gel state DPPC domains appear black while liquid crystalline state SDPC domains appear white [39].

[39]. Monolayers were made from 70 mol% DPPC (16:0, 16:0 PC, T_m = 41.3°C) and 29 mol% SDPC (18:0, 22:6 PC, T_m = −20°C) to which 1 mol% N-rhodamine-PE was added as the membrane imaging agent. Compressions were run and images taken at every 5 mN/m. At the experimental temperature, 23°C, DPPC is below its T_m and so would be in the gel (solid) state while SDPC is above its T_m and would be in the liquid crystalline (fluid) state. The N-rhodamine-PE fluorescent probe partitions preferentially into the fluid (SDPC) phase. The images clearly show phase separation into liquid crystalline (fluorescent, white) and gel (non-fluorescent, black) state domains that are affected by the lateral pressure.

In the late 1980s Haverstick and Glaser [40,41] investigated cation-induced phase separation of anionic phospholipids into domains. They employed a fluorescence microscope, a CCD camera, and a digital imaging processor to visualize lipid microdomains on the surface of very large (5—15 µM) lipid vesicles. The fluorophore NBD (4-nitrobenzo-2-oxa-1, 3-diazole) was attached to the *sn*-2 chain of PA, PS, PE, and PC. Vesicles were made from the bulk lipid DOPC (99 mol% of the total phospholipids) to which 1 mol% of the fluorescent probe (NBD-PA, NBD-PS, NBD-PE or NBD-PC) was added. It was demonstrated that Ca^{2+}, but not Mg^{2+}, Mn^{2+}, or Zn^{2+}, induced phase separation of the anionic fluorescent phospholipids NBD-PA and NBD-PS from DOPC (Figure 11.10, [40]). NBD-PE and NBD-PC did not phase separate. The conclusion that Ca^{2+} can induce formation of lipid microdomains composed of fluorescent anionic phospholipids was confirmed in erythrocytes, erythrocyte ghosts, and vesicles made from erythrocyte lipid extracts.

In a subsequent paper, Haverstick and Glaser extended their Ca^{2+} work to include the cationic peripheral protein cytochrome c (Figure 11.11) [41]. Vesicles were formed as in Figure 11.10 and observed by NBD fluorescence. Bar indicates 10 mm. (A) A vesicle viewed immediately after the addition of 100 mM cytochrome c. Note no domains are visible. (B) A similar vesicle after 30 min in the presence of 100 mM cytochrome c. Domains have formed. (C) A vesicle as in A viewed after 30 min in the presence of 10 mM cytochrome c. (D) A vesicle as in B (domain formation induced by 100 mM cytochrome c) further incubated for 30 min with 0.1 M NaCl. Fluorescent domains that are barely detectable in the presence of 10 mM cytochrome c are obvious with 100 mM cytochrome c and are further augmented with 0.1 M NaCl [41]. This experiment demonstrated that cytochrome c can function in a manner similar to Ca^{2+} by inducing the formation of NBD-PA domains in a DOPC membrane. The

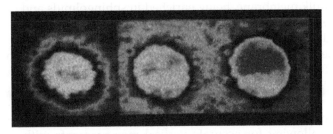

FIGURE 11.10   Fluorescence images of a Ca^{2+}-induced anionic phospholipid (PA) domain. Vesicles were made from the bulk lipid DOPC (99 mol% of the total phospholipids) to which 1 mol% of the fluorescent probe NBD-PA was added. Images were taken 2, 15, and 30 min after the addition of 2 mM $CaCl_2$. A faint fluorescent NBD-PA domain can be detected 2 min after addition of 2 mM Ca^{2+}. The domain grows and becomes more obvious after 15 and 30 min [40].

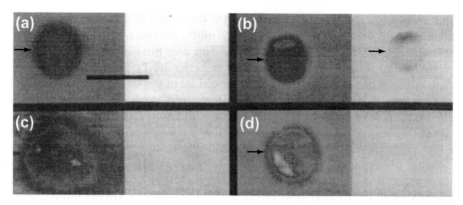

FIGURE 11.11   The cationic peripheral protein cytochrome c causes domain formation in phospholipid vesicles. Vesicles were formed containing 1 mol% NBD PA, 4 mol% DOPA and 95 mol% DOPC. Vesicle domains were observed by NBD fluorescence. Bar indicates 10 μm. (a) A vesicle viewed immediately after the addition of 100 μm cytochrome c. Note no domains are visible. (b) A similar vesicle after 30 min in the presence of 100 μm cytochrome c. (c) A vesicle as in (a) viewed after 30 min in the presence of 10 μm cytochrome c. (d) A vesicle as in (b) (domain formation induced by 100 μm cytochrome c) further incubated for 30 min with 0.1 M NaCl. Fluorescent domains that are barely detectable in the presence of 10 μm cytochrome c are obvious with 100 μm cytochrome c and are further augmented with 0.1 M NaCl [41].

experiment also showed that the cationic peripheral protein cytochrome c causes the trans-membrane protein gramicidin to be excluded from the PA-rich domain and to accumulate into the PC-rich domain. Also, measurement of the cytochrome c heme absorbance demonstrated that the peripheral protein location tracks with the PA domain. Therefore cytochrome c causes rearrangement of both lipid and protein components of a membrane. The Haverstick and Glaser imaging experiments are but representative examples of countless numbers of possible lipid and protein model membrane mixtures that await testing in the future.

The most relevant example of a lipid microdomain in a biological membrane is the highly controversial detergent-insoluble lipid raft. Lipid rafts were presented in Chapter 8 and their methods of isolation will be discussed in Chapter 13. Rafts are an example of a lipid micro-domain that may be partially isolated in crude form by extraction from the biological parent plasma membrane. In reality the isolated raft is defined by a lipid composition that is a little different than the bulk membrane, being enriched in sphingolipids and cholesterol. The raft fractions are also enriched in characteristic signaling proteins.

Since its early development in 1982 by Binnig and Rohrer [42], atomic force microscopy (AFM) has become a major imaging technique for investigating small biological structures, including lipid rafts. For their discovery, Binnig and Rohrer were awarded the 1986 Nobel Prize in Physics. AFM works by dragging a small, very sharp probe across the membrane surface while accurately measuring the vertical deflection of the probe due to changes in surface topography (Figure 11.12, [43]) [44]. The probe therefore 'feels' its way across the membrane surface allowing for accurate vertical measurements with resolution down to < 1 nm [45]. AFM also has the advantage of operating under water and so is applicable for use with biological membranes under near physiological conditions.

Since lipid rafts are enriched in saturated, long chain sphingolipids and cholesterol, they tend to be thicker than the surrounding non-raft membrane. Detecting small differences in

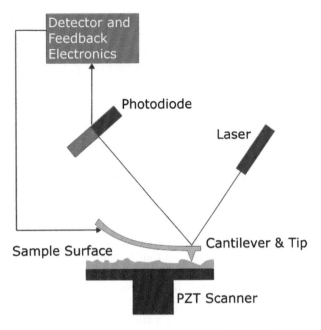

Detector and
Feedback
Electronics

Photodiode

Laser

Sample Surface

Cantilever & Tip

PZT Scanner

FIGURE 11.12   Schematic diagram of an atomic force microscope (AFM) [43].

surface height is exactly what AFM is best suited for [45]. As a result, by AFM the top of rafts appear lighter (they are higher) than the thinner and thus lower membrane domains. Figure 11.13 shows an example of lipid raft microdomains imaged by AFM. A mica-supported lipid bilayer was made from equimolar DOPC (18:1,18:1 PC) and brain SM to which the GPI-anchored protein placental alkaline phosphatase (PLAP) was added [46]. SM-rich lipid rafts are in gray and protrude above the black, DOPC-rich, non-raft domains by ~7 Å. PLAP are white spikes. The image demonstrates that PLAP is almost exclusively found in rafts.

Before the advent of AFM, lipid rafts had historically been imaged by fluorescence microscopy of raft-associated gangliosides that bind fluorescent cholera toxin [46]. Model bilayer

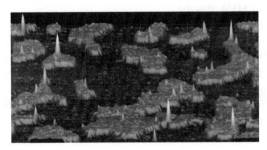

FIGURE 11.13   AFM image of lipid bilayer membrane composed of equimolar DOPC and brain SM to which the GPI-anchored protein placental alkaline phosphatase (PLAP) is added [46]. SM-rich lipid rafts are in gray, the DOPC-rich non-rafts in black, and PLAP are white spikes. The image demonstrates that PLAP is almost exclusively found in rafts. The width of the scan is ~2 μm [45].

membranes consisting of various raft (e.g. SM/cholesterol/ganglioside) and non-raft (e.g. various unsaturated phospholipids). With fluorescent cholora toxin, raft domains light up. If the same model bilayer membranes are imaged by AFM, overlapping domains are observed. Both techniques identify the same membrane patches as being raft domains. Many other examples simultaneously imaging rafts by fluorescence and AFM can be found in the literature (for example see [47]).

D. HOMEOVISCOUS ADAPTATION

Many of the essential membrane properties discussed in Chapters 9, 10, and 11 must be maintained within narrow constraints for life to continue. The well-studied examples of disease and aging are often associated with deleterious alterations of membrane physical properties. For warm-blooded (homeothermic) animals, membranes are naturally kept between a narrow, functional temperature range and are not exposed to harmful temperature fluctuations. However, most organisms on planet Earth have no internal temperature control mechanism (they are cold-blooded or poikilothermic) and are at the whim of the external environment. Examples of poikilothermic organisms are reptiles, fish, plants, fungi, bacteria, protists, and even to some extent hibernating mammals. Diving mammals face similar problems, as an increase in environmental pressure affects membranes in a similar way to a decrease in temperature. All organisms must be able to control environmental insults to assure proper membrane function. The effort to maintain proper membrane properties is known as *homeoviscous adaptation* (HVA), a general term where all temperature-dependent properties are lumped into 'membrane viscosity'. Viscosity is roughly the inverse of 'fluidity' (see Chapter 9).

The concept of HVA was first proposed in 1974 by Sinensky from membrane lipid studies on *Escherichia coli* [48]. HVA caught on quickly and was supported by many studies on arctic fish, plants, fungi, and bacteria. Of particular importance was a series of papers by A.R. Cossins [49,50]. Cossins assessed membrane order (a measure of 'fluidity') for a series of organisms of different body or habitat temperatures. The measured order parameters were found to directly follow the organism's temperature [Antarctic fish ($-1°C$) < perch ($15°C$) < convict cichlid ($28°C$) < rat ($37°C$) < pigeon ($42°C$)] [51]. These results indicated that 'evolutionary adaptation to cold environments produces membranes of significantly lower order' [51]. And this is at the heart of HVA theory.

In an excellent and thoughtful review, Jeffrey Hazel [51] agrees that lipids are indeed involved in HVA, but suggests their effect is far more complex than standard HVA theory would suggest. An abbreviated list of membrane properties that are hard to incorporate into conventional HVA theory include: membrane remodeling; microdomain heterogeneity; specificity of lipid—protein interactions and; proliferation of mitochondrial and sarcoplasmic reticular membranes. Several membrane properties, closely linked to fluidity and likely involved in HVA, are briefly discussed below.

Acyl Chain Length

Shorter chain lengths have lower T_ms than longer chain lengths (Chapter 4) and so accumulate in membranes of organisms exposed to lower temperatures.

Acyl Chain Double Bonds

Double bonds reduce fatty acid T_ms and so also accumulate in membranes of organisms exposed to lower temperatures. For example, in cold acclimation the proportion of di-unsaturated PC and PE in plants increases 50%. In bacteria, double bonds are often replaced by adding methyl branches or cyclopropyl rings (see Chapter 4). Phytol chains also contain methyl branches and so can interfere with tighter acyl chain packing associated with cold temperature.

Phospholipid Head Groups

Phospholipids with identical acyl chains often have widely different T_ms. Most important for homeoviscous adaptation are the two major membrane structural phospholipids, PC and PE. The T_ms for PEs are usually $> 20°C$ or higher than homologous PCs. Therefore another way to increase membrane 'fluidity' in colder temperatures is to increase the membrane PC/PE ratio.

Sterols

Cholesterol is known to fluidize gel state membranes and reduce the fluidity of fluid state membranes (Chapter 5 and Chapter 9). Therefore membranes exposed to low temperature may employ cholesterol or other similar sterols to fluidize their membranes.

Isoprene Lipids

Although far less abundant than membrane sterols, other isoprene lipids, such as vitamin E and D, could also serve to fluidize membranes at low temperatures.

Anti-freeze Proteins

Anti-freeze proteins (AFPs) were first isolated by Arthur De Vries from Antarctic fish in 1969 [52]. These proteins have since been found in a wide variety of organisms including certain vertebrates, plants, fungi, and bacteria that survive in sub-zero environments. AFPs do not function as ordinary anti-freeze solutes like ethylene glycol since they are effective at 300—500 times lower concentrations. They undoubtedly contribute to homeoviscous adaptation of membranes.

Divalent Metals

Divalent metals can induce anionic phospholipids to undergo isothermal phase transitions. By binding to the anionic head group, M^{2+}s impose order to the entire phospholipid, including the acyl chains. This can result in a substantial increase of the phospholipid's T_m, perhaps driving a liquid crystal state lipid into a gel state without any temperature change (isothermal phase transition). Table 11.3 shows the effect of Mg^{2+} and Ca^{2+} on the anionic phospholipid, DPPG. While 100 mM NaCl has no effect on the T_m of DPPG, 5 mM Mg^{+2} increases the T_m 11°C and 5 mM Ca^{2+} increases the T_m a substantial 25°C. Another option

TABLE 11.3   Effect of Divalent Metals on the T_m of the Anionic Phospholipid DPPG (16:0,16:0 Phosphatidylglycerol).

Added cation	T_m (°C)
NaCl, 100 mM	42
Mg^{2+}, 1 mM	50
Mg^{2+}, 5 mM	53
Ca^{2+}, 1 mM	57
Ca^{2+}, 1 mM	67

The unmodified T_m of DPPG is 42°C.

for poikilothermic organisms would therefore be to decrease their membrane anionic phospholipid content or to sequester their major divalent metals.

Don Juan Pond

An extreme example of life adjusting to a very harsh environment is the Don Juan Pond in the west end of the Wright Valley in Victoria Land, Antarctica [53] (Figure 11.14, [54]). Despite its provocative name, the pond is actually named after the two helicopter pilots, Lt Don Roe and Lt John Hickey, who discovered the pond in 1961. The Don Juan Pond is a small and very shallow hypersaline lake that almost never freezes despite temperatures that fall below −50°C. When it was first discovered, the pond was about 300 m long, 100 m wide and about 1 foot deep. Since then, the pond has shrunk considerably. The Don Juan Pond is defined as being hypersaline, as it is over 18 times more saline than the ocean and more than twice as saline as the Dead Sea. The major salt components of the pond are $CaCl_2$ (3.7 M) and NaCl (0.5 M). Don Juan Pond levels of NaCl have been reported to decrease the freezing point

FIGURE 11.14   The Don Juan Pond in Antarctica [54].

of water down to $-21°C$ and $CaCl_2$ levels down to $-50°C$, accounting for the failure of the Don Juan Pond to freeze.

What is most surprising is the fact that the Don Juan Pond supports a variety of life! Microflora of yeasts, blue-green algae, fungi, and bacteria have all been shown to inhabit the pond, although they may lack extensive capability for continuous carbon reduction. In a 1979 article in *Nature*, Siegel et al. [55] reported an extensive, irregular pellicle or mat-like structure 2−5 mm thick inhabiting the pond. Regardless of how abundant these Don Juan Pond organisms are, they certainly are a prime example of homeoviscous adaptation.

SUMMARY

Membrane properties must support resident protein activity. One measurable property is the 'packing free volume' that quantifies the required breathing space for enzyme activity. Several important membrane properties, including area/molecule, collapse point, 'lipid condensation', surface elasticity, and molecular 'squeeze out' have been assessed with lipid monolayers using a Langmuir Trough. It was demonstrated that lipids vary in their lateral compressibility. Cholesterol and di-saturated phospholipids are poorly compressible, while polyunsaturated phospholipids are highly compressible. The lateral pressure of a biological membrane is estimated to be ~30−35 mN/m. Freeze fracture EM has supplied direct evidence that proteins indeed reside inside membranes and, without cytoskeleton involvement, there is no long-range order. Observed order is only a few tenths of a micron and the membrane bilayer sea is very crowded. Membrane domains can be roughly divided into large, stable macrodomains (e.g. gap junctions, clathrin-coated pits, caveolae) and small, unstable lipid microdomains (e.g. lipid rafts). In general, membrane properties are maintained by homeoviscous adaptation.

Chapter 12 will discuss the methodologies used to homogenize a cell and then isolate, purify, and analyze the various cell membrane fractions.

References

[1] Slater SJ, Kelly MB, Taddio FJ, Ho C, Rubin E, Stubbs CD. The modulation of protein kinase C activity by membrane lipid bilayer structure. J Biol Chem 1994;269:4868−71.

[2] Epand RM. Lipid polymorphism and protein−lipid interactions. Biochim Biophys Acta 1998;1376:353−68.

[3] Straume M, Litman BJ. Influence of cholesterol on equilibrium and dynamic bilayer structure of unsaturated acyl chain phosphatidylcholine vesicles as determined from higher order analysis of fluorescence anisotropy decay. Biochemistry 1987;26:5121−6.

[4] Mitchell DC, Straume M, Litman BJ. Role of sn-1-saturated, sn-2-polyunsaturated phospholipids in control of membrane receptor conformational equilibrium: Effect of cholesterol and acyl chain unsaturation on the metarhodopsin I ↔ metarhodopsin II equilibrium. Biochemistry 1992;31:662−70.

[5] Litman BJ, Mitchell DC. A role for phospholipid polyunsaturation in modulating membrane protein function. Lipids 1996;31:S193−7.

[6] Mitchell DC, Gawrisch K, Litman BJ, Salem Jr N. Why is docosahexaenoic acid essential for nervous system function? Biochem Soc Trans 1998;26:365−70.

[7] Mitchell DC, Litman BJ. Molecular order and dynamics in bilayers consisting of highly polyunsaturated phospholipids. Biophys J 1998;74:879−91.

[8] Mitchell DC, Litman BJ. Effect of cholesterol on molecular order and dynamics in highly polyunsaturated phospholipids bilayers. Biophys J 1998;75:896−908.

[9]   Salem Jr N, Kim HY, Yergey JA. Docosahexaenoic acid: membrane function and metabolism. In: Simopoulos AP, Kifer RR, Martin R, editors. The Health Effects of Polyunsaturated Fatty Acids in Seafoods. New York: Academic Press; 1986. p. 319–51.

[10] Davies JT, Rideal EK. Interfacial Phenomena. 2nd ed. New York: Academic Press; 1963. p. 265.

[11] Girard-Egrot AP, Godoy S, Blum LJ. Enzyme association with lipidic Langmuir–Blodgett films: Interests and applications in nanobioscience. Adv Colloid Interface Sci 2005;116:205–25.

[12] Jain MK, Wagner RC. Introduction to Biological Membranes. New York: John Wiley and Sons; 1980. p. 61.

[13] Shaikh SR, Dumaual AC, Jenski LJ, Stillwell W. Lipid phase separation in phospholipid bilayers and monolayers modeling the plasma membrane. Biochim Biophys Acta 2001;1512:317–28.

[14] Bonn M, Roke S, Berg O, Juurlink LBF, Stamouli A, Muller M. A molecular view of cholesterol-induced condensation in a lipid monolayer. J Phys Chem B 2004;108:19083–5.

[15] Yeagle P. The roles of cholesterol in biology of cells. In: Yeagle P, editor. The Structure of Biological Membranes. Boca Raton, FL: CRC Press; 1992 [Chapter 7].

[16] Yeagle PL. The roles of cholesterol in the biology of cells. In: The Structure of Biological Membranes. 2nd ed. Boca Raton, FL: CRC Press; 2005 [Chapter 7].

[17] Finegold LX, editor. Cholesterol in Membrane Models. Boca Raton, FL: CRC Press; 1993.

[18] Smaby JM, Momsen MM, Brockman HL, Brown RE. Phosphatidylcholine acyl chain unsaturation modulates the decrease in interfacial elasticity induced by cholesterol. Biophys J 1997;73:1492–505.

[19] Boonman A, Machiels FHJ, Snik AFM, Egberts J. Squeeze-out from mixed monolayers of dipalmitoylphosphatidylcholine and egg phosphatidylglycerol. J Colloid Interface Sci 1987;120:456–68.

[20] Severs N. Freeze-fracture electron microscopy. Nat Protoc 2007;2:567–76.

[21] Structural Biochemistry/Lipids/Membrane Fluidity, En.wikibooks.org.

[22] Ellis RJ. Macromolecular crowding: obvious but unappreciated. TRENDS Biochem Sci 2001;26:597–604.

[23] Yu J, Branton D. Reconstitution of membrane particles in recombinants of erythrocyte protein band 3 and lipid: Effects of spectrin–actin association. Proc Natl Acad Sci USA 1976;73:3891–5.

[24] Edidin M. Shrinking patches and slippery rafts: Scales of domains in the plasma membrane. Trends Cell Biol 2001;11:492–6.

[25] Robertson JD. Ultrastructure of two invertebrate synapses. Proc Soc Exptl Biol Med 1953;82:219–23.

[26] Farquahr MG, Palade GE. Junction complexes in various epithelia. J Cell Biol 1963;17:375–412.

[27] Evans WH, Martin PE. Gap junctions: structure and function (Review). Mol Membr Biol 2002;19:121–36.

[28] Revel JP, Yee AG, Hudspeth AJ. Gap junctions between electronically coupled cells in tissue culture and in brown fat. Proc Natl Acad Sci USA 1971;68:2924–7.

[29] Goldstein JL, Brown MS. Receptor-mediated endocytosis: Concepts emerging from the LDL receptor system. Annu Rev Cell Biol 1985;1:1–39.

[30] Garland Publishing. 1998.

[31] Rappoport JZ. Focusing on clathrin-mediated endocytosis. Biochem J 2008;412:415–23.

[32] Gautier A, Bernhard W, Oberling C. Sur lexistence dun appareil lacunaire pericapillaire du glomerule de Malpighi, revele par la microscopie electronique. Comptes Rendus des Seances de la Societe de Biologie et de Ses Filiales 1950;144:1605–7.

[33] Yamada E. The fine structure of the gall bladder epithelium of the mouse. J Biophys Biochem Cytol 1955:445–58.

[34] Anderson RG. The caveolae membrane system. Annu Rev Biochem 1998;67:199–225.

[35] Li X, Everson W, Smart E. Caveolae, lipid rafts, and vascular disease. Trends Cardiovasc Med 2005;15:92–6.

[36] Wolf DE, Kinsey W, Lennarz W, Edidin M. Changes in the organization of the sea urchin egg plasma membrane upon fertilization: indications from the lateral diffusion rates of lipid-soluble fluorescent dyes. Dev Biol 1981;81:133–8.

[37] Starling AP, East JM, Lee AG. Effects of Gel Phase Phospholipid on the Ca^{2+}-ATPase. Biochemistry 1995;34:3084–91.

[38] Lee AG, Dalton KA, Duggleby RC, East JM, Starling AP. Lipid structure and (Ca^{2+})-ATPase function. (Review) Biosci Rep 1995;15:289–98.

[39] Dumaual AC, Jenski LJ, Stillwell W. Liquid crystalline/gel state phase separation in docosahexaenoic acid-containing bilayers and monolayers. Biochem Biophys Acta 2000;1463:395–406.

[40] Haverstick DM, Glaser M. Visualization of Ca^{2+}-induced phospholipid domains. Proc Natl Acad Sci USA 1987;84:4475–9.

[41] Haverstick DM, Glaser M. Influence of proteins on the reorganization of phospholipid bilayers into large domains. Biophys J 1989;55:677—82.

[42] Binnig G, Rohrer H, Gerber C, Weibel E. Phys Rev Lett 1982;1982(49):57.

[43] Wikipedia, the free encyclopedia. File: Atomic force microscope block diagram.svg

[44] Morris VJ, Kirby AR, Gunning AP. Atomic Force Microscopy for Biologists. 2nd ed. London: Imperial College Press; 2010.

[45] Henderson RM, Edwardson JM, Geisse NA, Saslowsky DE. Lipid rafts: feeling is believing. News Physiol Sci 2004;19:39—43.

[46] Saslowsky DE, Lawrence J, Ran X, Brown DA, Henderson RM, Edwardson IM. Placental alkaline phosphatase is efficiently targeted to rafts in supported lipid bilayers. J Biol Chem 2002;277:26966—70.

[47] Shaw JE, Epand RF, Epand RM, Li Z, Bitman R, Yip CM. Correlated fluorescence-atomic force microscopy of membrane domains: structure of fluorescence probes determines lipid localization. Biophys J 2006;90:2170—8.

[48] Sinensky M. Homeoviscous adaptation—A homeostatic process that regulates the viscosity of membrane lipids in Escherichia coli. Proc Natl Acad Sci USA 1974;71:522—5.

[49] Cossins AR, Prosser CL. Evolutionary adaptation of membranes to temperature. Proc Natl Acad Sci USA 1978;75:2040—3.

[50] Behan-Martin MK, Jones GR, Bowler K, Cossins AR. A near perfect temperature adaptation of bilayer order in vertebrate brain membranes. Biochim Biophys Acta 1993;1151:216—22.

[51] Hazel JR. Thermal adaptation in biological membranes: Is homeoviscous adaptation the explanation? Annu Rev Physiol 1995;57:19—42.

[52] DeVries AL, Wohlschlag DE. Freezing resistance in some Antarctic fishes. Science 1969;163:1073—5.

[53] Mitchinson A. Geochemistry: The mystery of Don Juan Pond. Nature 2010;464:1290.

[54] Goldstein R. A cool place to work. The Southern Illinoisan. College of Science. Microbiology: Southern Illinois University, http://www.micro.siu.edu/CoolPlace.html; 1999.

[55] Siegel BZ, McMurty G, Siegel SM, Chen J, LaRock P. Life in the calcium chloride environment of Don Juan Pond, Antarctica. Nature 1979;280:828—9.

[56] Girard-Egrot AP, Godoy S, Blum LJ. Enzyme association with lipidic Langmuir–Blodgett films: Interests and applications in nanobioscience. Advances in Colloid and Interface Science 2005;1—3:205—25.

[57] Hill M, UNSW Cell Biology: http://cellbiology.med.unsw.edu.au/units/science/lecture0808.htm

O U T L I N E

A. **Introduction**   239
  Microsomes   240
    Membrane Isolation Steps   241

B. **Breaking Open the Cell:**
  Homogenization   241
  Homogenization Methods   242
    1. Use of Enzymes   242
    2. Shear Force (Mortar and Pestle)   242
    3. Blenders   242
    4. Osmotic Gradients   244
    5. Sonication   245
    6. Bead Beaters   246
    7. Gas Ebullition (High Pressure
      'Bomb')   246
    8. French Press   247
    9. Freeze/Thaw   248

C. **Membrane Fractionation:**
  Centrifugation   248
  Differential Centrifugation   248
  Density Gradient Centrifugation   249

D. **Membrane Fractionation: Non-**
  Centrifugation Methods   253
  Affinity Chromatography   253
  Lectin-Affinity Chromatography   254
  Antibody-Affinity Chromatography   256
  Ligand-Receptor Affinity
  Chromatography   256
  Anion Exchange Chromatography   256
  Derivitized Beads   257
    Magnetic Beads   257
    Other Types of Microbeads   258
  Other Methodologies   258
    Two Phase Partitioning   258
    Silica Particles   259
    Separation by Size   259
    Membrane-Specific 'Tricks'   259

E. **Membrane Markers**   260

Summary   261

References   262

A. INTRODUCTION

Isolating biological membranes is as much an art form as it is a science. The procedures are tedious and require patience, precision, and organization. Since every membrane has its own

An Introduction to Biological Membranes
http://dx.doi.org/10.1016/B978-0-444-52153-8.00012-X

peculiarities, countless procedures have been published. This chapter will consider a few of the problems encountered in isolating membranes and some of the more common techniques that have been employed.

As discussed in Chapter 1, a eukaryotic cell is composed of an astonishing number of different membranes, all packed tightly into a very small volume. Membranes present a tiny and ever changing target for investigation, making their isolation particularly challenging. Every eukaryotic cell is surrounded by a relatively tough plasma membrane (PM) that encumbers countless more delicate internal membranes. This presents a dilemma. How does one break open the PM without severely impacting the tightly packed, interconnected and delicate internal membranes? An old analogy seems appropriate. You are given a bag of different types and sizes of watches that had been smashed with a sledgehammer into thousands of intermingled, broken parts. From this you are asked to reassemble the watches and determine how they measure time!

Instantly upon breaking open a cell, the internal membranes are relieved of curvature stress by taking new morphologies, often in the form of similar-sized vesicles called microsomes [1].

Microsomes

The concept of microsomes is critical to membrane studies. Microsomes are small sealed vesicles that originate from fragmented cell membranes (often the endoplasmic reticulum (ER)). These vesicles may be right-side-out, inside-out or even fused membrane chimeras. Microsomes may have unrelated proteins sequestered in their internal aqueous volumes or attached to their surfaces. Microsomes are therefore artifacts that arise as a result of cell homogenization and are very complex. By their very definition, microsomes *per se* are not present in living cells and are physically defined by an operational procedure, usually differential centrifugation (discussed below, [2]). In a centrifugal field, large particles including unbroken cells, nuclei, and mitochondria sediment out at low speed (<10,000g), whereas much smaller microsomes do not sediment out until much higher speeds (~100,000g). Soluble cellular components like salts, sugars, and enzymes remain in solution at speeds that pellet out microsomes.

Microsomes have been observed, if not understood, for a long time. In an early review from 1963, Siekevitz linked microsomes to remnants of the ER after cell homogenization [3]. Typically, discussions of microsome properties and isolation procedures are found buried in papers whose primary objective is to investigate a specific integral membrane protein (for example see [4]).

Once broken, all membrane fractions are instantly exposed to new osmotic stresses, divalent metal ions, unnatural pHs, and degradative enzymes including proteases, oxidases, lipases, phospholipases, and nucleases [2,5–7]. Historically, sucrose has been the major osmotic component (osmoticum) of cell homogenization buffers [8]. Sucrose is inexpensive, readily available in pure form, and is poorly permeable to most membranes. Importantly, it does not destroy enzymatic activity. In many contemporary membrane isolations, Ficoll™ (GE Healthcare companies) has replaced most or all of the sucrose. Ficoll is an uncharged, highly branched polymer formed by the co-polymerisation of sucrose and epichlorohydrin [9]. Due to its multiple (−OH) groups, Ficoll, like sucrose, is highly water-soluble. Often a little sucrose is added to the Ficoll to accurately control the density, viscosity, and osmotic strength of the buffer. Ficoll is extensively used in density gradient centrifugations (discussed below). In addition to the osmoticum (e.g. Ficoll/sucrose), many additional compounds are included in membrane isolation buffers. These include pH buffers, chelators to remove harmful divalent metals, antioxidants, and inhibitors of many degradative enzymes. Homogenization buffers are indeed a complex 'soup'.

The investigator must have many tools in his bag and work quickly, to identify and separate the various membrane fractions before serious artifacts arise. It is impossible to obtain an absolutely pure membrane; at best just an enriched fraction can be isolated.

Membrane Isolation Steps

The general sequence of steps involved in membrane isolation includes [5,6,8]:

- Identify the best cell to obtain the membrane of interest.
- Isolate and clean the cells.
- Break open the cell (homogenization).
- Separate large intact organelles (fractionation).
- Separate the small microsomal cell fractions.
- Identify the cell fractions by use of appropriate membrane markers (both positive and negative markers).

Possible additional steps may include:

- Isolate and purify a specific membrane protein.
- Isolate and purify membrane lipids.
- Reconstitute the membrane protein into lipid vesicles.
- Determine the mechanism of action of the reconstituted protein.

B. BREAKING OPEN THE CELL: HOMOGENIZATION

With the final objective in mind, the proper cell type is chosen and cleaned of unrelated debris such as veins. If the starting tissue is solid (e.g. liver, muscle etc), the target cells must be separated from neighboring cells. This process may involve use of proteases or other enzymes that disaggregate adjacent cells and may be aided by chelating agents that bind divalent cations. Disaggregation will alter some physical structures that reside on the cell plasma membrane outer surface (e.g. gap junctions and tight junctions). Obviously if the objective is to isolate and reconstitute rhodopsin for light receptor studies, liver, brain, and muscle tissue would be inappropriate. Once sufficient quantities of the cell are obtained, the cell must be broken open by as gentle a method as possible. This process is referred to as 'homogenization' [2,5,6,8]. Vigorous procedures result in damaged membranes and produce many small vesicular microsomes that are difficult to separate from one another. If the procedure is too gentle, many cells will remain intact. Since it is estimated that even the simplest, routine membrane isolation procedures take ~15% of an investigator's time [10], a variety of cell disruption methodologies have been developed for specific applications. For a brief comparison of the advantages and disadvantages of many of the methods discussed below see reference [11].

Homogenization Methods

1. Use of Enzymes

Cells that are encased in thick, tough cell walls (e.g. plant, fungi, and some bacteria) present a particularly challenging set of problems. The cell wall must be removed while doing as little damage as possible to the more delicate underlying plasma membrane. Once the cell wall is removed, the remaining cell wall-free protoplast can be homogenized. A variety of commercially available enzymes are used to digest the cell wall [7,12,13]. These enzymes include cellulase, pectinase, lysozyme, lysostaphin, zymolase, glycanase, mannase, mutanolysin, and many more. Unfortunately, it is almost impossible to remove a cell wall without doing considerable damage to the delicate molecules residing below, which are attached to the outer surface of the plasma membrane. Also, the enzymatic methodologies for cell wall degradation are not applicable to large-scale preparations, and the enzymes must be inactivated or removed before the protoplasts are to be homogenized.

2. Shear Force (Mortar and Pestle)

The most commonly employed cell homogenization technique involves use of a mortar and pestle. The basic concept involves forcing cells through a gap between two very close surfaces. The resulting shear force tears the cell apart [14]. The shear force will increase as the gap is narrowed or as the movement of the pestle relative to the fixed mortar is increased. Very accurate control of the shear force applied to the cells is essential. If the shear force is insufficient, cells will not be disrupted. If the force is too great, the entire cell, including its organelles, will be broken into many different microsomes. At extreme shear force, biochemicals may even be destroyed.

Several commercial mortar and pestle homogenizers are at the heart of most cell membrane studies. The most commonly used homogenizers are the Ten Broeck and Dounce homogenizers (Figure 12.1), both of which are glass mortar and pestles that are purchased with preset, controlled gap sizes. Cells are forced through the gap by manual manipulation of the pestle. The process is normally performed in ice to prevent heat-induced destruction of cell components. The addition of a motor driven Teflon pestle is part of the Potter-Elvehjem homogenizer. Although Van Rensselaer Potter (Figure 12.2) and Conrad Elvehjem are now best known for their homogenizers, both made major unrelated contributions to the life sciences. Potter pioneered and named the field of Bioethics while Elvehjem discovered Niacin.

3. Blenders

Another very commonly employed type of homogenizer is generally referred to as a 'blender'. Blenders can vary from simple, inexpensive household food blenders or juice extractors to sophisticated, expensive instruments that employ high-speed blades and specialized sample chambers.

The first simple food blender was introduced to the public in 1936 by big band leader and popular radio and television personality, Fred Waring, and was an immediate hit. However, Waring did not actually invent the blender. This was accomplished in 1922 by Stephen Poplawski and was later improved by Fred Osius in 1935. The Osius blender became the original Waring blender. In fact Fred Waring was only the shill that popularized the appliance. Originally Waring called his blender the Miracle Mixer but later changed it to the Waring

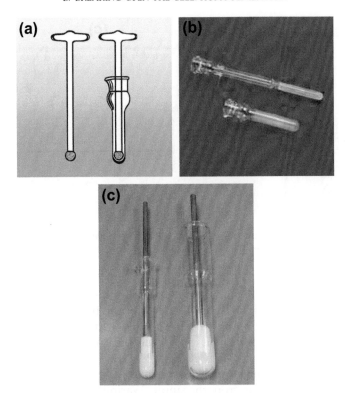

FIGURE 12.1 Commonly used mortar and pestle type cell homogenizers: (a) Dounce Homogenizer, *Courtesy of Canadawide Scientific Limited*; (b) Ten Broeck Homogenizer, *Courtesy of WHEATON* and; (c) Potter-Elvehjem Homogenizer [50].

FIGURE 12.2 Van Rensselaer Potter (1901–1962). *Courtesy of the American Association for Cancer Research*

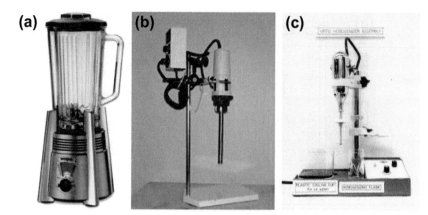

FIGURE 12.3   Types of blenders commonly used for cell homogenization: (a) Waring, *Courtesy of Tiki Bar Is Open.com*; (b) Brinkman Polytron Blender, *Courtesy of LabCommerce, Inc* and; (c) Virtis Tissue Homogenizer, *Courtesy of LabCommerce, Inc.*

Blendor (note 'or' not 'er'). By 1954 one million Waring Blendors had been sold and even today Waring blenders are considered the standard for the field. Soon after its introduction, the Waring Blendor became a commonly used scientific instrument for cell homogenization. Jonas Salk used one in developing his polio vaccine. A Waring Blendor is shown in Figure 12.3.

Although general food blenders are useful for many applications, they cannot efficiently disrupt microorganisms. A step up from the simple Waring type blenders are the high speed blade-type homogenizers including the Brinkman Polytron and the Virtis Tissue Homogenizer (Figure 12.3). These homogenizers employ a powerful electric motor to drive a shaft terminated by specially designed blades that spin at high speeds. As an example, the Virtis Cyclone IQ2 homogenizer has a 1/10HP motor mounted above a homogenizing flask/bottle. The blades rotate at between 5,000 to 30,000 rpm and the sample volume can vary from 0.2 to 2,000 ml.

4. Osmotic Gradients

As discussed in Chapter 2 the lipid bilayer component of a cell membrane is an excellent osmometer [15,16]. An osmometer is a device for measuring the osmotic strength of a solution. Membranes are osmometers because they are impermeable to most ions while allowing water to readily diffuse across. The osmotic strength of one solution relative to another can be defined by three basic classifications: hypertonic, hypotonic or isotonic. With regard to a cell, a hypertonic solution is one having a greater solute concentration outside the cell than is found in the cytosol. Water then leaves the cell and the cell shrinks (see Figure 2.1, Plasmolysis, in Chapter 2). In contrast, a hypotonic solution is one having a lesser solute concentration outside the cell than is found in the cytosol. Water then enters the cell and the cell swells. If the difference between the outside and cytosolic osmotic strength is sufficient, the cell may swell until it bursts. In the case of the erythrocyte, cell disruption releases internal hemoglobin in an easily followed process called hemolysis. As discussed previously, hemolysis played an important role in early membrane studies (see William Hewson in Chapter 2).

Isotonic solutions contain equal concentrations of impermeable solutes on either side of the membrane. The swelling or shrinking rate of an artificial lipid bilayer vesicle (called a 'liposome', discussed below) is an excellent osmometer [15,16].

In most cases, intracellular organelles are more easily extracted if the cell is slightly swollen, enhancing its breakage by other methods [17]. Cells are more susceptible to breakage when placed in hypo-osmotic buffers and are more stable in hyper-osmotic buffers. While there are countless homogenization buffers, historically the major osmoticum has been either ~250 mM sucrose or ~120 mM KCl, lightly buffered to pH 7.4.

5. Sonication

Ultrasonication employs high frequency sound waves, typically 20–50 kHz, to disrupt cells [18,19]. Ultrasonication can be used by itself for cell homogenization or more commonly in conjunction with other techniques. Light application of ultrasonication can even be used to separate cells from one another or to remove cells from a tissue culture substrate (glass or plastic).

There are three basic types of ultrasonication: tip, cup horn, and bath (Figure 12.4). In a tip sonicator, the high frequency is generated electronically and the energy transmitted directly to the sample via a metal titanium tip that oscillates at high frequency. The rapid tip movement results in alternating very low pressure areas (cavitation) and high pressure areas (impaction) that can tear the cell apart. The basic concept had its origins in the 1950's and technical advances have reduced required minimum sample volumes down to less than 200 μl. Although tip sonication remains the most popular for cell disruption, it does have some serious problems. As the titanium tip oscillates, it slowly disintegrates. As a result the tip must be replaced regularly and the generated titanium particles must be removed from the cell sample. In addition the rapid tip oscillation generates a lot of heat that must be dissipated quickly to prevent destruction of the delicate biological material.

To avoid problems associated with inserting the titanium tip probe directly into the sample, another method has been developed where the tip is placed in a water bath adjacent to a chamber that holds the sample. A commercial device that accomplishes this is referred to as a 'cup horn' (Figure 12.4). Since the sonic energy must pass through a water

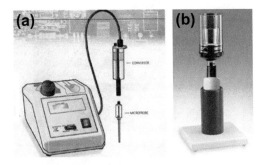

FIGURE 12.4 Common types of sonication homogenizers: (a) Tip Sonicater, *Courtesy of Giltron* and; (b) Cup Horn Sonicater, *Courtesy of Pontina Elettronica.*

bath and sample chamber wall before reaching the cell, this method is far gentler and less destructive than the direct tip method. Sonication is provided by a standard tip sonicater that is orientated upside down with the cup horn attached to the vertical sonicater shaft.

Another gentle, indirect sonication method involves use of a sonicating water bath, sometimes called a 'jeweler's bath' (Figure 12.4). No titanium tip is involved. Instead the walls of the bath supply the sonic energy and a vessel containing the cell sample is suspended in the bath. This is the least destructive of the three methods and has the advantage of providing accurate temperature control during sonication. However, its low energy limits its membrane applications. Sonic baths are often used to suspend lipids into water in the preparation of liposomes.

6. Bead Beaters

Another commonly employed homogenization procedure involves 'bead beaters' [20], an example of which is shown in Figure 12.5. These devices disrupt cells by violently shaking them in the presence of small (usually) glass beads. Sometimes ceramic, zirconium, or steel beads are used. Bead beating is done in sealed vials typically containing 100 µl to 1 ml samples and bead sizes are typically between 0.5 and 1.0 mm in diameter. Samples are agitated at 2,000 to 5,000 oscillations per minute in a specially designed clamp driven by a high energy electric motor.

7. Gas Ebullition (High Pressure 'Bomb')

Very rapid cell homogenization can be achieved by use of what is known as a 'high pressure bomb' [21]. The most commonly used pressure bomb for cell homogenization, a Parr Pressure Bomb, is shown in Figure 12.6. The basis of this method is similar to what divers experience as the bends (decompression sickness). The high pressures experienced by divers in deep water force gas to dissolve into their body fluids. If the diver returns to the surface too rapidly, the gasses (primarily nitrogen) leave the body fluids and return to the gaseous state

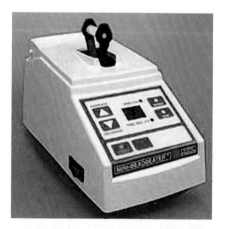

FIGURE 12.5   Bead Beater type of cell homogenizer. *Courtesy of Glen Mills*

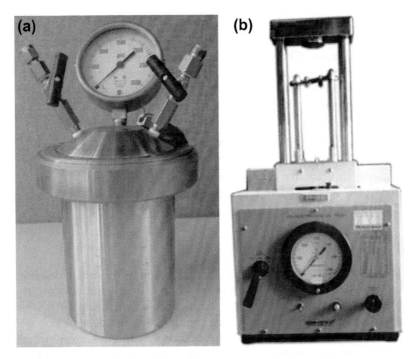

FIGURE 12.6   Parr Pressure 'Bomb' and French Press cell homogenizers: (a) Parr Pressure 'Bomb', *Courtesy of Glen Mills* and; (b) French Press, *Courtesy of Lab Recyclers Inc.*

as bubbles. The rapid formation of these bubbles results in a variety of painful abnormalities and even death. The first sign of pain is often in joints that are rendered stiff and cannot bend — hence the term 'bends'.

Cells are placed in a thick-walled chamber (Figure 12.6) and put under high pressure (usually nitrogen or another inert gas up to about 25,000 psi) and the pressure is rapidly released. The rapid pressure drop causes the dissolved gas to be released as bubbles that lyse the cell.

8. French Press

A less frequently used type of cell homogenizer is the French Press [22] (Figure 12.6). The device is named after its pioneer Stacy French, a former plant physiologist at the Carnegie Institute of Washington. This apparatus functions by forcing cells through a narrow needle valve under very high pressure. The cells are exposed to large shear forces that are capable of disrupting very tough cells, even stripping off thick cell walls. It has been reported that the French press preferentially produces inside-out microsomal vesicles that can have important biochemical applications. Disadvantages of the French Press are its large, heavy size and its high cost. French Presses are typically used for small samples (from 1 to 40 ml) and can reach pressures of 20,000–40,000 psi. By controlling the pressure and valve tension, the shear force can be regulated to optimize cell disruption.

9. Freeze/Thaw

The fact that ice occupies a larger volume than does an equal quantity of liquid water (water is denser than ice, see Chapter 3), can be used to disrupt cells [23]. Alternating rapid freezing in liquid nitrogen or in a slurry of dry ice and isopropanol, followed by thawing in warm water can homogenize cells. More often, freeze/thaw cycles are used as a preliminary step to be followed by another homogenization procedure.

C. MEMBRANE FRACTIONATION: CENTRIFUGATION

By far the major technique that has been employed to separate the various cell fractions is centrifugation [8,24–26]. Two basic types of centrifugation exist, differential and density gradient.

Differential Centrifugation

Differential centrifugation is a first crude step in separating major cell fractions whose physical properties, size, and density, are very different from one another. In fact at least one order of magnitude difference in the sedimentation coefficient (discussed below) is required to achieve separation, and even then separation is not complete.

After the tissue sample is homogenized the mixed cell contents are subjected to repeated centrifugations at increasing speed, a process known as differential centrifugation. After each centrifugation the pellet (heavier component) is removed and the remaining cell suspension (lighter components) is centrifuged at an increased speed. The following separations are achieved from biggest (heaviest) to smallest (lightest) fractions:

BIGGEST	Pellet 1	unbroken cells
	Pellet 2	nucleus
	Pellet 3	mitochondria
	Pellet 4	lysosome
	Pellet 5	microsomes
	Pellet 6	ribosomes
SMALLEST	Suspension	cytoplasmic proteins, sugars, amino acids, nucleotides, salts etc.

Centrifugation speeds in membrane studies vary over a wide range from ~1,000g for 5 min for unbroken cells up to 1,000,000g for hours to finally pellet cytosolic macromolecules. Each step in differential centrifugation only *enriches* the target membrane. It does not completely purify the membrane.

The limitation of differential centrifugation results from large, heavy particles initially residing at the top of the centrifuge tube pelleting at the same time as smaller, lighter particles that are initially near the bottom of the tube. For example, consider the case of a mixture of heavy particle A and light particle B. The initial centrifugation pellets ~95% of A, but also pellets ~20% of B. If the pellet is separated, re-suspended in fresh media and re-centrifuged at the same speed, ~85% of the initial amount of A is found in the pellet, but only ~1% of B. Therefore, differential centrifugation can only produce enriched fractions that may be further purified by repeated centrifugations. However, each centrifugation results in a significant loss of the target membrane. Final purification is often accomplished through density gradient centrifugation.

Density Gradient Centrifugation

Density gradient centrifugation is based on generating a centrifugation media that is densest at the bottom of the centrifuge tube and least dense at the top of the tube. The density gradient is used to separate particles on the basis of size and density. How fast something moves in a centrifugal field is defined by the sedimentation coefficient, s.

$$s = 2r^2(\rho_p - \rho_m)/9\eta$$

where:

s = sedimentation coefficient
r = radius of the particle
ρ_p = density of the particle
ρ_m = density of the media
η = viscosity of the media

From this equation it can be deduced that: bigger particles centrifuge faster than smaller particles; centrifugation rate is increased with increasing difference between density of the particle and density of the media; and centrifugation rate is slowed by an increase in media viscosity. Normally the sedimentation coefficient s has units of 10^{-13}. To avoid this cumbersome unit, the sedimentation coefficient is multiplied by 10^{13} generating the more commonly used term, S, the Svedberg Unit.'

$$S = s \times 10^{13}$$

Density gradients are of two basic types, discontinuous (or step, Figure 12.7a and continuous Figure 12.7b). In most cases the cell homogenate is placed on top of the medium in the centrifuge tube and the centrifugation is started. The cell fractions will migrate down the density gradient until $\rho_p = \rho_m$ and s drops to 0. At this point the centrifugation is stopped and the now separated cell fractions are individually isolated.

A discontinuous (or step) gradient is made by layering solutions of decreasing density on top of one another. The densest solution is placed at the bottom of the tube with successively lower density solutions on top (Figure 12.7a). The solutions are carefully layered by use of a pipette or syringe. To prevent unwanted mixing of the solutions during gradient formation, the centrifuge tube is tilted ~20 degrees from perpendicular as the solutions are layered. This method is inexpensive and requires no special 'gradient making' equipment. Since the solution volumes can vary from layer to layer and the densities need not decrease in a linear

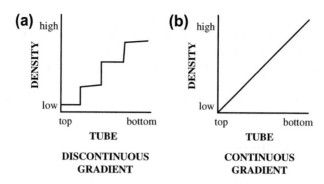

FIGURE 12.7 Density Gradient Centrifugation density profiles: (a) Discontinuous (or step) gradient and; (b) Continuous gradient.

fashion, it is possible to tailor a specific gradient fraction to match the density property of the membrane of interest and maximize its yield. Figure 12.8 shows a discontinuous gradient of various membrane fractions isolated from rat brain homogenate [27]. Densities of the five fractions were determined by the amount of Percoll (see below) in each fraction (0%, 3%, 10%, 15%, and 23%).

A continuous gradient displays a linear decrease in solution density from the bottom to the top of the centrifuge tube (Figure 12.7b). Gradients are generated by either a 'gradient maker' or can be self-generating during the centrifugation. Figure 12.9 demonstrates two

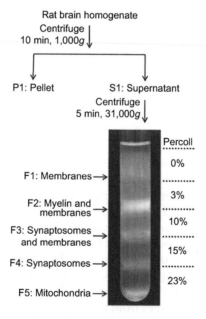

FIGURE 12.8 Discontinuous (step) gradient of various membrane fractions isolated from rat brain homogenate. Density of the 5 fractions was established by the amount of Percoll in each fraction (0%, 3%, 10%, 15%, and 23%) [27].

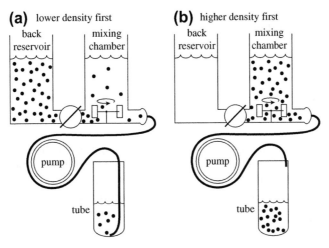

FIGURE 12.9 Formation of continuous density gradients using a density gradient maker. The device consists of two chambers, one of which contains a heavy (dense − black balls) solution and the other a light (less dense − clear) solution. The densities of the two solutions will eventually define the final solution density at the bottom and top of a centrifuge tube. The chamber on the right (the mixing chamber) has a mixer and an outlet where the mixed solution is pumped into the centrifuge tube. The chamber on the left holds a reservoir that is slowly added to the mixing chamber, but has no direct outlet to the centrifuge tube. Two types of density gradient systems are depicted, Lower Density First (Panel a) and Higher Density First (Panel b). Lower Density First: the lower density solution, initially in the mixing chamber, is first added to the bottom of the centrifuge tube. As the process continues the denser solution, initially in the back reservoir, is slowly added to the mixing chamber, increasing its density. Therefore the initial low density solution is continuously replaced from the tube bottom by a more dense solution, creating the gradient. Higher Density First: the solutions are reversed, with the denser solution in the mixing chamber. Therefore the first solution added to the tube is the most dense. In contrast to the low density first method, the solution is added from the top of the centrifuge tube. The gradient is made by adding less dense solutions from the top.

ways of making a continuous gradient using a gradient maker, the Lower Density First method (Figure 12.9a) and Higher Density First method (Figure 12.9b). Both methods are based on mixing two solutions of different density before adding them to the centrifuge tube. Details of the procedures are given in the legend for Figure 12.9. Gradients can be shallow (the two initial solutions have similar densities) or steep (the two initial solutions have very different densities). Shallow gradients are used to separate particles of similar physical properties. While separation of large, cell homogenate components (e.g. nuclei, mitochondria, chloroplasts etc) is relatively easy and can often be done with just differential centrifugation, separation of microsomes (small cell vesicles) is difficult. All microsomes have similar physical properties including size, shape, and density but may be separated using shallow density gradient centrifugation methodologies.

Some solutes have the ability to self-generate a density gradient during centrifugation. Examples include cesium chloride (CsCl), Percoll™, Ficoll™ and OptiPrep™. The major application for CsCl is in separating very dense molecules like nucleic acids. The density of DNA is about 1.7 g/ml and RNA about 2 g/ml while membrane vesicles fall between 1.1–1.3 g/ml, the range covered by Percoll, Ficoll and OptiPrep.

Densities of some major biomolecules (g/ml)	
Lipids	0.9–1.1
Proteins	1.25
DNA	1.7
RNA	2.0

In Figure 12.10, particle density (g/ml) is plotted against sedimentation coefficient in Svedberg Units (*S*) [28]. Clearly all membrane vesicle fractions (microsomes) are close together, making them very difficult to separate from one another. DNA and RNA, on the other hand, are much denser than the membrane fractions and can be easily separated on CsCl gradients.

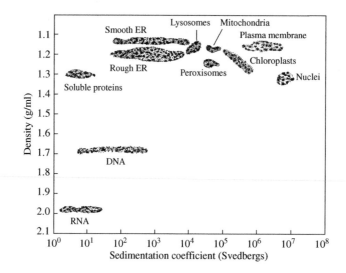

FIGURE 12.10   Particle density in g/ml is plotted against the log of the sedimentation coefficient in Svedberg Units. The values for DNA and RNA were determined from CsCl gradients [28].

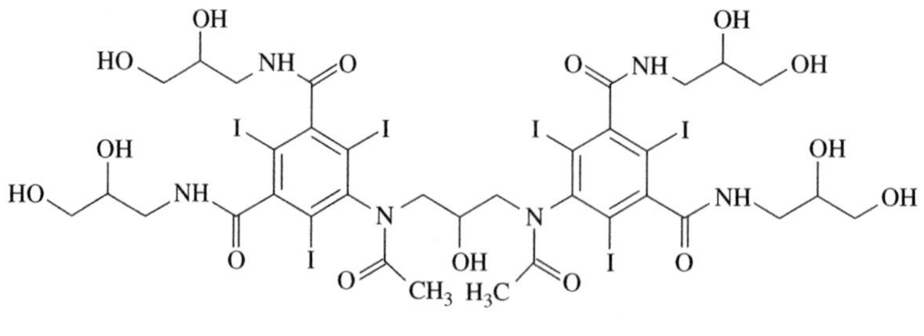

FIGURE 12.11   Iodixanol. *Courtesy of Focus Technology Co., Ltd.*

Percoll™ (GE Healthcare [29]) consists of silica particles (15–30 nm in diameter) coated with non-dialyzable polyvinylpyrrolidone (PVP). Percoll is non-toxic, chemically inert, easily adaptable to physiological ionic strength and pH, is of low viscosity and can be adapted to discontinuous or continuous gradients. Importantly it does not stick to membranes. It is its role in continuous gradient centrifugation that makes Percoll so attractive for membrane isolations (for example see [27]). A continuous Percoll gradient can be simply and quickly generated by moderate speed centrifugation (200–1,000g for a few minutes).

Ficoll™ (GE Healthcare) is an uncharged, highly branched polymer formed by the co-polymerization of sucrose and epichlorohydrin. Due to its many (−OH) groups, it readily dissolves in water. Like Percoll, Ficoll can be adapted to discontinuous or self-generating continuous gradients [9].

More recently a new self-generating density gradient medium, OptiPrep™ (Axis Shield [30]), has significantly impacted the field of membrane purification. OptiPrep is a solution of 60% iodixanol in water.

Iodixanol (Figure 12.11) is non-toxic to cells, is metabolically inert and has low viscosity and osmolality. OptiPrep avoids the high viscosity problems associated with sucrose and Ficoll gradients and the inconvenience of removing Percoll from membrane fractions after centrifugation. OptiPrep has been successfully applied in the isolation of plasma membranes and their domains, including lipid rafts.

Continuous density gradient centrifugations are of two basic types, isopycnic (or equilibrium) and rate zonal. In isopycnic, centrifugation is stopped when the particle no longer moves down the density gradient. At this point density of the particle equals density of the medium and $s = 0$. Sometimes two different particles may reach the same or very similar final location in the gradient and so cannot be separated after completion of the centrifugation. However, the particles may have different physical properties (e.g. size, shape) and so reach the same final location at different rates. In rate zonal, centrifugation is stopped before the particles reach final equilibrium. At this point, where density of the particles being separated is greater than density of the medium, the particles may be separated from one another.

D. MEMBRANE FRACTIONATION: NON-CENTRIFUGATION METHODS

Isolation of membrane fractions remains very much an art form and so there appears to be almost as many specific procedures as there are membranes to be isolated. While most isolation procedures are based on centrifugation, an almost limitless number of clever adaptations or 'tricks' have been employed for specific applications. Often these 'tricks' are combined with standard centrifugation methods to enhance final membrane purification. A brief sample of these techniques is discussed below.

Affinity Chromatography

Affinity chromatography is the most used of the non-centrifugation methods [31,32]. The most commonly employed applications for the isolation of membranes use lectins or antibodies. Fortunately, lectins and antibodies are among the most commercially available biochemicals and are ideally suited for column chromatography separations. Both applications work best for plasma membrane vesicles (microsomes). As discussed in Chapters 5 and 7, carbohydrates are attached to outer leaflet plasma membrane sphingolipids (cerebrosides, globosides and gangliosides). Most plasma membrane proteins are also heavily glycosylated (Chapter 6), with the sugars always facing outside of the cell. In fact, carbohydrates are a characteristic of the plasma membrane and so are obvious targets for membrane purification schemes.

Lectin-Affinity Chromatography

Lectins play a central role in membrane studies. Their history and function are thoroughly discussed in several excellent reviews by Nathan Sharon (for example see [33,34]). The term lectin is derived from the Latin word *legere*, 'to select', and indeed that is what lectins do. Lectins are proteins of non-immune origin that reversibly bind to specific sugars without modifying them. They are ubiquitous in nature, being found in animals, plants, insects, microorganisms, and even humans. Each lectin has two or more separate binding sites and so lectins are bifunctional reagents that have the ability to clump or agglutinate cells. It is this property that led to their 1888 discovery by Herrmann Stillmark (Figure 12.12) at the University of Dorpat in Estonia. In his Ph.D. thesis Stillmark described the agglutination of erythrocytes by the lectin ricin isolated from castor bean (*Ricinus communis*).

Countless lectins have now been identified making their classification difficult, but increasing their importance in membrane isolations. For example, the Sigma-Aldrich catalog offers almost 500 lectins from dozens of sources! Lectins vary from those that

FIGURE 12.12   Hermann Stilmark (1860–1923). *Courtesy of Tartu University*

TABLE 12.1   Specificity For Some Common Lectins.

Abbreviation	Name	Carbohydrate bound
Con A	Concanavalin A	Mannose, Glucose
GNA	Snowdrop Lectin	Mannose
RCA	*Ricinus communis*	Galactose, N-Acetyl Glucosamine Agglutin
PNA	Peanut Agglutin	Galactose, N-Acetyl Galactosamine
WGA	Wheat Germ	Sialic Acid, N-Acetyl Glucosamine Agglutin
UFA	*Aleuria europaeus*	Fucose Agglutin

bind to a specific simple sugar to many that bind to di-, tri-, and even polysaccharides. Table 12.1 lists a few common lectins with their carbohydrate specificity. This table is overly simplified as several of the lectins actually bind to a di- or tri-polysaccharide containing the listed sugars and not just the simple sugar itself. Lectins are bound to carbohydrates by many weak reversible non-covalent bonds [35,36], similar to binding of a substrate to the active site of an enzyme. Most lectins actually share little structural similarity as demonstrated by their amino acid sequence, molecular size, and other molecular properties. It is hard to comprehend how these widely varying structures can all bind carbohydrates.

Many free lectins are commercially available and several are even available conjugated to inert matrices that can be packed into columns for use in lectin-affinity chromatography. Conjugation to the inert matrix, often Sepharose or Agarose, does not affect the lectin's ability to bind to specific sugars, and so is useful for membrane isolations. Sugar binding is also not affected by detergents, compounds that are essential for membrane protein isolation (Chapter 13). Lectin-affinity chromatography consists of pouring the cell homogenate over a lectin column. Membranes that do not have the requisite sugar selected by the lectin simply pass through the column and are removed. The bound membranes are washed and then released from the column by a solution containing a high concentration of the simple sugar that the lectin recognizes. A single pass over an appropriate lectin-affinity column can greatly purify a mixed membrane population in one step. However, this procedure only works for membrane fractions that have sugars on their surfaces, mostly plasma membrane microsomes.

The fact that a simple sugar can block the attachment of a lectin to its normal polysaccharide target can have important medical implications. For example, a urinary tract infection can result from binding of *E. coli* to a resident urinary tract lectin [37]. It was noted that the *E. coli* infection can be greatly reduced by administrating mannose which dislodges the *E. coli*. This may explain why cranberry juice, rich in mannose, has historically been used to prevent bladder infections.

Antibody-Affinity Chromatography

Another important type of affinity chromatography employs columns with attached antibodies [32,38]. Antibody-affinity chromatography is more versatile and powerful than lectin-affinity chromatography. While use of lectins is limited to membranes possessing surface carbohydrates, appropriate antibodies can bind to any surface antigen including carbohydrates and proteins. Lipids and nucleic acids are antigenic only when bound to carbohydrates or proteins. In addition, antigen-antibody binding is much stronger than carbohydrate-lectin binding.

The term antigen was coined in 1899 by Hungarian microbiologist Ladislas Deutsch (he also had an alias, Laszio Detre). Antigen is a contraction of the original name 'antisomato-gen'. An antigen is simply any molecule or part of a molecule that is recognized by the immune system and can bind to the antigen-binding site of an antibody. Antibodies have the ability to recognize and bind specifically to an antigen in the presence of a vast solution of molecules that are often similar in structure. The precise specificity of an antibody for an antigen makes them ideal for membrane purification by affinity chromatography. Countless custom antibodies are now commercially available for attachment to column matrix material. A crude membrane homogenate is passed over the antibody column whereupon only the membrane fractions containing the appropriate surface antigen bind. Other membrane fractions pass through the column and are discarded. After washing the column, the bound membrane is released by altering the pH of the washing solution.

Ligand-Receptor Affinity Chromatography

Another type of affinity chromatography employs ligand-receptors. The outer surface of the plasma membrane has a large number of proteinaceous receptors whose function is to bind specifically to external molecules referred to as ligands. Included in a long list of ligands are small signaling peptides, hormones, neurotransmitters, vitamins, and toxins. An unusual application is employing zinc-affinity columns to isolate proteins that require zinc cofactors to function [39]. Ligands bind to, and dissociate from receptors according to the law of mass action. By this methodology, ligands are covalently bound to the column matrix and the cell homogenate is passed over the column. Membranes with the receptor for the ligand bind to the column and the non-bound membranes pass through the column and are discarded. Ligand-bound membrane fractions are released from the column by adding a large excess of free ligand. While ligand affinity columns have occasionally been used to separate membrane fractions, their major use has been in isolating and purifying receptors for signal transduction studies.

Anion Exchange Chromatography

Ion exchange chromatography [40] has been an important scientific tool since its discovery in about 1850. Anion exchange columns are made of resin beads, often agarose or cellulose, to which is covalently bound a cationic functional group. The stationary phase (the resin beads) interacts with anions dissolved in a mobile, aqueous phase. Anion exchange chromatography separates compounds based on their net negative surface charge. Since biological membranes are anionic, they are amenable to this technique. Membrane phospholipids are either neutral (they are zwitterions like PC) or they are anions (full anions like PA, PS, PI, PG, CL or a partial anion like PE). In addition, gangliosides by definition contain the anionic sugar sialic acid and so provide negative membrane surface charge. As a result, all membranes have varying degrees of negative charge density and will bind to anion exchange columns. The higher the negative charge density, the tighter the membrane binding. As with affinity chromatography, a cell homogenate is passed over the anion exchange column and washed with water. Membrane fractions are released from the column (stationary phase) with either increasing retention time in the presence of a constant anion concentration in the mobile phase (normally Cl^-) or by adding solutions of increasing anion concentrations. The poorest binding membrane fraction has the lowest negative charge density and is the first to be released. The most negative membrane fraction is the tightest bound and so is last to be released.

There are now many examples of commercially available anion exchange resins. The leader in the field is Dow Chemical Company's division Dionex Corporation (Dow-Ion Exchange). Membrane separations by anion exchange chromatography have serious limitations and do not have the specificity and precision of affinity chromatography.

Derivitized Beads

Magnetic Beads

Since the 1980s magnetic micro- and nano-beads have been used for a wide array of biological applications [41], including membrane separations. Unmodified magnetic beads would be of little or no value for these studies. However, magnetic beads have proved to be readily modified by several of the same molecules known to work so well with affinity chromatography (e.g. lectins and antibodies). In addition, magnetic beads have assumed a major role in molecular biology by being modified to bind to specific nucleic acid sequences, thus allowing separation of various nucleic acids [42] and isolation of DNA-binding proteins.

Separations using magnetic beads have proven to be not only versatile but also extremely gentle. The rougher techniques of column chromatography and centrifugation can often be completely avoided. The technique is quick, easy, sensitive, inexpensive, and amenable to automation.

The leader in the field of magnetic beads is Dynabeads™ (Invitrogen, Life Technologies [43]). Dynabeads are exceptionally uniform spheres ([43]) of highly defined composition. Importantly, they are superparamagnetic, meaning they have magnetic properties when placed in a magnetic field, but lose these properties completely when removed from the field. A uniform polymer shell surrounding the superparamagnetic core can be modified through a variety of bioreactive conjugates (e.g. lectins, antibodies) and separates the inert internal core from active surface components that protrude into the external aqueous environment. An appropriately modified Dynabead will attach specifically to a target membrane in a complex cell homogenate. A magnet placed on the outside of the tube containing the homogenate will attract the Dynabeads to the tube wall. The remaining, non-attached membrane fragments can then be simply washed away and discarded. After adding new media, the external magnet is removed and the desired membrane attached to the Dynabeads is left free in solution where it can be readily removed from the beads.

Other Types of Microbeads

There has been a recent explosion in the development and application of a wide variety of microbead technologies. One leading company, Phosphorex, Inc. (Fall River, MA), offers micro- and nanospheres with sizes ranging from 20 nm to 1,000 microns. The beads are made from polystyrene, poly(methyl methacrylate), poly(acrylic acid), nylon, poly(lactic acid), PLGA (poly(lactic-co-glycolic acid)), chitosan, and many other synthetic and natural polymers. They can be modified through attachment of several biomolecules including lectins and antibodies. These micro- and nanospheres have the additional property of being able to sequester molecules into their aqueous interiors, making them suitable as drug delivery vehicles. Membrane separations with modified micro- and nanospheres parallel methods discussed above for affinity chromatography.

Other Methodologies

Two Phase Partitioning

This unusual methodology is based on membranes partitioning between two co-existing but immiscible liquid phases. The phases are somewhat analogous to what is seen in some household novelty items that are commonly found in department stores. One is a slowly rocking device that generates what resembles highly colored ocean waves. The second is the more popular 'lava lamp' [44] The 'lava lamp' is a recent addition to pop culture, invented by Edward Craven-Walker (Figure 12.13) in 1963. The immiscible 'lava lamp' solutions are often water/glycerol and wax/carbontetrachloride. A heat source in the form of an incandescent light bulb decreases the wax phase density, causing it to rise. The lighter wax then rises as a blob and slowly cools, sinking back to the bottom in a lava-like display. For membrane separations, partitioning is between two high molecular weight immiscible

FIGURE 12.13   E. Craven-Walker (1918–2000).

liquids, often polyethylene glycol and dextran [45]. In one example Yoshida and Kawata [46] reported that 'mung bean tonoplast has a high partition coefficient for the polyethylene glycol-enriched upper phase and the smooth endoplasmic reticulum has a high partition coefficient for the Dextran-enriched lower phase'. Two phase separation of membrane fractions is at best poorly understood and is presently only a curiosity that will never likely receive wide application.

Silica Particles

Silica particles have been used to isolate plasma membrane (PM) fractions [47]. Affinity is provided by charge attraction as silica microbeads are positively charged while PM fractions are negatively charged. Silica microbeads are also dense and can be easily centrifuged away from other, non-bound and hence lighter, membranes. The procedure begins by binding whole cells to the dense silica particles. Upon cell lysis, large open sheets of PM remain bound to the silica and are isolated away from the unbound internal membranes by gentle centrifugation. Using this procedure, Chaney and Jacobson reported that PM from *Dictyostelium discoideum* was obtained in high yield (70–80%) and purified 10 to 17-fold [47].

Separation by Size

It is relatively easy to separate intact organelles from intact cells and even from one another on the basis of size and this is normally accomplished by differential centrifugation. However, separation on the basis of size can also be achieved by common filter techniques including Millipore Filters, Sephadex (size exclusion chromatography) and PAGE electrophoresis (polyacrylamide gel electrophoresis). These techniques have much more important applications in other aspects of bioscience, but have occasionally been used for membrane separations. For example, Millipore Filters are extensively used to remove contaminants from laboratory water and to size liposomes (Chapter 13), Sephadex to separate non-sequestered solutes from solutes sequestered inside liposomes (Chapter 13), and PAGE electrophoresis for protein purification.

Membrane-Specific 'Tricks'

Membrane purification is as much an art as it is a science. A large number of what could be best described as 'tricks' have been employed for very specific applications. These applications have no general use in membrane purification procedures. Three examples follow:

1. It is often difficult to separate microsomes derived from the rough endoplasmic reticulum (RER) from the smooth endoplasmic reticulum (SER). However, in the presence of 15 mM CsCl, RER microsomes aggregate, greatly facilitating the separation.
2. Rat liver lysosomes can be hard to separate from mitochondria and peroxisomes due to their similar sedimentation coefficients. Nobel Prize recipient Christian de Duve (in *Physiology or Medicine*, 1974) reported that if the detergent Triton WR 1339 was injected into the rat prior to organelle extraction, lysosomes accumulated the detergent making them lighter [48]. With detergent, lysosome density decreased from 1.21 to 1.12, facilitating their separation from the unmodified mitochondrial and peroxisomal fractions.
3. Finally, an ER separation has been reported based on the presence of an ER enzyme, glucose 6-phosphatase. Glucose 6-phosphate, the substrate for the enzyme, enters the ER

microsomes via a trans-membrane transporter protein. Once inside the vesicle, glucose 6-phosphate is cleaved yielding glucose and free phosphate. It was noted that if the membrane fractions were bathed in Pb^{2+}, insoluble lead phosphate $(Pb_3(PO_4)_2)$ accumulated in the ER microsome interior, significantly increasing their density. The ER microsomes could then be easily centrifuged away from the other microsomes.

E. MEMBRANE MARKERS

For each of the many steps in isolating and purifying a membrane, tests must be run not only to demonstrate that the target membrane is indeed present and being purified but also that possible contaminating membranes are being eliminated. For the initial run through a multi-step procedure this may require a tremendous amount of very tedious assays. Positive identification of a membrane is achieved by following two basic types of membrane markers, structural markers and marker enzymes [49]. Both types of markers must uniquely reside in the membrane being tested. To follow the outcome of each membrane purification step, relatively rapid and quantitative procedures, mostly enzyme assays, are preferred over slower and more qualitative procedures like electron microscopy. After the investigator has achieved what he hopes is high purity, a variety of confirming enzyme assays, structural measurements, and compositional analysis are preformed. Once the membrane purification scheme is shown to be accurate and reproducible, subsequent purifications need not be so detailed. Some of these markers are listed in Tables 12.2 (Membrane Structural Markers) and Table 12.3 (Membrane Marker Enzymes) (these tables are modified from Table 1.2 in [49].

TABLE 12.2   Partial list of Membrane Structural Markers.

Membrane	Structural marker
Nucleus	Chromosomes
	DNA (Feulgen Stain)
Plasma Membrane	Gap Junctions
	Brush Border Glycocalyx
	Microvilli
	Bind Lectins
Mitochondrial Inner	F_1-ATPase 'knob'
	Cardiolipin
RER	Ribosomes
Golgi	Stains well with Osmium
Thylakoid	Chlorophyll
Any Membrane	Specific Antibodies

TABLE 12.3   Partial list of Membrane Marker Enzymes.

Membrane	Marker enzyme
Plasma Membrane	5'-Nucleosidase
	Alkaline Phosphodiesterase
	Na^+/K^+ ATPase
	Adenylate Cyclase
	Amino Peptidase
Mitochondria Inner	Cytochrome c Oxidase
	Succinate-Cytochrome c Oxidoreductase
Mitochondria Outer	Monoamine Oxidase
Lysosome	Acid Phosphatase
	β-Galactosidase
Peroxisome	Catalase
	Urate Oxidase
	D-Amino Acid Oxidase
Golgi	Galactosyltransferase
	α-Mannosidase
Endoplasmic Reticulum	Glucose-6-Phosphatase
	Choline Phosphotransferase
	NADPH-Cytochrome c Oxidoreductase
(Cytosol)	Lactate Dehydrogenase

SUMMARY

Purifying a membrane is as much an art as it is a science. The first step in purifying a membrane is breaking open (homogenizing) the cell. Once broken, the many cell membrane components must be fractionated. Methods discussed include:

Homogenization Methods: Enzymes to break thick cell wall; Shear force (mortar and pestle); Food blenders; High speed blade-type homogenizers; Osmotic gradients; Sonicators; Gas ebullition; French Press; and Freeze/thaw.

Fractionation Methods: Centrifugation methods (differential centrifugation and density gradient centrifugation); Non-centrifugation methods (lectin-affinity chromatography, antibody-affinity chromatography, ligand-receptor affinity chromatography, anion exchange chromatography, magnetic beads, microbeads, two-phase partitioning, silica particles, size filtration and size exclusion chromatography).

After each step in membrane purification, tests must be run to assure that the step resulted in purification of the target membrane and that other membranes had been excluded. To achieve this, a large number of structural markers and marker enzymes characteristic of a particular membrane are employed.

Membrane reconstitution is often a necessary step in determining a membrane protein's mechanism of action. Chapter 13 will review the methodologies to purify membrane proteins and lipids and to reconstitute these components into a functional liposome. Lipid raft isolation will also be discussed.

References

[1]   Voet D, Voet JG. Biochemistry. 3rd ed. Wiley; 2004. p. 1309.
[2]   Evans WH. Isolation and characterization of membranes and cell organelles. In: Rickwood D, editor. Preparative Centrifugation — a Practical Approach. Oxford University Press; 1992. p. 233—70.
[3]   Siekevitz P. Protoplasm: Endoplasmic reticulum and microsomes and their properties. Annu Rev Physiol 1963;25:15—40.
[4]   Heinemann FS, Ozols J. Isolation and structural analysis of microsomal membrane proteins. (Review) Front Biosci 1998;3:483—93.
[5]   Graham JM. Homogenization of tissues and cells. In: Graham JM, Rickwood D, editors. Subcellular Fractionation — a Practical Approach. Oxford University Press; 1997. p. 1—29.
[6]   Garcia FAP. Cell Disruption and Lysis. Encyclopedia of Bioprocess Technology. John Wiley; 2002.
[7]   Garcia FAP. Cell Wall Disruption and Lysis. Encyclopedia of Industrial Bioprocess: Bioprocesses, Bioseparation and Cell Technology. 1—12. John Wiley; 2009.
[8]   Graham JM, Rickwood D. Subcellular Fractionation: A Practical Approach. Oxford, UK: Oxford University Press; 1997.
[9]   Amersham Bioscience. 2011. Ficoll PM 70, Ficoll PM 400. Cell Separation. Data file 18-1158-27 AA, pp 1—6.
[10] Camara C. Sample preparation for speciation. Anal Bioanal Chem 2005;381:277—8.
[11] Kido H, Micic M, Smith D, Zoval J, Norton J, Madou M. A novel, compact disk-like centrifugal microfluidics system for cell lysis and sample homogenization. Colloids Surf B Biointerfaces 2007;58:44—51.
[12] Hopkinson J, Mouston C, Charlwood KA, Newbery JE, Charlwood BV. Investigation of the enzymatic digestion of plant cell walls using reflectance fourier transform infrared spectroscopy. Plant Cell Rep 1985;4:121—34.
[13] Kalia A, Rattan A, Chopra P. A method for extraction of high-quality and high-quantity genomic DNA generally applicable to pathogenic bacteria. Anal Biochem 1999;275:1—5.
[14] Urban K, Wagner G, Schaffner D, Roglin D, Ulrich J. Rotor-stator and disc systems for emulsification processes. Chem Eng Technol 2006;29:24—31.
[15] de Gier J. Osmotic behaviour and permeability properties of liposomes. Chem Phys Lipids 1993;64:187—96.
[16] de Gier J. Liposomes recognized as osmometers. J Liposome Res 1995;5:365—9.
[17] Srinivastan PR, Miller-Faures A, Frrera M. Isolation techniques for HeLa-cell nuclei. Biochim Biophys Acta 1982;65:501—5.
[18] Diagonistics LLC. Sonication 2003;101.
[19] Chandler DP, Brown J, Bruckner-Lea CJ, Olson L, Posakony GJ, Stults JR, et al. Continuous spore disruption using radially focused, high-frequency ultrasound. Anal Chem 2001;73:3784—9.
[20] Burgmann H, Pesaro M, Widmer F, Zeyer J. A strategy for optimizing quality and quantity of DNA extracted from soil. J Microbiol Methods 2001;45:7—20.
[21] Goldberg S. Mechanical/physical methods of cell disruption and tissue homogenization. In: Methods in Molecular Biology, Vol 424. New York: Springer Publishing; 2008. p. 3—22.
[22] Rinker AG, Evans DR. Isolation of a chromosomal DNA from a methanogenic archaebacteria using a French pressure cell. Biotechniques 1991;11:612—3.
[23] Harju S, Fedosyuk H, Peterson KR. Rapid isolation of yeast genomic DNA: Bust n' grab. BMC Biotechnol 2004;4:8.
[24] Mikkelsen SR, Corton E. centrifugation methods. In: Bioanalytical Chemistry. Hoboken, NJ: John Wiley & Sons, Inc; 2004. p. 247—67 [Chapter 13].

[25] Rickwood D. Preparative Centrifugation: A Practical Approach. Oxford, UK: IRL Press at Oxford University Press; 1992.

[26] Graham JM. Biological centrifugation (The Basics). Oxford, UK: Bios Scientific; 2001.

[27] Dunkley PR, Jarvie PE, Robinson PJ. A rapid Percoll gradient procedure for preparation of synaptosomes. Nat Protoc 2008;3:1718−28.

[28] Price CA. Centrifugation in Density Gradients. New York: Academic Press; 1982.

[29] GE-Healthcare. Percoll™ www.gelifesciences.com.

[30] Shield Axis. Biological Separations. News Bulletin for Axis-Shield Density Gradient Media 2009; Issue 2. OptiPrepTM Preparation of gradient solutions.

[31] Hage, D.S. and Cazes, J., Eds. Handbook of Affinity Chromatography. 2nd ed. Chromatographic Science Series, Marcel Dekker; Taylor and Francis, 2006.

[32] Bailon P, editor. Affinity Chromatography: Methods and Protocols (Methods in Molecular Biology). Totowa, New Jersey: Humana Press; 2000.

[33] Sharon N. History of lectins: from hemagglutinins to biological recognition molecules. Glycobiology 2004;14:53R−62R.

[34] Sharon N, Lis H. Lectins. 2nd ed. Dordrecht, The Netherlands: Springer; 2007.

[35] Kennedy JF, Palva PMG, Corella MTS, Cavalcanti MSM, Coelho LCBB. Lectins, versatile proteins of recognition: a review. Carbohydr Polym 1995;26:219−30.

[36] Van Damme EJM, Peumans WJ, Pusztai A, Bardocz S. Handbook of Plant Lectins: Properties and Biomedical Applications. New York: John Wiley; 1998.

[37] Liu Y, Gallardo-Moreno AM, Pinzon-Arango PA, Reynolds Y, Rodriguez G, Camesano TA. Cranberry changes the physicochemical properties of *E. coli* and adhesion with uroepithelial cells. Colloids Surf B Biointerfaces 2008;65:35−42.

[38] Nisnevitch M, Firer MA. The solid phase in affinity chromatography: strategies for antibody attachment. J Biochem Biophys Methods 2001;49:467−80.

[39] Ogut O, Jin J-P. Expression, zinc-affinity purification, and characterization of a novel metal-binding cluster in Troponin T: Metal-stabilized α-helical structure and effects of the NH_2-terminal variable region on the conformation of intact troponin T and its association with tropomyosin. Biochemistry 1996;35:16581−90.

[40] Yamamoto S, Nakanishi K, Matsuno R. Ion Exchange Chromatography of Proteins (Chromatographic Science Series). New York: Marcel Dekker, Inc; 1988.

[41] Haukanes B-I, Kvam C. Application of magnetic beads in bioassays. Nat Biotechnol 1993;11:60−3.

[42] Levison PR, Badger SE, Dennis J, Hathi P, Davies MJ, Ian J, et al. Recent developments of magnetic beads for use in nucleic acid purification. J Chromatogr A 1998;816:107−11.

[43] Invitogen, Life Technologies. Dynabeads® Magnetic Separation technology. Carlsbad, CA.

[44] Bellis, M. The history of lava lamps: Craven Walker − How Lava Lamps Work, About.com Inventors, http://inventors.about.com/od/lstartinventions/a/lava_lamp.htm. New York Times; 2008.

[45] Persson A, Jergil B. Purification of plasma membranes by aqueous two-phase affinity partitioning. Anal Biochem 1992;204:131−6.

[46] Yoshida S, Kawata T. Isolation of smooth endoplasmic reticulum and tonoplast from mung bean hypocotyls (Vigna radiata [L.] Wilczek) using a Ficoll gradient and two-polymer phase partition. Plant Cell Physiol 1988;29:1391−8.

[47] Chaney LK, Jacobson BS. Coating cells with colloidal silica for high yield isolation of plasma membrane sheets and identification of transmembrane proteins. J Biol Chem 1983;258:10062−72.

[48] Leighton F, Poole B, Beaufay H, Baudhuin P, Coffey JW, Fowler S, et al. The large-scale separation of peroxisomes, mitochondria, and lysosomes from the livers of rats injected with triton WT-1339. J Cell Biol 1968;37:482−513.

[49] Gennis RB. Biomembranes. Molecular Structure and Function. (Table 1.2, p.20). New York, NY: Springer-Verlag; 1989.

[50] Frezza C, Cipolat S, Scorrano L. Organelle isolation: functional mitochondria from mouse liver, muscle and cultured filroblasts. Nature Protocols 2007;2:287−95.

[51] CONTEMPORARY JOE: http://contempjoe.blogspot.com/2008/10/edward-craven-walker-and-his-lava-lamp.html

Membrane Reconstitution

OUTLINE

A. **Detergents** 266
   Membrane Detergents 269

B. **Membrane Protein Isolation** 271

C. **Membrane Lipid Isolation** 275
   TLC 276
   HPLC 278
   Fatty Acid Analysis 279

D. **Model Lipid Bilayer Membranes —
Liposomes** 282
   Model Membranes 282
   Liposomes 284
   Unilamellar Vesicles 287
   SUV by Sonication 288
   Vesicles by Extrusion 288

Vesicles by Detergent Dilution 289
LUVs: Dilution Methods 290
Concentration Methods 291

E. **Membrane Reconstitution** 292
   Ca²⁺ ATPase 293
   Bacterial Proline Transporter 294

F. **Lipid Rafts** 295
   Experimental Support of Rafts 297
   Questions Concerning Rafts 298
   One Last Word on Rafts 299
   Lipid Raft Characteristics 299

Summary 300

References 300

Membranes are unimaginably complex. They consist of hundreds of different proteins floating in a bewildering and crowded sea of a thousand or more different lipids and this mixture is in constant flux. A membrane is not homogeneous but exists in fleeting patches that exhibit both lateral and trans-membrane asymmetry. Also, countless reactions occur simultaneously in and around the membrane. Can the biological membrane problem be simplified? In this chapter we will examine how one membrane protein can be isolated away from the others and reconstituted into a model lipid bilayer membrane in order to determine how the protein functions. Finally we will investigate the isolation and properties of an important membrane domain involved in cell signaling — the lipid raft.

An Introduction to Biological Membranes
http://dx.doi.org/10.1016/B978-0-444-52153-8.00013-1

A. DETERGENTS

It is hard to overstate the importance of detergents in life science studies, particularly those involving biological membranes. At present, detergents form the cornerstone of membrane studies at the molecular level. As a result, a large detergent industry has flourished, producing well over a thousand different products. Detergents are amphipathic molecules, simultaneously containing polar and non-polar ends (discussed in Chapter 3) [1]. The first detergent was also the first soap (see Chapter 5), namely the potassium salt of free fatty acids resulting from the saponification of animal fat.

Detergents are a class of molecules that have the ability to disrupt hydrophobic/hydrophilic interactions like those that drive the stability of membranes (see the hydrophobic effect, Chapter 3) [1]. Detergents are used to lyse cells and to extract membrane proteins and lipids for reconstitution experiments. Many additional functions have been assigned to detergents including protein crystallization, immunoassays, and electrophoresis. While detergents' amphipathic structure may grossly resemble that of membrane phospholipids, a closer examination shows that detergents are much more water soluble (Table 13.1). For example, solubility of the phospholipid DPPC is only ~4.7×10^{-7} mM, many orders of magnitude lower than for membrane detergents. This trait of resembling a phospholipid while being much more water soluble is the reason detergents can be successfully employed to extract and purify integral proteins from membranes.

Detergents that are used in membrane studies can be divided into 5 basic categories: anionic, cationic, non-ionic, zwitterionic, and bile salts. The structures of some of the more commonly used detergents in membrane studies are depicted in Figure 13.1. The first 4 of these classes are synthetic detergents and are classified by the net charge residing on their polar moiety. The 5th classification is based on it being a biologically produced type of

TABLE 13.1   Detergents Commonly Used in Membrane Studies with their HLB and CMC Values.

Type	Detergent	HLB	CMC (w/v%)
Anionic	SDS		0.17
Cationic	Cetylpyridinium Bromide	16.7	0.042
Non-Ionic	Brij-35	16.9	0.09
	Octyl-Glucoside	13.1	0.7
	Triton X-100	13.6	0.0155
	Tween 20	16.7	0.0987
Zwitterionic	CHAPS	13.1	0.5
Bile Salt	Cholate	18	0.388
	Deoxycholate	16	0.083
(phospholipid	DPPC	solubility	~3.5×10^{-7} g/l)

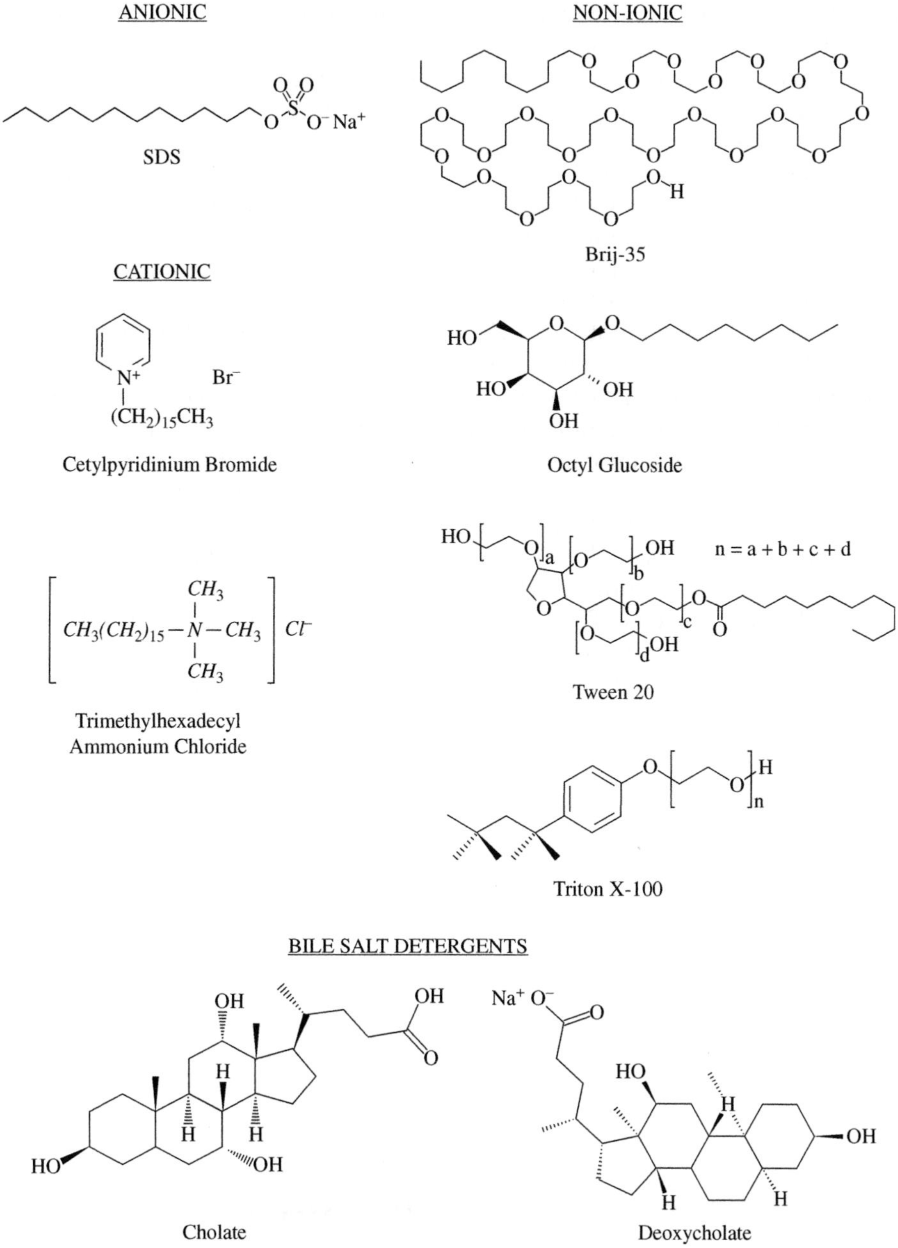

ANIONIC

SDS

CATIONIC

Cetylpyridinium Bromide

Trimethylhexadecyl
Ammonium Chloride

NON-IONIC

Brij-35

Octyl Glucoside

$n = a + b + c + d$

Tween 20

Triton X-100

BILE SALT DETERGENTS

Cholate

Deoxycholate

FIGURE 13.1 Examples of basic types of detergents most commonly used in studies of biological membranes.

detergent. Besides being separated into these five classes, each detergent is given two numbers that may be predictive of its use in membrane studies — the critical micelle concentration (CMC) and the hydrophile/lipophile balance (HLB). CMC is an actual, measured parameter while the HLB is an empirically derived number.

The CMC is the detergent (or any surfactant) concentration above which micelles are spontaneously formed [2]. Below the CMC each detergent molecule will either reside at the aqueous interface or, if water soluble enough, be solvated. Above the CMC additional surfactant will only increase the number of micelles in solution. At the CMC, micelles form, changing several basic parameters of the solution. The abrupt solution changes that occur at the CMC are commonly followed by an increase in solution turbidity (light absorbance), a change in electrical conductivity or a change in surface tension. Determination of CMC using surface tension measurements is shown in Figure 13.2 [3]. At low concentrations, the detergent will preferentially accumulate at the air/water interface greatly reducing the surface tension. As detergent accumulates at the interface, the increasingly crowded condition forces some detergent into the aqueous sub-phase, resulting in the generation of micelles. The detergent concentration where this occurs is the CMC. CMCs for some common membrane detergents are listed in Table 13.1.

A second number often used to characterize a detergent is the HLB (hydrophile/lipophile balance) [4,5]. The HLB is a number that is an empirical measure of how hydrophilic or hydrophobic the detergent is. There are several ways to calculate the HLB. The most commonly used method to calculate HLBs for membrane studies was developed by Griffin in papers published in 1949 and 1954 [4,5]. The Griffin method calculates the net HLB from

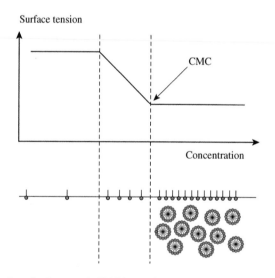

FIGURE 13.2   Determination of a detergent's CMC by surface tension measurements. At low concentrations, the detergent will preferentially accumulate at the air/water interface greatly reducing the surface tension. As detergent accumulates at the interface, the ever increasing crowded condition forces some detergent into the bathing solution generating micelles at the CMC. At this point there is no further decrease in surface tension with additional detergent [3].

combining the number of atoms comprising the hydrophilic and hydrophobic portions of the detergent.

$$HLB = 20 \times M_h/M$$

Where M_h is the molecular mass of the hydrophilic portion of the molecule and M is the molecular mass of the entire molecule. By this method HLBs can range from 0 to 20 with 0 being a molecule that is completely hydrophobic and 20 a molecule that is completely hydrophilic. Since HLB is entirely empirical, other formulations can span over different ranges. A high HLB indicates that the detergent will partition into the aqueous phase while a low HLB indicates the detergent will prefer remaining in the membrane and so will be very difficult to extract from the membrane. Detergents used for membrane studies generally have HLBs between 13 and 15. Structures of several detergents commonly used in membrane studies are depicted in Figure 13.1 and their CMCs and HLBs listed in Table 13.1. It should be mentioned that CMCs and HLBs can vary significantly with pH and solvent ionic strength.

Membrane Detergents

The most suitable detergents for extraction of functional membrane proteins are classified as non-ionic, zwitterionic or bile salts (detergent structures are shown in Figure 13.1). A neutral detergent that has received a great deal of use in isolating integral membrane proteins is octyl glucoside (n-octyl-β-D-glucoside). Variations of this basic detergent are commercially available with mannose replacing glucose and with decyl (10-carbon) or lauryl (12-carbon) chains replacing octyl (8-carbon). Octyl glucoside is hydrophilic enough to be readily removed from extracted membrane proteins. CHAPS (3-[(3-cholamidopropyl)dimethylammonio]-1-propanesulfonate) is a popular zwitterionic detergent used to solubilize integral membrane proteins without causing denaturation. Brij-35 (a polyoxyethyleneglycol lauryl ether) is another non-ionic detergent involved in membrane integral protein extractions. Brij-35 is also used to prevent nonspecific protein binding to gel filtration and affinity chromatography supports. A major drawback to Brij-35 is its difficulty in being removed by dialysis. However, it has been reported that Brij-35 can be removed by Extracti-Gel D Detergent Removing Gel. Triton X-100 is also a non-denaturing, non-ionic detergent. However, its application in membrane studies is generally different than octyl glucoside or Brij-35. The major use for Triton X-100 is to lyse or permealize cells. It is not as useful in solubilizing functional membrane proteins and has difficulty breaking protein–protein interactions. Triton X-100 is also hard to remove from cell membranes. Anyone who has ever tried to pipette Triton X-100 will remember the unpleasant experience. Triton X-100 is *very* viscous! Often Triton X-100 is used in conjunction with zwitterionic detergents (such as CHAPS) that are better suited for membrane protein extractions.

Cholate is the major bile salt comprising ~80% of all bile salts produced in human liver. This detergent is a natural product synthesized from cholesterol. Another closely related bile salt that is popular as a membrane protein extraction detergent is deoxycholate whose structure differs from cholate only by lacking a single (−OH) (Figure 13.1). Deoxycholate is considered to be a secondary bile acid and is a metabolic byproduct of intestinal bacteria. Both cholate and deoxycholate have been proposed to function in a similar fashion

(Figure 13.3). Each has a smooth hydrophobic surface that interacts with acyl chains of the membrane phospholipids. The other surface is polar, being dominated by (−OH) groups (3 for cholate and 2 for deoxycholate). A cylinder is formed with the membrane integral protein forming the core, surrounded by a ring of phospholipids and finally a second ring

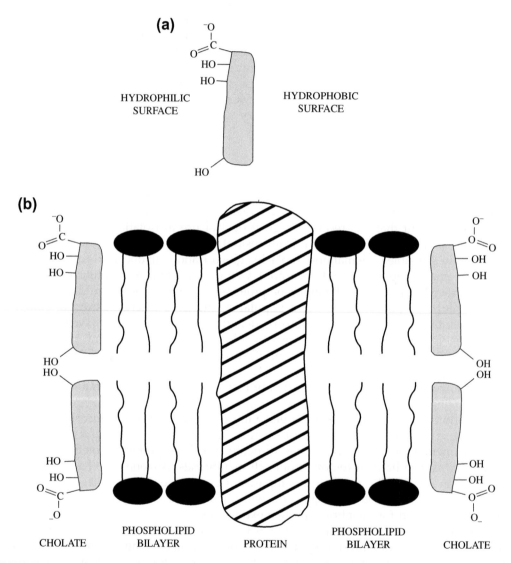

FIGURE 13.3 Mechanism by which cholate solubilizes an integral membrane protein. (The structure of cholate is shown in Figure 13.1.) Panel (a) is a cartoon drawing of cholate demonstrating its hydrophilic surface and its hydrophobic surface. Panel (b) shows how cholate may extract a plug from the membrane that contains an integral protein and phospholipids.

of cholate or deoxycholate. The outer ring is stabilized by polar (−OH) interactions with water. The charged carboxylate on the detergents further stabilizes the cylinder.

SDS (sodium dodecylsulfate) is by far the best known of a family of related anionic detergents that bind to and denature proteins. Similar detergents can have different alkyl chain lengths and different anionic head groups (e.g. sulfonate, carboxylate, phosphate etc.). Below the CMC, SDS monomers bind to both membrane and cytosolic proteins in an equilibrium-driven process. Binding is cooperative meaning the binding of a first SDS facilitates binding of a second and so forth. Upon SDS binding, the protein structure is altered into rigid rods whose length is proportional to its molecular weight. SDS binds to the protein chain at a ratio of one SDS for every two amino acid residues. Since the SDS head is negatively charged, net charge on the denatured protein is proportional to the protein's molecular weight. SDS-laden proteins can then be separated in an electric field generated on a stable support. The process is referred to as SDS-PAGE (polyacrylamide gel electrophoresis). The function of SDS in membrane studies is therefore to determine the number and size of membrane proteins. Since proteins are denatured by SDS, this detergent is of little value in reconstituting functional proteins into membranes. SDS has many more applications outside the world of science. It is a powerful surfactant that in high concentrations is the major ingredient in many industrial cleaners and at low concentrations is found in toothpaste, shampoo, shaving cream, bubble bath formulations, and even laxatives.

B. MEMBRANE PROTEIN ISOLATION

Methodologies for the isolation and purification of integral membrane proteins have always lagged far behind those for water-soluble cytosolic proteins. A water-soluble protein can often be precipitated by ammonium sulfate (salting-out) in relatively pure form. In sharp contrast, integral membrane proteins must first be extracted from the membrane and come attached to membrane lipids, making them water-insoluble goos that defy purification.

An early attempt to isolate membrane proteins involved what is known as 'acetone powders'. Tissues are ground up and the aqueous fraction removed by filtration or centrifugation. The non-soluble fraction is then mixed with a large excess of cold acetone that dissolves the membrane lipids. The solution is then filtered to remove the insoluble proteins. The filtered protein powder is washed several times with cold acetone and dried producing the 'acetone powder'. Unfortunately, most proteins comprising the residual protein powder are denatured and are therefore useless for almost all functional studies. This method of producing 'acetone powders' is just too harsh for delicate biological membrane material. However, acetone powers have been successfully employed in some cytoskeletal protein isolations [6].

A gentler method to isolate membrane proteins is by use of chaotropic agents [7]. This method has also been shown to have unwanted complications, thus limiting its application. Chaotropic agents act by disrupting inter-molecular interactions including hydrogen bonding, van der Waals forces, and hydrophobic effects that are responsible at the molecular level for most biological structure. Chaotropic agents include urea, guanidinium chloride, lithium perchlorate, Na thiocyanate, NaBr and NaI. When dissolved in water at very high levels (~4−8 M) chaotropic agents cause membranes to fall apart by eliminating the stabilizing hydrophobic effect driven by water structure (Chapter 3). Unfortunately, chaotropic agents can also denature proteins, and the enormous concentrations employed makes their complete removal difficult, limiting their usefulness.

The methods discussed in Chapter 12 for isolation and purification of membrane fractions are also applicable for the isolation and purification of membrane integral proteins. Many of the protein isolation methods are based on centrifugation and affinity chromatography. The major difference is that extraction of an integral protein from its membrane in non-denatured form generally involves the use of mild detergents. Obtaining a non-contaminated membrane fraction in high yield greatly enhances the chance of subsequently obtaining a particular protein in pure form. For example, if the objective is to extract and purify cytochrome c oxidase, a first step will involve obtaining an intact mitochondria by simple differential centrifugation.

Moderate concentrations of mild detergents are used to compromise membrane integrity and facilitate the extraction of integral proteins in water-soluble, non-denatured form. The extracted proteins are found in mixed micelles composed of the protein, membrane phospholipids, and the detergent. The effect of detergent levels on membranes is concentration-dependent.

Detergent/Lipid ratio	Effect on membrane
0.1:1 to 1:1	Bilayer remains intact. Some proteins extracted.
1:1 to 2:1	Bilayer solubilized into mixed micelles.
10:1	All lipid—protein interactions are replaced by detergent—protein interactions.

The membrane solubilization buffer should contain enough detergent to provide greater than one micelle per membrane protein in order to assure that individual proteins are isolated in separate micelles. The required amount of detergent will depend on the detergent's physical properties (CMC, HLB, aggregation number) as well as the aqueous buffer pH, salt concentration, and temperature.

The complex solution containing many membrane proteins (mostly located in mixed micelles), can then be passed over a lectin or antibody affinity column (discussed in Chapter 12). Only the selected protein binds to the column and the remaining, un-bound proteins are discarded. The bound protein can then be removed by addition of excess free sugar for a lectin affinity column or by altering the pH or salt concentration for an antibody affinity column.

If the desired protein is hard to isolate or is present in low amounts, recombinant DNA technology may be employed. The protein may be expressed in large quantities in another cell type. Also the newly expressed protein can be tagged by something, often a poly-histidine tag (His-tag) that can greatly facilitate the protein's isolation.

The His-tag technique usually involves engineering a sequence of 6—8 histidines to either the C- or N-terminal of the protein [8,9]. The poly-histidine tag binds strongly to divalent metal ions like nickel or cobalt. Therefore, if a complex mixture of proteins is passed through a column containing immobilized nickel, only the His-tag protein binds and all other proteins can be discarded. The His-tag protein can be eluted from the column by an imidazole-containing solution. Imidazole, the side chain of histidine, competes for the poly-histidine binding site. The His-tag protein can also be released from the nickel column by lowering the pH to ~4.5. It should be mentioned that any protein with a high affinity for divalent metals can be separated by an immobilized-nickel column without the need of a special His-tag.

Another approach to purification of a protein through use of an engineered tag involves addition of an antigen peptide tag. Separation is achieved using an appropriate antibody-attached resin column. This methodology is highly specific to a single protein. After the desired protein is isolated, the engineered tags can be easily removed by a protease.

After the desired protein is obtained in pure form, it is usually so dilute it must be substantially concentrated for subsequent reconstitution studies. The most commonly used concentration technique is lyphilization or freeze-drying. The sample is frozen and the water content reduced by sublimation under vacuum. Another procedure increases the protein concentration by ultrafiltration. Selectively permeable membranes allow water and small molecules to cross the membrane while retaining the much larger protein. The solution is forced through the filter by mechanical or gas pressure or by centrifugal force in a specially designed centrifuge tube.

At each step in the multi-step protein purification process, both the total amount of all proteins present and the activity of the desired protein must be determined. Although the total protein and total activity of the desired protein will decrease after each step, the specific activity will increase if any purification is achieved (Figure 13.4). Enzyme activity is equal to the moles of substrate converted to product per unit of time. The standard unit of activity is the katal where:

$$1 \text{ katal} = 1 \text{ mol converted/sec}$$

However a katal is such a large unit, specific activity is more usually expressed as:

$$1 \text{ enzyme unit } (U) = 1 \ 10^{-6} \text{ mol substrate converted/min}$$

Therefore, 1 U corresponds to 16.67×10^{-9} katals.

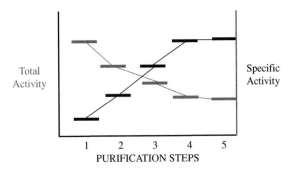

FIGURE 13.4 With each step in the purification process, the total enzyme activity decreases while specific activity increases until the enzyme is pure.

FIGURE 13.5   Oliver Lowery (r) (1910–1996). *Courtesy of the National Library of Medicine*

FIGURE 13.6   J. Folch-Pi (1911–1979). *Courtesy of Global Contents*

The specific activity (SA) is the activity of an enzyme per mg of total protein.

$$SA = 10^{-6} \text{ mol converted mg}^{-1} \text{min}^{-1}$$

The SA of a pure enzyme is constant and is a characteristic of the enzyme. Therefore, the purity of an enzyme at any step in the purification, expressed as % purity, can be determined by:

$$\% \text{ Purity} = \text{measured SA/SA of pure enzyme} \times 100$$

Until purity is achieved, the measured SA will always be less than that of the pure enzyme because other proteins besides the desired enzyme are still present in the solution.

In addition to enzyme activity, total protein must be determined at each purification step. Many methods to accurately quantify total protein are currently in use. The simplest is to measure solution absorbance at 280 nm. This method is not sensitive and suffers from many non-protein compounds in the purification buffers that can also absorb at 280 nm. A revolutionary colorimetric procedure to quantify proteins was published by Oliver Lowery (Figure 13.5) in 1951 [10]. For decades this paper was the most cited in all scientific literature. A more sensitive method to quantify proteins was described by Bradford in 1976 [11] and further improved by Zor and Selinger in 1996 [12]. A Bradford Protein Assay kit, commercially available from Bio-Rad, is perhaps the most popular protein assay in use today. Another very popular protein assay is the Pierce BCA (bicinchoninic) Protein Reagent Assay that is reputed to be the best detergent-compatible protein assay available.

C. MEMBRANE LIPID ISOLATION

During the early years of biochemistry, knowledge of lipids far exceeded that of proteins. Isolation of the first protein was not until 1925 when Sumner reported his work on urease from jack bean [13]. In comparison, the early pioneering lipid extractions by Chevreul were done in the early 19th century [14]. Modern membrane polar lipid extraction procedures had their beginnings with the classic 1957 paper by J. Folch (Figure 13.6) (Folch, Lees, and Stanley [15]). With slight modifications, this method is still in use today. Lipid extraction requires that the tissue must be fresh and devoid of lysophospholipids, oxidized products, large quantities of mono- or di-acyl glycerol, and free fatty acids. The tissue is homogenized with chloroform/methanol (2:1) at a ratio of 1 g tissue to 20 ml solvent. The tissue naturally contains a small amount of water that adds to the mixture but often must be further augmented to improve phase separation. After filtration or centrifugation to remove solid cellular debris, the liquid mixture is removed and washed with 0.2 vol of 0.9% NaCl in water. Low speed centrifugation separates the mixture into two phases. The upper phase, primarily methanol and water, is enriched in very polar lipids like gangliosides, while the chloroform-rich lower phase contains the phospholipids and non-polar lipids. This phase is isolated and evaporated under vacuum and further extracted with cold acetone. The acetone extraction removes the non-polar lipids leaving the membrane phospholipids in the acetone-insoluble pellet. Recently, dichloromethane has occasionally been used in place of chloroform to avoid adverse health effects. Another very similar procedure involving extraction with chloroform/methanol/water was later proposed by Bligh and Dyer in 1959 [16].

After extraction by the classic Folch method, lipids are distributed between 3 fractions: the methanol/water upper phase containing gangliosides and other very polar lipids; the lower chloroform phase containing the membrane phospholipids; and the acetone-extracted phase containing non-polar lipids like triacylglycerols and cholesterol. Each of these fractions can be dried and subjected to additional fractionation. Although there are a variety of methodologies used to separate membrane lipid fractions, the two most common are thin layer chromatography (TLC) and high-performance liquid chromatography (HPLC) [17].

TLC

TLC was discovered by Ukranian scientists N.A. Izmailov and M.S. Shraiber in 1938 [18]. At first, TLC was difficult to use since the TLC plates all had to be home made. Now, a wide variety of pre-made TLC plates are commercially available. TLC is so fast, simple, and inexpensive it is routinely employed and is considered to be 'a poor man's HPLC' (discussed below). TLC plates consist of a supporting sheet of glass, aluminum or plastic, to which is adhered a thin coating of a solid adsorbent, usually silica or alumina. Aluminum or plastic plates have the advantage of being easily cut to a desired shape by a pair of scissors. A small amount of a volatile solvent containing the dissolved membrane lipid mixture is spotted near the bottom of the plate [19,20]. The spot is dried and the plate then placed inside a sealed developing chamber containing a shallow pool of a developing solvent known as the eluent. The eluent constitutes the mobile phase and percolates up through the stationary phase (the adsorbant that is used for lipid separations is usually silica), by capillary action. The spotted lipids are bound to a different extent to the stationary silica depending on their physical properties of polarity and hydrophobicity. As the eluent passes through the bound lipids, an equilibrium is established for each lipid in the mixture between the stationary silica and the mobile liquid phase. The lipids that are most soluble in the mobile phase will spend more time in that phase and move farther up the TLC plate. When the mobile solvent front approaches the end of the plate, the plate is removed from the developing chamber and dried. The separated lipids are then visualized by a number of agents that can be sprayed on the plate [21]. Table 13.2 lists some of the most commonly used sprays to visualize lipids on TLC plates. The relative distance the lipid travels from the original spot (the origin) is quantified by its R_f value (retention factor). R_f is defined as the distance the lipid travels up the plate divided by the total distance the eluent travels (from the origin to the front). Determination of R_f values is depicted in Figure 13.7 [20].

Lipids can be separated by 1-dimensional or 2-dimensional TLC as shown in Figure 13.8 [22]. Sequential 1-dimensional TLC is shown in the left panel. The polar lipids were first separated by two successive runs with chloroform-methanol-water (60:30:5 by volume). In the third run, the non-polar lipids were separated using hexane-diethyl ether-acetic acid (80:20:1.5 by volume). In the first two runs the non-polar lipids ran at the front and were not separated. In the third run, the polar lipids did not migrate, but the non-polar lipids did, resulting in their separation. The right panel of Figure 13.8 demonstrates a 2-dimensional TLC lipid separation. In the first direction, the eluent was chloroform/methanol/concentrated ammonia (65/35/4, by volume). This was followed by a second eluent, 2, butanol/acetic acid/water (60/20/20, by volume) run at right angles to the first. Better

TABLE 13.2 Visualization Methods For Identifying Lipids on TLC Plates [21].

Reagent	Reacts with	Lipid Identified
Ultraviolet Light	non-specific	all lipids
H_2SO_4 heat	carbon charring	all lipids
Iodine	double bonds	lipids with double bonds
Primuline	non-specific	all lipids
Rhodamine 6G	non-specific	all lipids
Molybdate	phosphate	all phospholipids
Malachite green	phosphate	all phospholipids
Ninhydrin	primary amines	PE, PS
Dragendorf Reagent	choline	PC
2,4-dinitrophenylhydrazine		plasmalogens
Resorcinol	sugars	gangliosides
Ferric Chloride		cholesterol

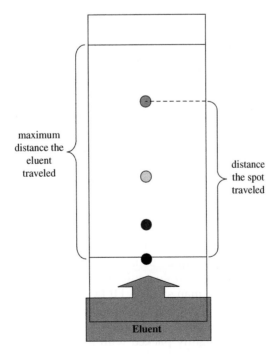

FIGURE 13.7 Determination of R_f values [20]. R_f is the distance the spot travels divided by the maximum distance the eluent travels. *Courtesy of Instructables*

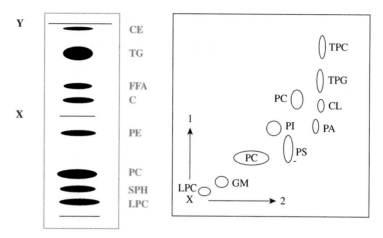

FIGURE 13.8   Lipid separation by one dimensional (left panel) and two dimensional (right panel) TLC [22]. Left Panel: One dimensional TLC with three successive solvent systems. The first two solvents (up to X, chloroform-methanol-water (60:30:5 by volume)) are used to separate the polar lipids, while the third system (up to Y, hexane-diethyl ether-acetic acid (80:20:1.5 by volume)) separates the non-polar lipids. Abbreviations: CE, cholesterol esters; TG, triacylglycerols; FFA, free fatty acids; C, cholesterol; PE, phosphatidylethanolamine; PC, phosphatidylcholine; SPH, sphingomyelin; LPC, lysophosphatidylcholine. Right Panel: Two dimensional TLC. First dimension (1, chloroform/methanol/conc ammonia (65/35/4, by volume). Second dimension (2, butanol/acetic acid/water (60/20/20, by volume). Abbreviations: PI: phosphatidylinositol, PS: phosphatidylserine, SM: sphingomyelin, PC: phosphatidylcholine, PE: phosphatidylethanolamine, PDME: dimethyl phosphatidylethanolamine, PMME: monomethyl phosphatidylethanolamine.

separation is normally achieved by 2-dimensional TLC. Lipid spots on the TLC plates can be directly quantified by scanning densitometry.

Several manufacturers (e.g. Camag, Shimadzu, X-Rite, Tobias etc.) offer computer-controlled densitometers. Digital camera pictures can also be used to quantify the spots on TLC plates. In addition, the lipid components can be removed from the TLC sheet using an eluent solvent that results in a high R_f for the lipid (i.e. the lipid is very soluble in the eluent). The isolated lipid can then be subjected to further analysis like determining the fatty acid content of a phospholipid (see below). Since the number of possible eluent solutions is almost infinite, it is theoretically possible to separate almost any lipids by TLC.

HPLC

In recent years, HPLC has supplanted TLC as the method of choice for membrane lipid separations [23–25]. The basic principles behind HPLC are essentially the same as TLC. While TLC does not involve any elaborate equipment, HPLC uses expensive instrumentation. Therefore HPLC is the 'rich man's' TLC. HPLC uses a variety of stationary phases and mobile phases that are selected to identify and separate the desired analyte (a substance undergoing analysis). The mobile phase is pumped at extremely high pressure (up to 400 atmospheres) through a column tightly packed with the stationary phase. Also required is a sensitive detector. The analytes dissolved in the mobile phase are injected into the HPLC column at time zero. The time required for an analyte to produce its maximum peak as

TABLE 13.3   Relative Retention Times in RP-HPLC of Various Phospholipids in The Most Commonly Used Mobile Phases Acetonitrile/Water and Propan-2-ol/water. Retention Times Increase from Top to Bottom.

Acetonitrile/water	Propan-2-ol/water
PA	CL
CL	PE
PI	PI
PS	PS
PE	PA
PC	PC
SM	SM
Lyso PC	Lyso PC

indicated by the detector is referred to as the retention time and is characteristic of the analyte under conditions of the experiment (e.g. stationary and mobile phase, column length, pressure, and temperature). Initially the HPLC stationary phase was silica gel and the separation process resembled that described above for TLC. The more polar analytes bind better to the silica gel while the more non-polar analytes partition more readily into the non-polar mobile phase and so reach the detector quicker and have a shorter retention time.

While silica gel HPLC is termed 'normal phase' HPLC, it is not the most commonly used procedure. Reversed phase HPLC (RP-HPLC or RPC) is by far the most popular type of HPLC used in biological membrane studies [25–27]. RP-HPLC employs a non-polar stationary phase and a polar mobile phase that contains water. The RP-HPLC stationary phase is often silica gel that has been modified by RMe_2SiCl where R is either a C-8 or C-18 saturated hydrocarbon chain. Therefore the RP-HPLC stationary phase is polar silica with an oily, hydrophobic coat. Opposite to 'normal phase' HPLC, with RP-HPLC retention time is longer for non-polar analytes and can be further increased by adding more water to the mobile phase and decreased by adding more organic solute (e.g. methanol, acetonitrile). Analytes with a large number of non-polar C-H or C-C bonds have longer retention times while those with polar −OH, -NH$_2$ or -COO$^-$ groups have short retention times. Similarly, retention times decrease with double bond content (C-C single bond > C=C double bond). Table 13.3 presents relative retention times in RP-HPLC of various phospholipids in the most commonly used mobile phases acetonitrile/water and propan-2-ol/water.

Fatty Acid Analysis

As discussed above, the various membrane lipid and phospholipid classes can be isolated, primarily by TLC and RP-HPLC. Most membrane lipids are classified as 'complex lipids' (see Chapter 4) that contain esterified fatty acids. Often the nature of these fatty acids constitutes essential information as they comprise the bulk of a membrane's hydrophobic interior. The thin oily center of a lipid bilayer is responsible for basic membrane properties and stability. It constitutes the diffusion barrier and provides the proper environment for membrane protein activity (see Chapter 6). Determining membrane fatty acid composition is therefore an essential component of membrane studies and their analysis is usually achieved by gas chromatography (GC) or more properly gas-liquid chromatography (GLC).

Gas chromatography (GC) is used to separate and identify compounds that can be vaporized without decomposing [28]. GC was first described by the Russian scientist Mikhail Semenovich Tswett in 1903 and was fully developed as gas-liquid chromatography in 1950 by the 1952 Nobel Laureate A.J.P. Martin (Figure 13.9), the pioneer of liquid—liquid and paper chromatography. The basic principle for GC is analogous to that of TLC and HPLC with a few very important differences. In GC the mobile phase is a carrier gas and the analytes partition between an immobile liquid stationary phase and a mobile gas phase. This accounts for the more correct term gas-liquid chromatography. The carrier gas is usually an inert gas like helium or an unreactive gas like nitrogen. The stationary phase is a thin coating of various materials lining the inside of a long thin tube referred to as a column. The most common stationary phase used in fatty acid analysis is Carbowax (polyethylene

FIGURE 13.9   A.J.P. Martin (1910—2002). *Courtesy of International Autograph Auctions Ltd.*

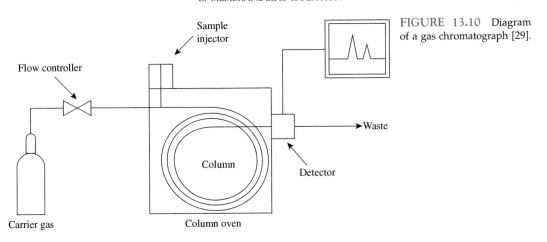

FIGURE 13.10 Diagram of a gas chromatograph [29].

glycol, $H\text{-}(OCH_2CH_2)_n\text{-}OH))$. The analytes partition to a different extent between the mobile gas phase and the stationary phase. Analytes that favor the gas phase leave the column faster and so have shorter retention times. The column is located in an oven to assure accurate, programmable temperature control. It is an elevated temperature that volatizes the fatty acid methyl esters and makes GC feasible. The fatty acids must be methylated to make them volatile. The final components of a functional GC, depicted in Figure 13.10 [29], are an injection port and a detector. There are many types of detectors used in GC, the most common being the flame ionization detector (FID) and the thermal conductivity detector (TCD).

Free fatty acids, as well as fatty acids esterified to other biochemicals (e.g. acyl glycerols, cholesterol esters, phospholipids), are not volatile and if heated to a high enough temperature would disintegrate. However, methyl esters of fatty acids are volatile and can be used in GC analysis [30]. Preparation of fatty acyl methyl esters is perhaps the most commonly used reaction in lipid analysis. Fatty acid methyl esters can be made from free fatty acids that have been hydrolyzed (saponified) from their biological ester source (see Chapter 5) [31]. However, it is not necessary to actually obtain free fatty acids *before* methylation. Fatty acid methyl esters can be directly obtained from biological esters by a process known as trans-esterification [32]. A method that can trans-esterify can also directly methylate free fatty acids. Methylation can be achieved in either acidic (by methanolic hydrogen chloride or boron trifluoride in methanol) or basic (sodium methoxide in methanol) anhydrous conditions. An emphasis must be placed on *anhydrous* as even traces of water can be detrimental to esterification. The methoxide method is extremely fast (complete in a few minutes) and is accomplished at room temperature [33].

Figure 13.11 shows a fatty acid profile of liver phospholipids determined by GC [34]. Each fatty acid is identified by three numbers: number of carbons, number of double bonds, and the series (omega-family). Elution time is on the abscissa and the detector intensity on the ordinate. The area under each curve represents the relative amount of each fatty acid in the sample. A fatty acid methyl ester database has been established [35]. It contains retention times and mass spectral information on ~100 fatty acids.

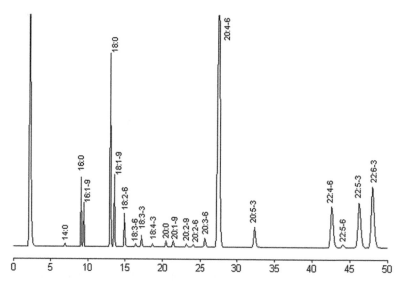

FIGURE 13.11   Fatty acid profile of liver phospholipids determined by GC on a Carbowax stationary phase column (Alltech) [34]. The mobile gas phase was helium and the temperature profile was 100°C for 2 min, 195°C for 33 min and 220°C up to the end. Temperature ramp was 40°C/min. Each fatty acid is identified by three numbers; number of carbons, number of double bonds, and the series (omega-family). Elution time is on the abscissa and the detector intensity on the ordinate. The area under each curve represents the relative amount of each fatty acid.

D. MODEL LIPID BILAYER MEMBRANES — LIPOSOMES

In previous sections of this chapter the importance of detergents in membrane studies was discussed and how membrane integral proteins and polar lipids (including phospholipids, gangliosides, and cholesterol) can be isolated. The next step involves reconstituting a functional membrane using detergents, polar lipids, and integral proteins. But first, model, protein-free, stable lipid bilayer vesicles, commonly known as liposomes, will be described.

Model Membranes

Biological membranes are so complex that they often defy comprehensive analysis. For this reason a variety of relatively simple model membrane systems have been developed. The simplest model consists of membrane lipids dissolved in organic solution. Obviously this model is extremely limited as it has no three dimensional structure that in any way resembles a membrane. Its primary application is in following very short distance molecular interactions, mainly lipid–lipid associations. The oldest model membrane is the lipid monolayer, discussed in Chapter 2. Lipid monolayers have their origins in a very distant, murky past. Although monolayers are only one half of a bilayer, they formed the foundation for modern membrane biophysical studies. The first stable lipid bilayer was reported by Mueller et al. in 1962 [36]. This model membrane, introduced in Chapter 9, was made by spreading an organic solution of membrane polar lipids across a small (~1mm) hole in a teflon partition separating two aqueous chambers (Figure 13.12) [37]. Initially a thick glob of lipid in organic

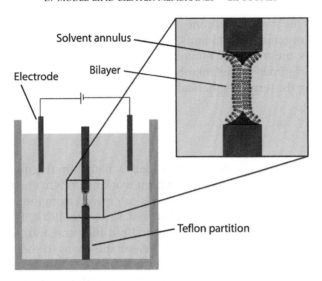

FIGURE 13.12 The planar bimolecular lipid membrane is also referred to as a black lipid membrane or BLM. It is formed by spreading an organic solution of membrane polar lipids across a small hole in a Teflon partition separating two aqueous chambers [37].

FIGURE 13.13 Alec D. Bangham (1921–2010). *Reprinted with permission* [80].

solution occupied the hole. This rapidly thinned as the organic solvent was squeezed out leaving a thin membrane that occupied most of the hole. Electrical and 90° reflected light interference measurements proved this suspended membrane was indeed a lipid bilayer. It was therefore called a bimolecular lipid membrane (or black lipid membrane, due to total destructive interference when viewed at 90° from incident light) or simply BLM. Although a BLM is well suited for trans-membrane electrical measurements, it is very prone to vibrations and even slight organic contaminants causes its demise. Also, proteins often denature at BLM-water interfaces and most trans-membrane solute flux measurements are not possible. A more realistic and rigorous model membrane that could reconstitute integral proteins was required. This came in the form of spherical lipid bilayer vesicles, more commonly referred to as 'liposomes'.

Liposomes

The discovery of liposomes is attributed to Alec D. Bangham (Figure 13.13) in 1961 (published in 1964, [38]). For years, Bangham had been working on what he referred to as 'multilamellar smectic mesophases'. The word 'smectic' is derived from the Greek word for soap. Although smectic mesophases had been observed since the 1930s, it was Bangham who recognized them for what they were, concentric lipid bilayers, and so initially they were called Bangosomes. A few years later, Gerald Weissmann changed their name to liposomes, and it stuck. Alec Bangham is considered to be the 'Father of Liposomes'. It is amazing how easy it is to prepare multilamellar liposomes. Phospholipids are dissolved in a volatile organic solvent and transferred to a round-bottom flask. The solvent is evaporated using nitrogen or argon gas to prevent lipid oxidation. As the flask is rotated during the evaporation process, a thin layer of phospholipid coats the flask's inside wall. The flask is then placed under a vacuum overnight to eliminate traces of the solvent. An aqueous buffer at a temperature above the phospholipid's melting point (T_m, Chapter 5) is added to the flask with a few glass beads to aid in dispersion. As the solution is vortexed, it immediately becomes milky, indicating the formation of multilamellar liposomes. An electron micrograph of a family of multilamellar vesicles (MLVs) is shown in Figure 13.14. Depending on the type of

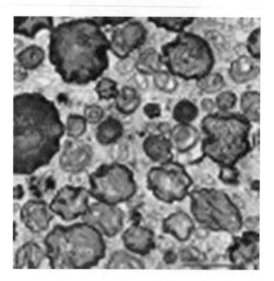

FIGURE 13.14   Electron micrograph of a family of multilamellar vesicles (MLVs) made by the method of Bangham [38].

phospholipid, the process is complete in approximately one to two minutes. The process is much slower with poorly hydrated lipids such as PE and with phospholipids containing long saturated chains with high T_ms. Liposome versatility makes them important for many types of scientific investigations. Liposomes can sequester water-soluble solutes in their aqueous interiors and house membrane integral proteins in their bilayers. Since liposomes can also be made from non-immunogenic membrane lipids, they have obvious potential as drug delivery vehicles (see Chapter 15), and indeed thousands of examples of such vehicles can be found in the medical literature [39,40]. Besides the very simple method for making multilamellar liposomes described above, there are a wide variety of alternate methodologies used in the production of other types of liposomes. Before we proceed to some of these methodologies, a brief description of size exclusion chromatography (also known for life science applications as gel filtration chromatography) is appropriate.

One important use for liposomes is in measuring trans-membrane solute diffusion. Since lipid bilayers are poorly permeable to most solutes, solute gradients can be readily established across the liposomal membranes. One important way to create solute gradients is by use of size exclusion (gel filtration) chromatography, a process that allows rapid separation of big from small particles (Figure 13.15). The technique was first described in the mid 1950s by Lathe and Ruthven [41,42] and requires at least a 10% difference in size between the two particles. This criterion is easily met when very large liposomes are separated from very small solutes [43]. The most commonly used stationary phase for gel filtration is the resin Sephadex™. Sephadex is a trademark for cross-linked dextran gels made by Pharmacia since 1959. Sephadex comes in several discrete bead sizes (Sephadex G-10, G-15, G-25, G-50, G-75, and G-100) with properties altered by the extent of cross-linking. The stationary phase is

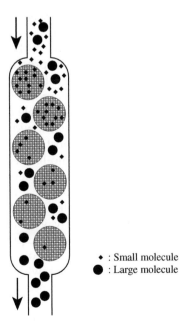

♦ : Small molecule
● : Large molecule

FIGURE 13.15   Separation of small from big particles using gel filtration (size exclusion) chromatography [81].

riddled with tiny pores that are large enough to be penetrated by a solute, but small enough to exclude a liposome. As a result, as the solute/liposome mixture passes down the gel filtration column, the solute enters the pores and so travels a longer, more convoluted path to reach the end of the column than does the large, excluded liposomes. Liposomes leave the column before the non-liposome-sequestered solute. Of course, solutes initially sequestered inside the liposomes pass through the gel and exit the column with the liposome fraction. The solute 'sees' ~80% of the total column volume, while the liposome only 'sees' ~35% of the volume. As an example, liposomes could be made in a 100 mM KCl buffer and passed through the gel filtration column using a large volume of a 100 mM NaCl buffer. The liposomes exiting the column would have an interior concentration of 100 mM KCl and 0 mM NaCl, while the extra-liposomal solution would be 0 mM KCl and 100 mM NaCl. The established trans-membrane gradients of K^+ and Na^+ can be made to parallel those of living cells.

Another method that can accomplish many of the same objectives as gel filtration chromatography is filtration using dialysis tubing. Like gel filtration, dialysis tubing can be used to separate large from small particles. Dialysis tubing is a thin semi-permeable sheet made of cellulose or cellophane (a form of cellulose). The sheet has microscopic holes that allow small solutes to pass through while retaining large particles. Dialysis tubing can be obtained of different overall size and with different cut-off pore diameter. For example, liposomes are made in a buffer containing a solute (A) and sealed in a dialysis tube. The tube is floated in a large excess of buffer devoid of (A), but containing solute (B) at the same osmotic strength. During dialysis (A) goes down its concentration gradient through the dialysis tube pores into the bathing solution where it is readily washed away. Liposomes containing (A) cannot squeeze through the pores and can then be removed from the dialysis tube for further experimentation. Similar to the example described for gel filtration, liposomes made in 100 mM KCl can be sealed in a dialysis tube and floated in a large excess of the same buffer but containing 100 mM NaCl in place of 100 mM KCl. Ideally, upon completion of dialysis, the tube contains liposomes with 100 mM KCl sequestered inside and 100 mM NaCl outside. Although gel filtration and dialysis theoretically result in the same final product, dialysis is many, many times slower. While complete dialysis may take days, gel filtration is complete in minutes. During extended dialyses, lipids can oxidize and hydrolyze, producing products that are deleterious to lipid bilayer structure and, in addition, the entire sample is ripe for bacterial growth.

Measurement of leakage rates from liposomes has a wide variety of applications and as a result many methods are available. One of the most common and sensitive methods uses ^{14}C-glucose (or any other small, water-soluble, radio-labeled molecule) as the solute. After the initial dialysis that produces the ^{14}C-glucose sequestered liposomes, the final product is then sealed into a fresh dialysis tube and placed in a new bathing solution devoid of ^{14}C-glucose. As ^{14}C-glucose leaks out of the liposomes, it can be detected at low levels outside the dialysis bag by a scintillation counter. Similarly, gel filtration can be used to measure ^{14}C-glucose leakage. Aliquots of the final ^{14}C-glucose-sequestered liposomes can be taken at different times and run through a small gel filtration column. The ^{14}C-glucose still sequestered in the liposomes comes off the column first while the leaked (non-sequestered) ^{14}C-glucose elutes in a later fraction. The ratio of non-sequestered/sequestered ^{14}C-glucose as a function of time represents the leakage rate.

Gel filtration and dialysis account for the production of most solute-sequestered liposomes. However, in some specific examples the second dialysis or gel filtration steps used to monitor solute leakage can be avoided. For example, carboxyfluorescein (CF^-) leakage can be directly monitored by fluorescence in a cuvette [44]. This technique works because at very high concentration (~160 mM) CF^- is self-quenching and appears as dark brown, but at low concentrations fluoresces a brilliant yellow. CF^- is sequestered inside the liposomes at 160 mM where it appears brown and starts to fluoresce yellow as the solute leaks out, where it first appears at low concentration in the external solution. Therefore liposome leakage is directly proportional to CF^- fluorescence intensity.

In another example, glucose leakage is followed by a linked series of enzymatic reactions in a cuvette followed by light absorbance at 340 nm [45]. Liposomes are made with sequestered glucose. The bathing solution contains ATP, Mg^{2+}, NADP, and the enzymes hexokinase and glucose-6-phosphate dehydrogenase. As glucose leaks out of the liposomes, the following reactions occur in the bathing solution:

$$Glucose + ATP + Mg^{2+} \xrightarrow{(Hexokinase)} glucose\text{-}6\text{-}phosphate + ADP + Mg^{2+}$$

$$Glucose\text{-}6\text{-}Phosphate + NADP^+ \xrightarrow{(Glucose\text{-}6\text{-}Phosphate\ Dehydrogenase)} 6\text{-}Phosphoglucanate + NADPH$$

NADPH absorbs at 340 nm while $NADP^+$ does not. Therefore glucose leakage from the liposomes is proportional to an increase in absorbance of NADPH at 340 nm. Since no samples are taken for analysis, glucose permeability can be continuously monitored.

Unilamellar Vesicles

The original liposome preparation method of Bangham produces multilamellar vesicles (MLVs) of varying size [38]. MLVs have several important shortcomings that limit their use. They are not of uniform size and a large portion of their total lipid is sequestered inside the onion-like vesicles. This severely limits the internal aqueous space available to solutes and makes it impossible to accurately assess the amount of lipid comprising the external membrane surface. In addition, evaluating the net leakage rate is very complex as solutes must traverse several bilayers before making their way through to the external bathing solution. Therefore a different type of liposome, one having only a single surrounding bilayer and of uniform size, was required. This type of liposome is generally referred to as a unilamellar vesicle (ULV) in contrast to a multilamellar vesicle (MLV). ULVs are further subdivided by size. The following list covers the most common types of liposomes:

Abbreviation	Name
MLV	Multilamellar Vesicle
ULV	Unilamellar Vesicle
SUV	Small Unilamellar Vesicle
LUV	Large Unilamellar Vesicle
GUV	Giant Unilamellar Vesicle

SUV by Sonication

There are three basic ways to make SUVs: by sonication, by extrusion, and by use of detergents. The sonication methods discussed in Chapter 12 for cell homogenization can also be used to convert MLVs into SUVs. High energy, tip-type sonicators are preferred. MLVs are a mixed population of large (0.1 to 5 μm), partially collapsed onion-like structures that in solution have an opaque, milky appearance. Upon sonication a striking change in the solution appearance occurs. The MLVs are broken down into SUVs that appear translucent with a slight bluish sheen. The transition indicates that the sonication is complete. Sonicated SUVs are ~200 Å in diameter. After sonication, the vesicles are single walled with two-thirds of the lipids in the outer bilayer leaflet and one-third comprising the inner leaflet. These vesicles cannot get any smaller and continued sonication will only lead to lipid oxidation and hydrolysis. Sonication is a vigorous procedure, producing local heating that must be dissipated by sonicating on ice with an on/off duty cycle. Sonication will also generate unwanted titanium particles and will denature any proteins that may have been reconstituted into the starting MLVs. The very tight radius of curvature also makes small sonicated vesicles highly susceptible to fusion, thus altering their size over time. Extensively fused sonicated vesicles can then be used as a starting population to make larger single lamellar vesicles (LUVs).

Vesicles by Extrusion

By an imprecise definition, SUVs range between 200–500 Å in diameter. Unilamellar vesicles larger than this are either defined as LUVs or, if they are very large, GUVs. SUVs and perhaps some smaller LUVs can be made by extruding MLVs several times through a series of Nuclepore™ filters. Nuclepore filters are a plastic (polycarbonate) filter that is perforated with perfectly round, uniform-sized holes. The holes are made by exposing the plastic filter disc to radiation by a process called Nuclear Particle Track Etching. The round holes can be made of almost any size. Nuclepore filters can be obtained with holes of 10, 8, 5, 3, 2, 1, 0.8, 0.6, 0.4, 0.2, 0.1, 0.05, and 0.015 μm. MLVs are forced through these filters under pressure, normally 10–20 times for each chosen pore size. Pressure is provided through devices known as extruders. Two types of extruders are available. With large pore filters it is easy to force MLVs through the filter and a manual dual syringe-type extruder will suffice (e.g. Avanti Mini-Extruder). In order to force the MLVs through small pore filters, very high pressures are required, far more than can be obtained by a manual syringe extruder. High pressure extruders (Figure 13.16, Lipex Extruder) force the MLVs through the Nuclepore filters under extremely high nitrogen pressure where a special gas regulator is even required. Usually the MLVs are first passed through a larger pore filter, then a medium pore filter, and finally a small pore filter. Approximately 20 passes are run for each filter. The final product should be a uniform-sized population of unilamellar vesicles. Unfortunately, unless the vesicles are very small, they may be of uniform size and much smaller than the parent MLVs, but still contain some sequestered lipid bilayers. A 2007 paper by Lapinski et al. [46] compared the basic physical properties of LUVs made by sonication versus extrusion.

FIGURE 13.16   High pressure extruder (Lipex Extruder). *Courtesy of ATS Engineering Inc*

Vesicles by Detergent Dilution

Detergents are essential for membrane studies. They are amphipathic molecules that in some ways resemble membrane polar lipids but are orders of magnitude more water soluble. If added to membranes at high enough concentration, they can replace the resident membrane lipids while maintaining normal protein conformation and activity. Detergents make functional protein isolation and reconstitution into proteoliposomes possible, but also provide another path to unilamellar vesicle construction. If membrane polar lipids are hydrated in a buffer containing a large excess of detergent, they become solubilized into detergent micelles. The water-soluble detergents can then be easily removed by dialysis or by gel filtration. When the detergent/membrane lipid mixture is sealed in a dialysis tube, the detergent diffuses down its concentration gradient, eventually crossing the dialysis tubing and entering the external bathing solution. As the detergent concentration decreases inside the dialysis tube, the membrane polar lipids fall out of solution as unilamellar lipid vesicles that are too large to fit through the dialysis tubing pores. These vesicles do not have sequestered lipids and so are unilamellar. Detergents may also be eliminated rapidly by passing the detergent/membrane lipid mixture through a gel filtration column. As the mixture passes down the column, the detergent is extracted into the small spaces in the gel matrix, while the polar membrane lipids come out of solution as unilamellar vesicles that are excluded from the gel. The newly formed unilamellar vesicles leave the column before the free detergent and so can be isolated largely devoid of detergent. Vesicle size is dependent on the detergent used and can be partially altered by changing the dialysis or gel filtration rate. Effect of detergents on vesicle size can be dramatic [47]. For example:

Detergent	Vesicle size
Sodium Cholate	SUV (small)
n-octyl (or hexyl or heptyl) β-D-glucopyranoside	LUV (medium)
POE4 (lauryl dimethylamine oxide)	GUV (very large)

Unfortunately it is very difficult to remove all of the detergent from new liposomes prepared by dialysis or gel filtration. This is particularly true for detergents with a low CMC (e.g. Triton X-100). Another detergent-removal process, extraction using hydrophobic adsorption onto non-polar polystyrene beads [48], has proven to be very effective and has become commonly used in unilamellar vesicle preparations. The best known polystyrene product is Bio Beads™ from Bio-Rad. Polystyrene beads can be used to clean up unilamellar vesicles made by any detergent process or can be used in lieu of dialysis or gel filtration. Polystyrene beads are responsible for major advances in membrane integral protein reconstitutions into proteoliposomes discussed below.

LUVs: Dilution Methods

LUVs can be made by variations of methods that are all based on a similar theme — rapid dilution of a concentrated membrane lipid solution. By these methods membrane polar lipids are dissolved in a small volume of an organic solvent. Solvents include ethanol, ether, Freon ($CHFCl_2$) or a detergent-in-water solution. The concentrated membrane lipid solution is injected into a large excess of a rapidly mixing aqueous bathing solution. The organic solvent or detergent is instantly dispersed into a large excess of the bathing solution as the water-insoluble membrane polar lipids separate into LUVs. Ethanol works well as the lipid solvent since it is infinitely miscible in water. Another method, known as 'ether vaporization', is demonstrated in Figure 13.17 [49]. While ether is totally insoluble in water, it does have a high vapor pressure and can readily be removed from water by vacuum. The membrane lipids are dissolved at high concentration in ether and slowly injected into an aqueous bath that is both warmed and under vacuum. The ether is instantly vaporized and pumped away while the membrane lipids are left behind as LUVs. The Freon method is similar to the 'ether vaporization' method. Membrane lipids are dissolved at high concentration in Freon and slowly injected into an aqueous bath maintained at 37°C. Since Freon boils at ~9°C, it is instantly vaporized leaving behind LUVs. Finally, membrane lipids, dissolved in an aqueous detergent solution, can be injected into a large aqueous bath where the water-soluble detergent is rapidly dispersed leaving LUVs. The detergent dilution method has a very important advantage over the ethanol, ether vaporization and Freon methods. Functional integral proteins, isolated in detergent/membrane lipid micelles, can be incorporated into LUVs as they form, a process known as 'membrane reconstitution'.

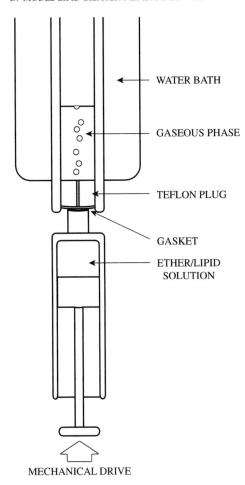

WATER BATH

GASEOUS PHASE

TEFLON PLUG

GASKET

ETHER/LIPID
SOLUTION

MECHANICAL DRIVE

FIGURE 13.17   Ether vaporization method for production of LUVs [49].

Concentration Methods

The dilution methods for making LUVs have one major disadvantage – the LUVs are formed in very dilute solution and must be concentrated. This is routinely done using two types of filters, the best known being Amicon concentrators (Figure 13.18). For both systems pressure must be applied to the diluted LUV sample to force the buffer through small holes crossing the filter. LUVs are too large to traverse the pores and so remain behind at ever increasing concentration. In one type of concentrator high pressure is applied by nitrogen gas from the top (Figure 13.18, Left) while the second type of concentrator relies on centrifugation to drive the buffer through the filter pores (Figure 13.18, Right).

Using dilution methodologies, liposome size, and hence sequestered volume, can be altered by changing the injection rates. This becomes very important if the LUVs are to be

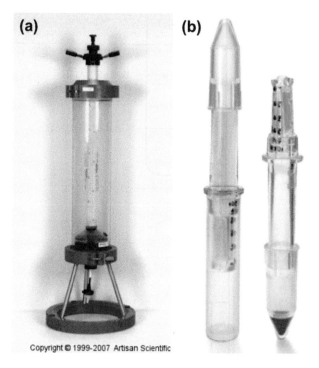

FIGURE 13.18   Two types of Amicon concentration filter devices. (a) high pressure is applied by nitrogen gas from the top. *Courtesy of Artisan Scientific Corporation.* (b) centrifugation drives the buffer through the filter pores. *Courtesy of EMD Millipore Corporation*

used for drug delivery. The sequestered volume can vary tremendously from one type of liposome to another. Although MLVs are large, their available interior volumes are surprisingly small, due to the sequestered multi-lipid bilayers.

Liposome type	Available sequestered volume
MLV	1.8 µg/nmol lipid
SUV (sonication)	0.8 µg/nmol lipid
LUV (ether vaporization)	14.0 µg/nmol lipid

E. MEMBRANE RECONSTITUTION

Previously in this chapter it was discussed how integral membrane proteins can be isolated into detergent micelles, while maintaining their correct (active) conformation. It was also shown that membrane polar lipids can be isolated and dissolved into detergent micelles, often using the same detergent used for protein isolation. If these two populations of detergent micelles are mixed and the detergent removed by dialysis, gel filtration or preferably by binding to non-polar polystyrene beads (Bio-Beads), proteoliposomes can be directly produced [50,51]. In addition to this method, a variety of other methods rely on adding the protein/detergent micelles to pre-made LUVs. These methods depend on first partially destabilizing the LUV bilayers by either detergent or brief sonication before adding the protein/detergent micelles. Partial bilayer disruption is necessary to prevent protein denaturation at the surface of tightly packed lipid bilayers. Several detergents have been used as destabilizing agents, primarily low levels of Triton X-100. Tip sonication must be very brief to prevent protein denaturation or may be replaced by gentle bath sonication.

Ideally, the proteoliposomes should be nearly detergent-free LUVs that contain the active integral protein spanning the membrane lipid bilayer. However, one potential problem concerns determining the protein's orientation in the membrane. Four orientations of the newly reconstituted protein are possible. Suppose in the biological membrane the transmembrane protein exists with an A-end exposed to the aqueous cell interior and a B-end exposed to the aqueous cell exterior. The possible orientations for the reconstituted proteoliposome are:

1. A-end inside, B-end outside (correct orientation).
2. B-end inside, A-end outside (reverse orientation).
3. A-end inside, B-end inside (bent orientation).
4. A-end outside, B-end outside (bent orientation).

During the process of reconstitution some orientations are more favorable than others. For example, if the protein is rigid like those with multiple span helices, the protein will not be flexible enough to exhibit the two bent conformations (3 and 4). Many plasma membrane proteins have an extensive sugar component that in a living cell will always face the external aqueous environment (the B-end in the example). If the sugar component is sufficiently large, steric hindrance in the small LUV interior will favor proteoliposomes with the sugars on the external surface (1 and 4). A sugar attachment containing sialic acid presents an additional problem — high negative charge density. Multiple negative charges require more spacing between the sugar components, favoring an external orientation. It is also possible to isolate and reconstitute a protein with a very large antibody attached at one end. This end will exclusively favor external orientations. Of course, once reconstituted, the antibody can be removed, thus restoring function.

There are now countless numbers of integral membrane proteins that have been successively reconstituted into proteoliposomes. Single protein proteoliposomes allow for close inspection of any lipid specificity and molecular mode of action that may be difficult to determine in complex, multi-component biological membranes. A few examples follow.

Ca^{2+} ATPase

Ca^{2+} ATPase is one of the most studied of all integral membrane proteins [52]. This transport protein (pump) was first discovered in 1966 and successfully reconstituted into proteoliposomes by the early 1970s [53]. The pump comes in two basic types, the plasma membrane Ca^{2+} ATPase (PMCA) and the sarcoplasmic reticulum Ca^{2+} ATPase (SERCA). Both pumps belong to the large family of enzymes known as P-Type ATPases because they form a phos-phorylated intermediate during transport. Their job is to pump Ca^{2+} from inside to outside the cell using ATP as the energy source. The pumps maintain constant, low intercellular Ca^{2+} levels that are required for proper cell signaling. In 1972 Efraim Racker [53] reported a successful reconstitution of the sarcoplasmic reticulum Ca^{2+} ATPase into soybean phos-pholipid proteoliposomes. Importantly, the reconstituted vesicles pumped Ca^{2+} at the expense of ATP and were sensitive to chlorpromazine, a known inhibitor of calcium ATPases. Two years later, Warren et al. [54] reconstituted the SERCA into proteoliposomes where > 99% of the native membrane lipids had been replaced by synthetic dioleoyl phosphatidyl-choline (DOPC). However, full expression of the capacity to accumulate Ca^{2+} into the proteo-liposomes required the presence of internal oxalate (Figure 13.19), a Ca^{2+} chelator. The explanation of these observations is that a truly functional calcium ATPase had been recon-stituted into a proteoliposome devoid of native phospholipids but comprised of a synthetic PC. However, both the right side out and inside out protein conformations were found in the proteoliposomes. ATP caused Ca^{2+} to be pumped into the vesicle interior by one conforma-tion, whereupon the opposite conformation ATPase would pump the Ca^{2+} back out, thus limiting net Ca^{2+} uptake. In the presence of internal oxalate, Ca^{2+} that was pumped into the vesicle was chelated as water-insoluble calcium oxalate (the major component of kidney stones) and so remained sequestered, resulting in a large increase in net Ca^{2+} accumulation.

Bacterial Proline Transporter

Reconstitution studies with the bacterial proline transporter into proteoliposomes demon-strate how a variety of biochemical 'membrane tricks' can be used to deduce the transporter's molecular mechanism [55,56].

PROTEOLIPOSOME PRODUCTION

The bacterial proline transporter was isolated using a 17 amino acid polyhistidine tag attached to the C-terminal and a nickel-affinity column [55]. The purified transporter was reconstituted into preformed detergent (Triton X-100) destabilized LUVs. Triton X-100 was removed by polystyrene beads. Proline transport was supported by PE and PG, but not PC and CL.

ENERGETICS

Experiments with reconstituted proteoliposomes proved the mechanism of action for the bacterial proline transporter [56]. Proline uptake was shown to be an example of active sym-port (a form of co-transport, see Chapter 14). The bacterial proline transporter was reconsti-tuted into proteoliposomes in a KCl buffer and diluted into a NaCl buffer. The proteoliposomes were therefore 'K$^+$-loaded' with a K$^+$ gradient that was high inside and low outside and a Na$^+$ gradient that was high outside and low inside. Upon addition of the K$^+$-ionophore valinomycin, K$^+$ moved down its concentration gradient generating a net negative potential in the proteoliposome interior. The negative interior drove Na$^+$ down its gradient from outside to inside, carrying with it proline. This mechanism is active sym-port. Formation of the trans-membrane electrical gradient was followed by fluorescence of the lipophilic cationic fluorescent dye carbocyanine. As carbocyanine entered the proteo-liposome it became concentrated, resulting in fluorescenc self-quenching. The decrease in net fluorescence was a direct indicator of the generation of a net negative potential in the proteo-liposome interior. Inherent leakage of K^+ was very slow indicating the proteoliposomes were surrounded by an intact, non-leaky membrane. Upon addition of valinomycin, K^+ leakage was dramatically enhanced. When ^{14}C-proline was added to the NaCl bathing solution, the addition of valinomycin corresponded with an uptake of ^{14}C-proline. If Na^+ was not included in the external solution or the Na^+ gradient was collapsed by a sodium ionophore (e.g. gramicidin D), ^{14}C-proline was not taken up into the proteoliposomes. A pH gradient was also ineffective at driving proline uptake. The stoichiometry of the symport system was approximately one Na+ for one proline.

The following is a summary of the mechanism of the bacterial proline transporter:

Need a non-leaky membrane.
Need a negative interior potential.
Need a Na^+ gradient, outside to inside.
The transporter is an example of active symport.

F. LIPID RAFTS

Although the basic concept of a lipid raft is compelling, it does not necessarily mean it is correct. It has been known for decades that biological membranes are not homogenous dispersions of lipids and proteins, but instead exist in a bewildering array of patches known as micro-domains [57]. Lipid rafts are an attempt to identify one of these lipid–protein patches and to assign it an essential biochemical function, namely cell signaling [58]. But do lipid rafts actually exist in biological membranes, or are they just a figment of membranol-ogists' fertile imaginations?

Lipid rafts can be defined by two basic operational definitions [59]. First, they are frac-tions of a plasma membrane that are enriched in cholesterol, and (predominantly) satu-rated phospholipids and sphingolipids. These lipids, when reconstituted into lipid bilayer vesicles, prefer being in a tightly packed, liquid ordered (l_o) state. Raft fractions also contain a characteristic set of proteins that are important in cell signaling [60]. Second, lipid rafts can be dispersed by extraction of cholesterol, usually with β-cyclodextran (Chapter 10). Intuitively it makes sense to compartmentalize a related set of functions into distinct membrane domains (e.g. lipid rafts and cell signaling) that are separate from the remaining membrane that surrounds them. The implication has been that lipid–lipid affinities create the lipid raft that then attracts the appropriate membrane proteins. However, there are many unsolved problems associated with the raft model [61,62]. Michael Edidin [59] has made a strong argument that it is more likely raft proteins select the appropriate lipids from the surrounding lipid sea than vice versa. Perhaps the actual answer is that the lipid–protein associations responsible for raft formation and stability are the result of *simultaneous* lipid–lipid, lipid–protein, protein–lipid and protein–protein interactions. A complex set of such interactions would be very hard to observe in biological or model membranes.

One of the major problems with the concept of lipid rafts is getting a handle on their size and stability [63]. In model membranes lipid rafts can be very large (microns in diameter).

However, the size of lipid rafts in biological membranes has yet to be fully determined, but it is estimated to be very small (1 to 1,000 nm in diameter). The estimated size of biological lipid rafts continues to decrease as imaging methodologies improve [63]. Many of the more recent studies have estimated lipid raft size at the lower end of the range, ~5 nm in diameter. The vanishingly small size of lipid rafts in biological membranes raises the possibility that even if rafts do exist, they may not be stable enough (they are just too small) to have any biological function. Rafts may not exist for a long enough stretch of time to attract the appropriate signaling proteins.

Lipid rafts are of two basic types, planar rafts and invaginated, flask-shaped rafts called caveolae [61,64]. Planar rafts and caveolae can be easily distinguished. Although both have essentially the same typical raft lipid composition, planar rafts have flotillin proteins, while caveolae have a characteristic flask-shape and caveolin proteins. Flotillins and caveolins both function in recruiting signaling molecules into lipid rafts and thus provide the raft skeletal frame. In a parallel fashion it has been proposed that cholesterol is the 'molecular glue' that holds the raft lipids together. It has been known for decades that cholesterol has a strong preference for sphingolipids over other phospholipids (see Chapter 10) [65]. Cholesterol also has a stronger affinity for saturated acyl chains and avoids polyunsaturated chains [66,67]. Finally, a hallmark of lipid rafts is their dissociation upon removal of cholesterol by β-cyclodextrin [68].

The earliest support for lipid rafts came from cold temperature detergent extractions, but it is not certain what these experiments actually mean. At 4°C, the non-ionic detergent Triton X-100 (later success was also reported with Brij-98 and CHAPS) solubilized much of the membrane, while the insoluble components, cholesterol, (predominantly) saturated phospholipids, sphingolipids, and characteristic signaling proteins, were left behind as lipid rafts [69]. The insoluble fraction is often referred to as detergent resistant membranes (DRMs) or sometimes detergent-insoluble glycolipid-enriched complexes (DIGs). Later advances eliminated the need for detergents, reducing possible artifacts. There are now countless published methods to isolate lipid rafts, most of which are slight variations on a basic theme. Below is an abbreviated outline of one of the most commonly used non-detergent procedures, known as the 'Smart Prep' after its inventor Eric Smart of the University of Kentucky [70]. The procedure is for the isolation of caveolae, a type of lipid raft.

BASIC STEPS IN THE SMART PREP [70]

Wash cells in 250 mM sucrose (isotonic) buffer.
Low speed centrifugation.
Homogenize cells with Dounce homogenizer (20x) and pass through a 21 gage needle (20x) in sucrose.
Low speed centrifugation to remove debris and obtain a post nuclear supernatant.
High speed centrifuge on 30% Percoll self-generating gradient.
Band at middle is the plasma membrane.
Sonicate the plasma membrane fraction.
High speed centrifugation on Optiprep to isolate caveolae.

All steps done at 4°C.

It is now clear that if non-caveolae lipid rafts exist, they must be extremely small and fleeting. The process of isolating rafts takes many orders of magnitude longer than the probable raft lifetime. Perhaps the membrane components that are isolated together as lipid rafts only share similar physical properties but never actually exist together in the intact membrane.

Experimental Support of Rafts

There is considerable experimental support for the existence of lipid rafts, primarily phase separations in model membranes, cholesterol depletion experiments, and multiple component experiments linking DRM analysis, cholesterol depletion, and microscopy. Micron-sized domains have been visualized on the surface of GUVs (giant unilamellar vesicles) using fluorescent membrane lipid tags that preferentially partition into l_o (raft) or l_d (non-raft) phases [71]. In one report, Dietrich et al. [72] demonstrated phase separation into l_o and l_d domains using fluorescence microscopy. Importantly, Triton X-100 was shown to selectively solubilize the l_d non-raft domain, while not perturbing the l_o raft domain. Similar model membrane studies demonstrated that phase separation could be achieved at both cold temperature (4°C) and physiological temperature (37°C), supporting the possibility of rafts existing in living cells.

Hammond et al. [73] reported a study using a complex model membrane that supported the feasibility of lipid rafts. GUVs were made from DOPC, DOPG, SM, cholesterol, the ganglioside GM1, and the rhodamine-labeled LAT trans-membrane peptide. This lipid mixture formed a single phase that was poised close to a separation boundary (Figure 13.20, panel a). Upon addition of the cholera toxin B subunit, a dramatic phase separation occurred (Figure 13.20, panel b). The fluorescent rhodamine-peptide preferentially partitioned into the l_d phase. Phase transition was induced by addition of the pentavalent cholera toxin B subunit that binds to and cross-links the GM1s. Panel (a) is before the addition of cholera toxin, while panel (b) is after cholera toxin addition. The dark regions in panel (b) are domains in the l_o (raft) phase.

One of the hallmarks of lipid rafts in biological membranes is a loss of physiological function, primarily related to signal transduction or membrane trafficking, upon cholesterol depletion [74]. By far the major agent used to extract cholesterol from membranes is β-cyclodextrin, although the polyene antibiotics filipin and nystatin (Chapter 14) can effectively remove cholesterol from rafts through complexation in the membrane. One of the strongest pieces of evidence in support of lipid rafts is a strong correlation between DRM-associated proteins, cholesterol depletion, loss of physiological function, and microscopy. Although it now appears that DRMs are not the same as pre-existing rafts [75], a surprisingly large number of examples of these correlations exist in the lipid raft literature.

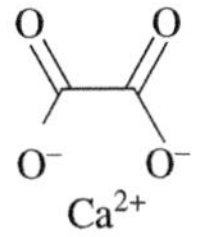

FIGURE 13.19   Calcium Oxalate.

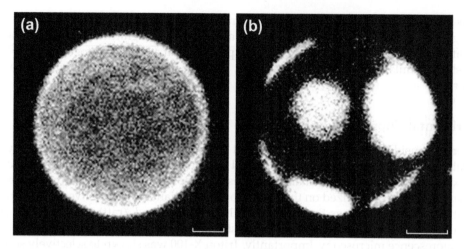

FIGURE 13.20   Fluorescence imaging of micron-sized domains on the surface of GUVs [73]. The model bilayer membranes were composed of DOPC, DOPG, SM, cholesterol, the ganglioside GM1, and a rhodamine-labeled LAT trans-membrane peptide. This lipid mixture formed a single phase (Panel a) that was poised close to a separation boundary. Upon addition of the cholera toxin B subunit, a dramatic phase separation occurred (Panel b). The fluorescent rhodamine-peptide preferentially partitioned into the l_d phase.

Questions Concerning Rafts

Many unanswered questions concerning the nature and even existence of lipid rafts abound [61]. The major trouble concerns the extremely small apparent size of lipid rafts in biological membranes (perhaps as small as 5nm and still shrinking) while lipid rafts in model membranes can be in the micron range. How small is too small for lipid rafts to have any biological significance? GPI-anchored proteins are routinely found in DRMs, indicating that they are common raft components, yet by fluorescence resonance energy transfer (FRET) GPIs are shown to be evenly distributed on the cell surface or are present in at most very small (nanoscale) clusters containing only a few proteins. Also, lateral diffusion measurements using fluorescence recovery after photobleaching (FRAP) show that lipid raft components (GPI-anchored proteins, other raft proteins, and raft lipids) do not diffuse as a single large unit in the plasma membrane. If raft components diffuse at different rates, how can they be found together in a raft? It remains to be shown if lateral membrane diffusion rates are sufficient to allow proteins time to diffuse into lipid rafts. The preponderance of lipid raft isolations into DRM fractions use cold temperatures (4°C) and non-ionic detergents, both of which can produce misleading artifacts. Using model membranes it is hard to access the potential importance of lipid asymmetry and cytoskeletal involvement in rafts. Also, it is not certain if intracellular rafts even exist.

Reputed raft proteins present another group of problems. Since raft lipids exist in the l_o state where they are extended, membrane lipid raft domains should be thicker than the surrounding non-raft l_d state regions. Indeed this has been confirmed by atomic force microscopy in model membranes [76]. Therefore one would expect that raft proteins should exhibit a longer hydro-phobic match surface than non-raft proteins. However, to date it has yet to be shown that a protein's hydrophobic span length is related to its preferential location in rafts. The rough surface associated with the membrane-spanning portion of an integral protein should exclude trans-membrane proteins from l_o state raft lipids. Indeed, most trans-membrane proteins do associate poorly with DRMs. However, rafts are known to accommodate a few characteristic trans-membrane proteins. The question remains, what makes raft integral proteins different from non-raft integral proteins and why are they concentrated in rafts?

There are many reports where polyunsaturated fatty acyl chains including arachidonic acid ($20:4^{\triangle 5,8,11,14}$) and docosahexaenoic acid ($22:6^{\triangle 4,7,10,13,16,19}$) have been found at significant levels among raft lipids [77]. This seems contrary to the basic precept that rafts exist in the tightly packed l_o state. How does a raft accommodate a highly contorted polyunsaturated fatty acid? To justify this contradiction it has been proposed for $PI(4,5)P_2$, a common component of rafts, that the sn-2 arachidonic chain fits into a groove on a protein's surface while the saturated stearic sn-1 chain inserts into the bilayer [78]. The effect of cholesterol depletion on trans-membrane signaling functions has become a cornerstone in defining rafts. Yet even this has its problems. Cholesterol levels can affect membrane structure and function by ways that are unrelated to raft structure. For example, Pike and Miller [79] have made a strong case that cholesterol depletion affects both raft structure and signaling through $PI(4,5)P_2$.

It is now clear that DRMs are not exactly the same as lipid rafts that are resident in the plasma membrane [75]. The current lipid raft model is in flux and will continue to be adjusted to match new experimental findings.

One Last Word on Rafts

The fundamental importance of the lipid raft model was to propose a type of domain that could serve as a simple paradigm for general membrane structure/function. Unfortunately, the original concept of a lipid raft has continued to evolve, becoming ever more complex to where it is now hard to define exactly what a lipid raft is and is not. Raft dynamics is poorly understood and fixed rules remain elusive. Listed below are some recently suggested lipid raft characteristics:

Lipid Raft Characteristics

1. Rafts are highly regulated and may occur only in response to stimulation.
2. In unstimulated cells, rafts are very small and unstable — if they exist at all.
3. Lipid rafts have a high affinity for ordered lipids.
4. Plasma membrane lipid composition is maintained close to that required for phase separation so that small changes in protein or lipid can have a large influence on lipid rafts.
5. Caveolins and flotillins probably organize rafts into specialized membrane domains.
6. Some raft proteins are linked to the actin cytoskeleton network.
7. The myristate/palmitate motif on lipoproteins appears to be an especially efficient raft-targeting signal.

8. Particularly important is palmitoylation for raft signaling.
9. Polyunsaturated fatty acyl chains are hidden in clefts on a raft protein surface.
10. It will be essential to determine the actual structure of lipid rafts as they exist in a biological membrane.

SUMMARY

Since it is so difficult to study a single protein in a complex biological membrane, methods have been devised to purify and reconstitute it into a liposome of known composition. These methods rely on detergents. Detergents are amphipathic molecules that resemble membrane polar lipids, but are orders of magnitude more water soluble. Hundreds of detergents are used to lyse cells, extract membrane proteins and lipids, and reconstitute them into membranes.

After isolation, the purified membrane is dissolved in an appropriate detergent solution, and the target integral protein purified, usually by some form of affinity chromatography. After purification, the protein is concentrated and mixed with a desired lipid that is also solubilized by the same detergent. Upon rapid removal of the detergent (by dialysis, size-exclusion chromatography, rapid dilution or polystyrene beads), polar lipids and integral proteins fall out of solution as proteoliposomes of known composition.

Recently the hot topic in membrane structure/function has become lipid rafts, either planar rafts or caveolae. It is now clear that if planar lipid rafts exist, they must be extremely small (<5nm) and fleeting.

Chapter 14 will discuss how solutes enter and leave a cell. Included will be discussions of: osmosis, simple passive diffusion, facilitated passive diffusion, active transport, Gap Junctions, receptor-mediated endocytosis, phagocytosis, pinocytosis, exocytosis, and apoptotic membrane blebbing.

References

[1] Helenius A, Simons K. Solubilization of membranes by detergents. Biochim Biophys Acta 1975;415:29–79.
[2] Rosen MJ. Surfactants and Interfacial Phenomena. New York: John Wiley; 1989.
[3] Kunjappu JT. (March). Chemistry World. RSC. Advancing the Chemical Sciences; 2003.
[4] Griffin WC. Classification of surface-active agents by 'HLB'. J Soc Cosmet Chem 1949;1:311.
[5] Griffin WC. Calculation of HLB values of non-ionic surfactants. J Soc Cosmet Chem 1954;5:259.
[6] Wilson L. Methods in Cell Biology: The Cytoskeleton. Cytoskeletal Proteins, Isolation and Characterization, vol. 24. New York: Academic Press; 1982. Part 1. p. 275.
[7] Hatefi Y, Hanstein G. Solubilization of particulate proteins and nonelectrolytes by chaotropic agents. Proc Natl Acad Sci USA 1969;62:1129–36.
[8] Hochuli E, Bannwarth W, Döbeli H, Gentz R, Stüber D. Genetic approach to facilitate purification of recombinant proteins with a novel metal chelate adsorbent. Bio Technology 1988;6:1321–5.
[9] Hengen P. Purification of His-Tag fusion proteins from *Escherichia coli*. Trends Biochem Sci 1995;20:285–6.
[10] Lowery OH, Rosebrough NJ, Fan AL, Randall RJ. Protein measurement with the Folin phenol reagent. J Biol Chem 1951;193:265–75.
[11] Bradford MM. Rapid and sensitive method for the quantitation of microgram quantities of protein utilizing the principle of protein-dye binding. Anal Biochem 1976;72:248–54.
[12] Zor T, Selinger Z. Linearization of the Bradford protein assay increases its sensitivity: theoretical and experimental studies. Anal Biochem 1996;236:302–8.

[13] Sumner JB. The isolation and crystallization of the enzyme urease. J Biol Chem 1926;69:435—41.

[14] The AOCS Lipid Library. Giants of the Past. Michel Eugène Chevreul 2010:1786—889.

[15] Folch J, Lees M, Stanley GHS. A simple method for the isolation and purification of total lipids from animal tissues. J Biol Chem 1957;226:497—509.

[16] Bligh EG, Dyer WJ. A rapid method of total lipid extraction and purification. Can J Biochem Physiol 1959;37:911—7.

[17] St. John LC, Bell FP. Extraction and fractionation of lipids from biological tissues, cells, organelles and fluids. BioTechniques 1989;7:476—81.

[18] Berezkin, V.G. History of Analytical Chemistry: Contributions from N.A. Izmailov and M.S. Schraiber to the development of thin-layer chromatography (On the 70th anniversary of the publication of the first paper on thin-layer chromatography). J Anal Chem 63, 400—4.

[19] Sherma J, Fried B. Practical Thin-Layer Chromatography: A Multidisciplinary Approach. Boca Raton, FL: CRC Press; 1996.

[20] www.instructables.com. 2011. Preparing your own thin layer chromatography plates (and then using them).

[21] Vaskovsky VE, Svetashev VI. Phospholipid spray reagents. J Chromatogr 1972;65:451—3.

[22] Christie, W The AOCS Lipid Library. Thin-Layer Chromatography of Lipids. The American Oil Chemists Society.

[23] Hax WMA, Geurts van Kessel WSM. High-performance liquid chromatographic separation and photometric detection of phospholipids. J Chromatogr 1977;142:735—41.

[24] Bell MV. Separations of molecular species of phospholipids by high-performance liquid chromatography. In: Christie WW, editor. Advances in Lipid Methodology — Four. Dundee, Scotland: Oily Press; 1977. p. 45—82.

[25] Dong MW. Modern HPLC for Practicing Scientists. Hoboken, NJ: John Wiley; 2006.

[26] Biosciences Amersham. Reversed Phase Chromatography. Piscataway, NJ: Principles and Methods; 1999.

[27] Corran PH. Reversed-phase chromatography of proteins. In: Oliver RWA, editor. HPLC of Macromolecules, A Practical Approach. IRL Press: Oxford Press; 1989. p. 127—56.

[28] Grob RL, Barry EF. Modern Practice of Gas Chromatography. New York: John Wiley; 1995.

[29] Chromatographer. Resolution Matters. Gas Chromatography.

[30] Christie WW. Gas Chromatography and Lipids: A Practical Guide. Dundee, Scotland: The Oily Press; 1989.

[31] Christie WW. Preparation of ester derivatives of fatty acids for chromatographic analysis. In: Christie WW, editor. Advances in Lipid Methodology — Two. Dundee, Scotland: Oily Press; 1993. p. 69—111.

[32] Christie WW. A simple procedure for rapid transmethylation of glycerolipids and cholesteryl esters. J Lipid Res 1982;23:1072—5.

[33] Bannon CD, Breen GJ, Craske JD, Hai NT, Harper NL, O'Rourke KL. Analysis of fatty acid methyl esters with high accuracy and reliability: III. Literature review of and investigations into the development of rapid procedures for the methoxide-catalysed methanolysis of fats and oils. J Chromatogr A 1982;247: 71—89.

[34] Leray C. Separation of fatty acids by GLC. Cyberlipid Center. Resource Site for Lipid Studies, www .cyberlipid.org/fattyt/fatt0003.htm 2011.

[35] Bicalho B, David F, Rumplel K, Kindt E, Sandra P. Creating a fatty acid methyl ester database for lipid profiling in a single drop of human blood using high resolution gas chromatography and mass spectrometry. J Chromatogr A 2008;1211:120—8.

[36] Mueller P, Rudin DO, Tien HT, Westcott WC. Reconstitution of Excitable Cell Membrane Structure in Vitro. Circulation 1962;26:167—1171.

[37] Tien HT. Bilayer Lipid Membranes (BLM): Theory and Practice. New York: Marcel Dekker; 1974.

[38] Bangham AD, Horne RW. Negative staining of phospholipids and their structural modification by surface active agents. J Mol Biol 1964;8:660—8.

[39] Lasic DD, Papahadjopoulos, editors. Medical Applications of Liposomes. New York: Elsevier; 1998.

[40] Florence AT. Liposomes in Drug Delivery (Drug Targeting and Delivery). Harvard Academic; 1993.

[41] Lathe GH, Ruthven CR. The separation of substances on the basis of their molecular weights, using columns of starch and water. Biochem J 1955;60(4):xxxiv.

[42] Lathe GH, Ruthven CR. The separation of substances and estimation of their relative molecular sizes by the use of columns of starch in water. Biochem J 1956;62:665—74.

[43] Mori S, Barth HG. Size Exclusion Chromatography. New York: Springer; 1999.

[44] Weinstein JN, Blumenthal R, Klausner RD. Carboxyfluorescein leakage assay for lipoprotein-liposome interabtion. Methods Enzymol 1986;128:857—68.

[45] Gould RM, London A. Specific interaction of central nervous system myelin basic protein with lipids. Effects of basic protein on glucose leakage from liposomes. Biochim Biophys Acta 1972;290:200—18.

[46] Lapinski MM, Castro-Forero A, Greiner AJ, Ofoli RY, Blanchard GJ. Comparison of liposomes formed by sonication and extrusion: Rotational and translational diffusion of an embedded chromophore. Langmuir 2007;23:11677—83.

[47] Rhoden V, Goldin SM. Formation of unilamellar lipid vesicles of controllable dimensions by detergent dialysis. Biochemistry 1979;18:4173—6.

[48] Rigaud JL, Levy D, Mosser G, Lambert O. Detergent removal by non-polar polystyrene beads. Eur Biophys J 1998;27:305—19.

[49] Deamer D, Bangham AD. Large volume liposomes by an ether vaporization method. Biochim Biophys Acta 1976;443:629—34.

[50] Silvius JR. Solubilization and functional reconstition of biomembrane components. Annu Rev Biophys Biomol Struct 1992;21:323—48.

[51] Rigaud J-L. Membrane proteins: functional and structural studies using reconstited proteoliposomes and 2-D crystals (review). Brazilian J Med Biol Res 2002;35:753—66.

[52] Carafoli E. Calcium pump of the plasma membrane (review). Physiol Rev 1991;71:129—53.

[53] Racker, E. Reconstitution of a calcium pump with phospholipids and a purified Ca^{++}adenosine triphosphatase from sarcoplasmic reticulum. J Biol Chem 247, 8198—200.

[54] Warren FB, Toon PA, Birdsall NJM, Lee AG, Metcalf JC. Reconstitution of a calcium pump using defined membrane components. Proc Nat Acad Sci USA 1974;71:622—6.

[55] Jung H, Tebbe S, Schmid R, Jung K. Unidirectional reconstition and characterization of purified Na^{+}/proline transporter of Escherichia coli. Biochemistry 1998;1998(37):11083—8.

[56] Chen CC, Wilson TH. Solubilization and functional reconstition of the proline transport system of Escherichia coli. J Biol Chem 1986;261:2599—604.

[57] Karnovsky MJ, Kleinfeld AM, Hoover RL, Klausner RD. The concept of lipid domains in membranes. J Cell Biol 1982;94:1—6.

[58] Simons K, Ikonen E. Functional rafts in cell membranes. Nature 1997;387:569—752.

[59] Edidin M. The state of lipid rafts: from model membranes to cells. Annu Rev Biophys Biomol Struct 2003;32:257—83.

[60] Brown DA. Lipid rafts, detergent-resistant membranes, and raft targeting signals. Physiology 2006;21:430—9.

[61] Pike LJ. The challenge of lipid rafts. J Lipid Res 2009;(Suppl. 50):S323—8.

[62] Shaw AS. Lipid rafts: now you see them, now you don't. Nat Immunol 2006;(7):1139—42.

[63] Edidin M. Shrinking patches and slippery rafts: Scales of domains in the plasma membrane. Trends Cell Biol 2001;11:492—6.

[64] van Meer G. The different hues of lipid rafts. Science 2002;296:855—7.

[65] Estep TN, Mountcastle DB, Barenholz Y, Biltonen RL, Thompson TE. Thermal behavior of synthetic sphingomyelin-cholesterol dispersions. Biochemistry 1979;18:2112—7.

[66] Shaikh SR, Cherezov V, Caffrey M, Soni S, Stillwell W, Wassall SR. Molecular organization of cholesterol in unsaturated phosphatidylethanolamines: X-ray diffraction and solid state ^{2}H NMR studies. J Am Chem Soc 2006;128:5375—83.

[67] Wassall SR, Shaikh SR, Brzustowicz MR, Cherezov V, Siddiqui RA, Caffrey M, et al. Interaction of polyunsaturated fatty acids with cholesterol: A role in lipid raft phase separation. Talking About Colloids; 2005.

[68] Ilangumaran S, Hoessli DC. Effects of cholesterol depletion by cyclodextrin on the sphingolipid microdomains of the plasma membrane. In: Danino D, Harries D, Wrenn SP, editors. Biochem J 335, 433—40. Macromolecular Symposia, vol. 219. Wiley-VCH; 1998. p. 73—84.

[69] Brown DA. Lipid rafts, detergent-resistant membranes, and raft targeting signals. Physiology 2006;21:430—9.

[70] Smart EJ, Ying Y-S, Mineo C, Anderson RGW. A detergent-free method for purifying caveolae membrane from tissue culture cells. Proc Natl Acad Sci 1995;92:10104—8.

[71] Schroeder R, London E, Brown D. Interactions between saturated acyl chains confer detergent resistance on lipids and glycosylphosphatidylinositol (GPI)-anchored proteins: GPI-anchored proteins in liposomes and cells show similar behavior. Proc Natl Acad Sci USA 1994;91:12130–4.

[72] Dietrich C, Bagatolli LA, Volovyk ZN, Thompson NL, Levi M, Jacobson K, et al. Lipid rafts reconstituted in model membranes. Biophys J 2001;80:1417–28.

[73] Hammond AT, Heberle FA, Baumgart T, Holowka D, Baird B, Feigenson GW. Crosslinking a lipid raft component triggers liquid ordered-liquid disordered phase separation in model plasma membranes. Proc Natl Acad Sci USA 2005;102:6320–5.

[74] Kabouridis PS, Janzen J, Magee AL, Ley SC. Cholesterol depletion disrupts lipid rafts and modulates the activity of multiple signaling pathways in T lymphocytes. Eur J Immunol 2000;30:954–63.

[75] Brown DA. Lipid rafts, detergent-resistant membranes, and raft targeting signals. Physiology 2006;21:430–9.

[76] Henderson RM, Edwardson JM, Geisse NA, Saslowsky DE. Lipid Rafts: Feeling is Believing. News Physiol Sci 2004;19:39–43.

[77] Li Q, Wang M, Tan L, Wang C, Ma J, Li N, et al. Docosahexaenoic acid changes lipid composition and interleukin-2 receptor signaling in membrane rafts. J Lipid Res 2005;46:1904–13.

[78] Hope HR, Pike LJ. Phosphoinositides and phosphoinositide-utilizing enzymes in detergent-insoluble lipid domains. Mol Biol Cell 1996;7:843–51.

[79] Pike LJ, Miller JM. Cholesterol depletion delocalizes phosphatidylinositol bisphosphate and inhibits hormone-stimulated phosphatidylinositol turnover. J Biol Chem 1998;273:22298–304.

[80] Watts G. Alec Douglas Bangham. The Lancet 2010;375:2070.

[81] John W. Kimball: http://home.comcast.net/~john.kimball1/BiologyPages/

Membrane Transport

OUTLINE

A. **Introduction** 305
   Fick's First Law 307
   Osmosis 307

B. **Simple Passive Diffusion** 309

C. **Facilitated Diffusion** 311
   Glucose Transporter 311
   Potassium Channels 312
   Sodium Channel 313
      Tetrodotoxin 314
      Saxitoxin 314
      Solute Equilibrium 315
      Aquaporins 316

D. **Active Transport** 317
   Primary Active Transport 319
   Na$^+$/K$^+$ ATPase 319
      Mechanism of the Na$^+$/K$^+$ ATPase 321
   Secondary Active Transport 321

Bacterial Lactose Transport 322
Vectorial Metabolism 323

E. **Ionophores** 324
   Valinomycin 325
   2,4-Dinitrophenol (DNP) 326
   Crown Ethers 326
   Nystatin 327

F. **Gap Junctions** 328

G. **Other Ways to Cross the Membrane** 329
   Receptor Mediated Endocytosis 329
   Pinocytosis and Phagocytosis 330
      Pinocytosis 331
      Phagocytosis 332

Summary 334

References 335

A. INTRODUCTION

Life depends on a membrane's ability to precisely control the level of solutes in the aqueous compartments bathing the membrane. The membrane determines what solutes enter and leave a cell. Trans-membrane transport is controlled by complex interactions between membrane lipids, proteins, and carbohydrates. How the membrane accomplishes these tasks is the topic of Chapter 14.

A biological membrane is semi-permeable, meaning it is permeable to some molecules, most notably water, while being very impermeable to most solutes (various biochemicals and salts) found in the bathing solution. This very important concept of unequal trans-membrane distribution and hence permeability between water and other solutes came out of the pioneering work of Charles Overton in the 1890s (see Chapter 2). How does a biological membrane accomplish semi-permeability? The barrier to solute movement is largely provided by the membrane's hydrophobic core, a very thin (~ 40 Å thick), oily layer. The inherent permeability of this core varies from membrane to membrane. Generally, the more tightly packed the lipids comprising the bilayer, the lower its permeability will be. Lipid bilayers are very impermeable to most solutes because of their tight packing. Figure 14.1 depicts the membrane permeability of a variety of common solutes [1]. Note the data is presented as a log scale of solute permeability (P in cm/s) and ranges from $Na^+ = 10^{-12}$ cm/s to water $= 0.2 \times 10^{-2}$ cm/s, spanning almost 10 orders of magnitude!

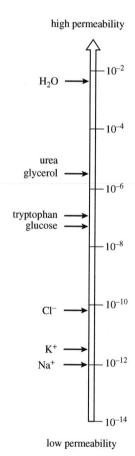

FIGURE 14.1   Log of the permeability (P in cm/s) across lipid bilayer membranes for common solutes ranging from Na^+ (10^{-12} cm/s) to water (0.2×10^{-2} cm/s). This range spans almost 10 orders of magnitude [1].

For example, LUVs made from DPPC (16:0, 16:0 PC) have a sharp phase transition temperature, T_m, of 41.3°C. At temperatures well below T_m, the LUVs are in the tightly packed gel state and permeability is extremely low. At temperatures well above T_m, the LUVs are in the loosely packed liquid disordered state (l_d, also called the liquid crystalline state) and permeability is high. However, maximum permeability is not found in the l_d state, but rather at the T_m [2]. As the LUVs are heated from the gel state and approach the T_m, domains of l_d start to form in the gel state. Solutes can then pass more readily through the newly formed l_d domains than the gel domains, resulting in an increase in permeability. At T_m there is a maximum amount of coexisting gel and l_d state domains that exhibit extremely porous domain boundaries. It is through these boundaries that most permeability occurs. As the temperature is further increased, the LUVs pass into the l_d state and the boundaries disappear, reducing permeability to that observed for the single component l_d state. Thus maximum permeability is observed at the T_m.

Fick's First Law

The tendency for solutes to move from a region of higher concentration to one of lower concentration was first defined in 1855 by the physiologist Adolf Fick (Figure 14.2). His work is summarized in what is now the very well known Fick's Laws of Diffusion [3]. The laws apply to both free solution and diffusion across membranes. Fick developed his laws by measuring concentrations and fluxes of salt diffusing between two reservoirs through tubes of water.

Fick's First Law describes diffusion as:

$$\text{Diffusion rate} = -DA\, dc/dx$$

Where:
D = diffusion coefficient (bigger molecules have lower Ds)
A = cross sectional area over which diffusion occurs
dc/dx is the solute concentration gradient (diffusion occurs from a region of
higher concentration to one of lower concentration)
The relationship between a solute's molecular weight and its diffusion coefficient is shown in Table 14.1. Large solutes have low diffusion coefficients and therefore diffuse more slowly than small solutes. The diffusion rate for a particular solute under physiological conditions is a constant and cannot be increased. This defines the theoretical limit for an enzymatic reaction rate and also limits the size of a cell. If a solute starts at the center of a bacterial cell, it takes $\sim 10^{-3}$ s to diffuse to the plasma membrane. For this reason, typical cells are microscopic (see Chapter 1). At about 3.3 pounds and the size of a cantaloupe, the largest cell on Earth today is the ostrich egg. However, a dinosaur egg in the American Museum of Natural History in New York is about the size of basketball. Since an egg's only function is to store nutrition for a developing embryo, its size is many orders of magnitude larger than a normal cell.

Osmosis

Osmosis is a special type of diffusion, namely the diffusion of water across a semipermeable membrane. Water readily crosses a membrane down its potential gradient

FIGURE 14.2   Adolf Fick (1829–1901) [61].

TABLE 14.1   The Relationship Between a Solute's Molecular Weight and its Diffusion Coefficient, D.

Compound	O_2	Acetyl choline	Sucrose	Serum albumin
D (cm^2/s × 10^6)	19.8	5.6	2.4	0.7
Molecular Weight	32	182	342	69,000

from high to low potential (Figure 14.3) [4]. Osmotic pressure is the force required to prevent water movement across the semi-permeable membrane. Net water movement continues until its potential reaches zero. An early application of the basic principles of osmosis came from the pioneering work on hemolysis of red blood cells by William Hewson in the 1770s (see Chapter 2). It has also been discussed that MLVs (multilamellar vesicles, liposomes) behave as almost perfect osmometers, swelling in hypotonic solutions and shrinking in hypertonic solutions (see Chapter 3) [5,6]. Liposome swelling and shrinking

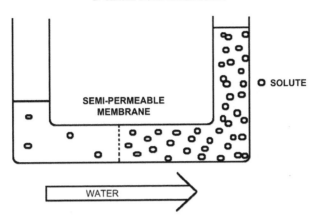

FIGURE 14.3   Osmosis and osmotic pressure. Water is placed in a U-shaped tube where each of the tube arms is separated by a semi-permeable membrane with pores of a size that water can easily pass through but a solute cannot. Upon addition of the solute to the tube's right arm, water diffuses from left to right (high water potential to low). The column of water in the tube's right arm (the one containing the solute) rises until the extra weight of the column equals the osmotic pressure caused by the solute. A pump could then be used to counter the osmotic pressure whereupon the solution columns in the right and left arms of the tube are made the same. The pump pressure required to equalize the height of the two columns is the osmotic pressure [4]. Note a small amount of solute leaks from right to left since no filter is perfect.

can be easily followed by changes in absorbance due to light scattering using a simple spectrophotometer. Therefore, osmosis has been investigated for many years using common and inexpensive methodologies and a lot is known about the process.

Membranes are rarely, if ever, perfectly semi-permeable. Deviation from ideality is defined by a reflection coefficient (σ). For an ideal semi-permeable membrane where a solute is totally impermeable, $\sigma = 1$. If a solute is totally permeable (its permeability is equal to water), $\sigma = 0$. Biological membranes are excellent semi-permeable barriers with $\sigma = 0.75$ to 1.0.

B. SIMPLE PASSIVE DIFFUSION

Movement of solutes across membranes can be divided into two basic types, passive diffusion and active transport [7]. Passive diffusion requires no additional energy source other than what is found in the solute's electrochemical (concentration) gradient and results in the solute reaching equilibrium across the membrane. Passive diffusion can be either simple passive diffusion, where the solute crosses the membrane anywhere by simply crossing the lipid bilayer, or facilitated passive diffusion, where the solute crosses the membrane at specific locations where diffusion is assisted by solute-specific facilitators or carriers. Active transport requires additional energy, often in the form of ATP, and results in a non-equilibrium, net accumulation (uptake) of the solute on one side of the membrane. The basic types of membrane transport, simple passive diffusion, facilitated diffusion (by channels and carriers) and active transport are summarized in Figure 14.4 [8]. There are countless different examples of each type of membrane transport process [7]. Only a few representative examples will be discussed here.

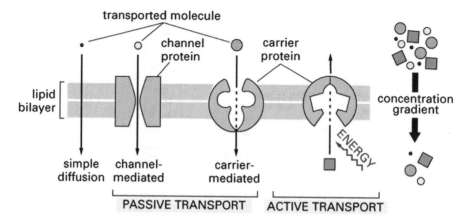

FIGURE 14.4   Basic types of membrane transport: simple passive diffusion; facilitated diffusion (by channels and carriers); and active transport [8].

Even for simple passive diffusion it requires energy to cross a bilayer membrane. In order to cross a membrane the solute must first lose its waters of hydration, diffuse across the membrane, and then regain its waters on the opposite side. The limiting step involves the energy required to lose the waters of hydration. Table 14.2 shows the relationship between the waters of hydration (proportional to the number of −OH groups on a homologous series of solutes) and the activation energy for trans-membrane diffusion. As the number of waters of hydration increases from glycol < glycerol < erythritol, the activation energy for diffusion also increases. The activation energy compares very well with the energy of hydration.

However, water diffusion does not fit this model. Water permeability is just too high. Several possibilities have been suggested to account for the abnormally high membrane permeability of water:

1. Water is very small and so just dissolves in bilayers better than larger solutes.
2. Due to its size, water can enter very small statistical pores (~4.2Å in diameter) more readily. Statistical pores result from the simultaneous lateral movement of adjacent

TABLE 14.2   Relationship Between the Waters of Hydration (Related to the Number of −OH Groups on a Homologous Series of Solutes) and the Activation Energy for Trans-Membrane Diffusion.

Solute	−OH groups	Activation energy (KJ/mol)
Glycol (HO-CH$_2$-CH$_2$-OH)	2	60
Glycerol (HO-CH$_2$-CH(OH)-CH$_2$-OH)	3	77
Erythritol (HO-CH$_2$-CH(OH)-CH(OH)-CH$_2$-OH)	4	87

membrane phospholipids in opposite directions. Statistical pores have only a fleeting existence and cannot be isolated or imaged.

3. Passage down water chains.
4. Water can be carried down kinks in acyl chains that result from acyl chain melting (see lipid melting in Chapter 9).
5. Water may rapidly cross membranes through non-lamellar membrane patches (e.g. micelle, cubic or H_{II} phase).
6. High water permeability will occur at locations of packing defect (e.g. surface of integral proteins, boundary between membrane domains).
7. Through pores or channels used to conduct ions.
8. Through specific water channels known as aquaporins.

The only molecules that can cross a membrane by simple passive diffusion are water, small non-charged solutes, and gasses. Charged or large solutes are virtually excluded from membranes and so require more than just simple passive diffusion to cross a membrane.

C. FACILITATED DIFFUSION

Facilitated diffusion (also known as carrier-mediated diffusion) is, like simple passive diffusion, dependent on the inherent energy in a solute gradient. No additional energy is required to transport the solute and the final solute distribution reaches equilibrium across the membrane. Facilitated diffusion, unlike simple diffusion, usually requires a highly specific trans-membrane integral protein to assist in the solute's membrane passage. Facilitators come in two basic types, carriers and gated channels. Facilitated diffusion exhibits Michaelis-Menton saturation kinetics (Figure 14.5), indicating the carrier has an enzyme-like active site. Like enzymes, facilitated diffusion carriers recognize their solute with exquisite precision, easily distinguishing chemically similar isomers like D-glucose from L-glucose and exhibit saturation kinetics. Figure 14.5 compares simple passive diffusion to facilitated diffusion. The figure is not to scale, however, as facilitated diffusion is orders of magnitude faster than simple passive diffusion.

Glucose Transporter

A well-studied example of a facilitated diffusion carrier is the glucose transporter or GLUT [9]. From the activation energy for trans-membrane simple passive diffusion of glycol, glycerol and erythritol presented in Table 14.2, it can be estimated that the activation energy for glucose should be well over 100 KJ/mol, but instead it is only 16 KJ/mol. This large discrepancy is attributed to the presence of a glucose facilitated diffusion carrier. Figure 14.6 demonstrates the mode of action of one of these transporters, GLUT-1, from the erythrocyte [10]. GLUTs occur in nearly all cells and are particularly abundant in cells lining the small intestine. GLUTs are but one example in a superfamily of transport facilitators. GLUTs are integral membrane proteins whose membrane-spanning region is composed of 12 α-helices. GLUTs function through a typical membrane transport mechanism [10]. Glucose binds to the

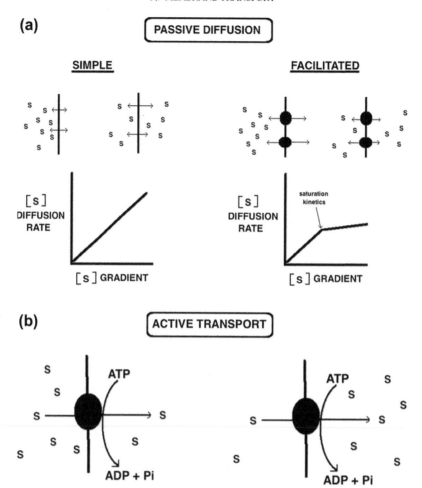

FIGURE 14.5   (a) Simple passive diffusion (Top, Left) and (b) Facilitated diffusion (Top, Right). Passive diffusion results in a final equilibrium distribution of a solute across the membrane. For a non-charged solute, the final distribution of the solute would find equal amounts of S on both sides of the membrane. Facilitated diffusion employs a specific transporter and exhibits Michaelis-Menten saturation kinetics. Active transport (Bottom) utilizes energy, often in the form of ATP, to drive solute uptake against its gradient resulting in a net accumulation of the solute.

membrane outer surface site causing a conformational change associated with transport, releasing glucose to the inner side of the membrane where it enters into the internal aqueous solution (Figure 14.6).

Potassium Channels

In virtually all organisms there exists a wide variety of ion channels, the most widely distributed being potassium channels [11]. There are four basic classes of potassium

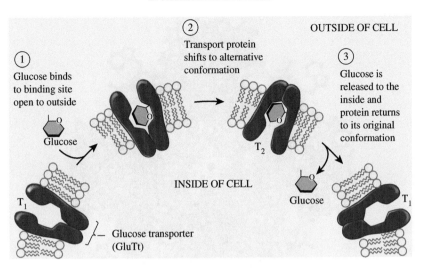

FIGURE 14.6   Glucose facilitated diffusion transporter GLUT-1 [10].

channels, all of which provide essential membrane-associated functions including setting and shaping action potentials and hormone secretion:

1. Calcium-activated potassium channel.
2. Inwardly rectifying potassium channel.
3. Tandem pore domain potassium channel.
4. Voltage-gated potassium channel.

Potassium channels are composed of four protein subunits that can be the same (homotetramer) or closely related (heterotetramer). All potassium channel subunits have a distinctive pore-loop structure that sits at the top of the channel and is responsible for potassium selectivity [12]. This is often referred to as a selectivity or filter loop. The selectivity filter strips the waters of hydration from the potassium ion. Further down the structure is a 10 Å diameter trans-membrane, water filled central channel that conducts potassium across the membrane. Elucidating the three-dimensional structure of this important integral membrane protein by X-ray crystallography (Figure 14.7) [12] was a seminal accomplishment in the field of membrane biophysics. For this work, from 1998, Rod MacKinnon of Rockefeller University (Figure 14.8) was awarded the 2003 Nobel Prize in Chemistry. Until the potassium channel work, just obtaining the structure of non-water soluble proteins was next to impossible. MacKinnon's work not only elucidated the structure of the potassium channel but also its molecular mechanism. It has served as a blueprint for determining the structure of other membrane proteins and has greatly stimulated interest in the field.

Sodium Channel

In some ways Na^+ channels [13] parallel the action of K^+ channels. They are both facilitated diffusion carriers that conduct the cation down the ion's electrochemical gradient. In excitable cells such as neurons, myocytes, and some glia, Na^+ channels are responsible for

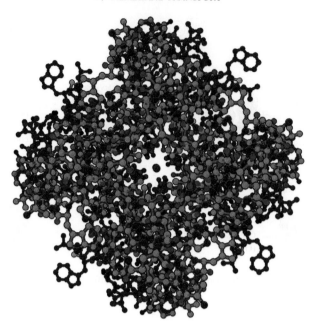

FIGURE 14.7   Three-dimensional structure of the potassium channel [12]. The channel itself is the clear opening in the center of the structure and a single K^+ is shown in the center of the channel.

the rising phase of action potentials. Therefore, agents that block Na^+ channels also block nerve conduction and so are deadly neurotoxins. There are two basic types of Na^+ channels, voltage-gated and ligand-gated. The opening of a Na^+ channel has a selectivity filter that attracts Na^+. From there the Na^+s flow into a constricted part of the channel that is ~3–5 Å wide. This is just large enough to allow the passage of a single Na^+ with one attached water. Since the larger K^+ cannot squeeze through, the channel is selective for Na^+. Of particular interest are two extremely potent biological toxins, tetrodotoxin (TTX) and saxitoxin (STX) [14], that have killed and injured many humans. Both toxins shutdown Na^+ channels by binding from the extra-cellular side.

Tetrodotoxin

Tetrodotoxin (TTX) is encountered primarily in puffer fish but also in porcupine fish, ocean sunfish, and triggerfish. TTX is a potent neurotoxin (Figure 14.9) that blocks Na^+ channels while having no effect on K^+ channels. Puffer fish are the second most poisonous vertebrate in the world trailing only the Golden Poison Frog. In some parts of the world puffer fish are considered to be a delicacy, but must be prepared by chefs that really know their business, as a slight error can be fatal. Puffer poisoning usually results from consumption of incorrectly prepared puffer soup, and TTX has no known antidote!

Saxitoxin

Saxitoxin (STX) is a neurotoxin produced by some marine dinoflagellates that can be accumulated in shellfish during toxic algal blooms known as Red Tide. Saxitoxin is one of the

FIGURE 14.8   Rod MacKinnon (1956–) [62].

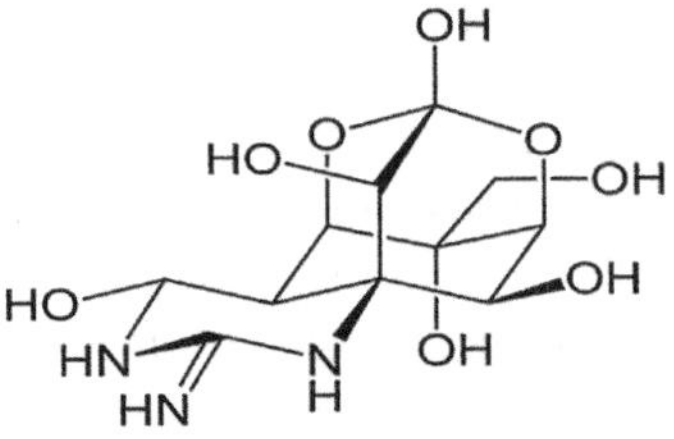

FIGURE 14.9   Tetrodotoxin.

most potent natural toxins (Figure 14.10) and it has been estimated that a single contaminated mussel has enough STX to kill 50 humans! STX's toxicity has not escaped the keen eye of the United States military, which has weaponized the toxin and given it the designation TZ.

Solute Equilibrium

The driving force for trans-membrane solute movement by simple or passive diffusion is determined by the free energy change ΔG.

$$\Delta G = RT \ln[s_o']/[s_o] + ZF\Delta\Psi$$

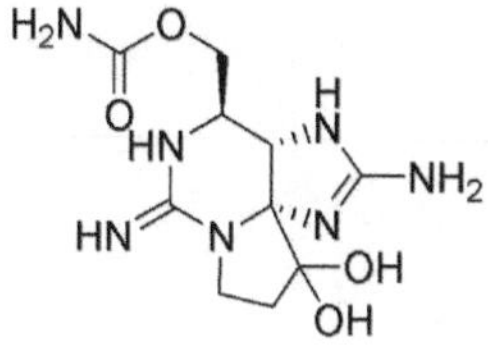

FIGURE 14.10   Saxitoxin.

Where:

ΔG is the free energy change

$[s_o']$ is the solute concentration on the right side of a membrane

$[s_o]$ is the solute concentration on the left side of a membrane

R is the gas constant

T is the temperature in °K

Z is the charge of the solute

F is the Faraday

$\Delta \Psi$ is the trans-membrane electrical potential

Solute movement will continue until $\Delta G = 0$. If ΔG is negative, solute movement is left to right (it is favorable as written). If ΔG is positive, solute movement is right to left (it is unfavorable in the left to right direction) or energy must be added for the solute to go from left to right The equation has two parts: a trans-membrane chemical gradient ($[s_o']$ / $[s_o]$); and a trans-membrane electrical gradient ($\Delta \Psi$). The net movement of solute is therefore determined by a combination of the solute's chemical gradient and an electrical gradient inherent to the cell. If the solute has no charge, $Z = 0$ (as is the case for glucose) and the right hand part of the equation ($ZF\Delta \Psi$) drops out. Therefore, the final equilibrium distribution of glucose across the membrane will have the internal glucose concentration equal to the external glucose concentration and is independent of $\Delta \Psi$, the electrical potential. At equilibrium for a non-charged solute, $\Delta G = RT \ln [s_o']$ / $[s_o]$ and ΔG can only be equal to zero if $[s_o'] = [s_o]$.

The situation for a charged solute like K^+ is more complicated. The net ΔG is determined by both the chemical gradient ($[s_o']$ / $[s_o]$) and electrical gradient ($\Delta \Psi$). The $\Delta \Psi$ results from the sum of all charged solutes on both sides of the membrane, not just K^+. Therefore, even if the K^+ concentration is higher inside the cell than outside (the chemical gradient is unfavorable for K^+ uptake), the $\Delta \Psi$ may be in the correct direction (negative interior) and of sufficient magnitude to drive K^+ uptake against its chemical gradient.

Aquaporins

Aquaporins are also known as water channels and are considered to be 'the plumbing system for cells' [15,16]. For decades it was assumed that water simply leaked through biological membranes by numerous processes described above. However, these methods of water permeability could not come close to explaining the rapid movement of water across some cells. Although it had been predicted that water pores must exist in very leaky cells, it wasn't until 1992 that Peter Agre (Figure 14.11) at Johns Hopkins University identified a specific trans-membrane water pore that was later called aquaporin-1. For this accomplishment Agre shared the 2003 Nobel Prize in Chemistry with Rod MacKinnon for his work on

FIGURE 14.11   Peter Agre (1949–). *Courtesy of The Johns Hopkins University, The Johns Hopkins Hospital, and Johns Hopkins Health System*

the potassium channel. Aquaporins are specific for water permeability, excluding the passage of other solutes. A type of aquaporin known as aqua-glyceroporins can also conduct some very small uncharged solutes such as glycerol, CO_2, ammonia, and urea across the membrane. However, all aquaporins are impermeable to charged solutes. Water molecules traverse the aquaporin channel in single file (Figure 14.12) [17].

D. ACTIVE TRANSPORT

A characteristic of all living membranes is the formation and maintenance of trans-membrane gradients of all solutes including salts, biochemicals, macromolecules, and even water. In living cells large gradients of Na^+ and K^+ are particularly important. Typical cell concentrations are:

Cell Interior: 400 mM K^+, 50 mM Na^+
Cell Exterior: 20 mM K^+, 440 mM Na^+

Living cells will also have a $\Delta\Psi$ from -30 to -200mV (negative interior) resulting from the uneven distribution of all ionic solutes including Na^+ and K^+. The chemical and electrical gradients are maintained far from equilibrium by a multitude of active transport systems. Active transport requires a form of energy (often ATP) to drive the movement of solutes against their electrochemical gradient, resulting in a non-equilibrium distribution of the solute across the membrane. A number of non-exclusive and overlapping terms are commonly used to describe the different types of active transport. Some of these are depicted in Figure 14.13 [18].

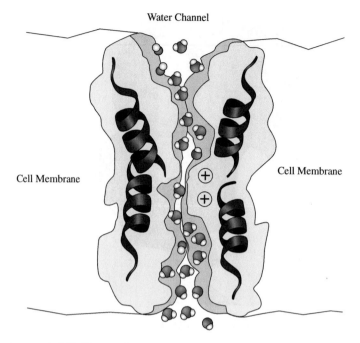

FIGURE 14.12 Aquaporin [17]. Water molecules pass through the aquaporin channel in single file.

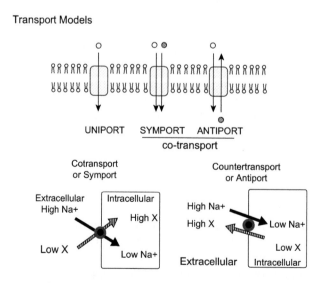

FIGURE 14.13 Basic types of active transport [18].

TABLE 14.3   The Four Types of ATP-utilizing Primary Active
Transport Systems.

ATP-utilizing primary active transport systems	Example
P-Type	Na^+/K^+ ATPase
	Ca^{2+} pump
	H^+ acid pump
F-Type	Mitochondrial ATP synthase
	Chloroplast ATP synthase
V-Type	Vacuolar ATPase
	ABC
(ATP binding cassette transporter)	Many

Primary Active Transport

Primary active transport is also called direct active transport or uniport. It involves using energy (usually ATP) to directly pump a solute against its electrochemical gradient.

The most studied example of primary active transport is the plasma membrane Na^+/K^+ ATPase discussed below. Other familiar examples of primary active transport are the redox H^+-gradient generating system of mitochondria, the light-driven H^+-gradient generating system of photosynthetic thylakoid membranes and the ATP-driven acid (H^+) pump found in the epithelial lining of the stomach. There are 4 basic types of ATP-utilizing primary active transport systems (Table 14.3).

Na^+/K^+ ATPase

A crucial active transport protein is the plasma membrane-bound Na^+/K^+ ATPase. This single enzyme accounts for one-third of human energy expenditure and is often referred to as the 'pacemaker for metabolism'. As a result, the Na^+/K^+ ATPase has been extensively studied for more than fifty years. The enzyme was discovered in 1957 by Jens Skou (Figure 14.14) who, 40 years later, was awarded the 1997 Nobel Prize in Chemistry. As is often the case in biochemistry, a serendipitous discovery of a natural product from the jungles of Africa has been instrumental in unraveling the enzyme's mechanism of action. The compound is ouabain, a cardiac glucoside, first discovered in a poison added to the tip of Somali tribesmen's hunting arrows (Figure 14.15). In fact the name ouabain comes from the Somali word *waabaayo* which means 'arrow poison'. The sources of ouabain are ripe seeds and bark of certain African plants and it is potent enough to kill a hippopotamus with a single arrow. For decades after its

FIGURE 14.14   Jens Skou (1918e). *Copyright to Scanpix Danmark*

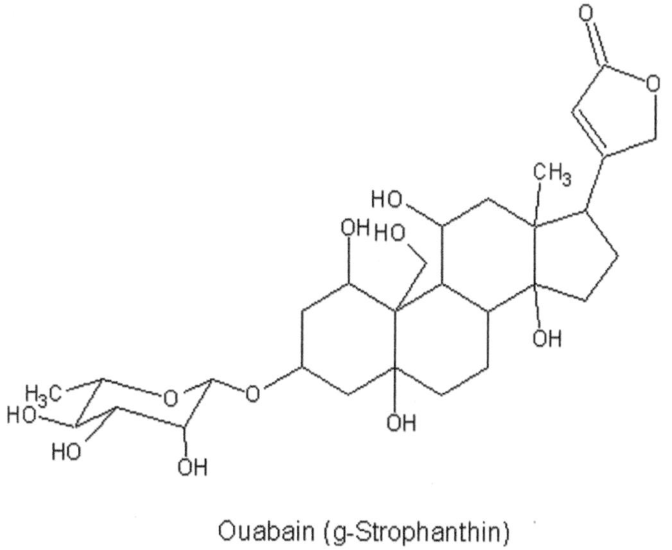

Ouabain (g-Strophanthin)

FIGURE 14.15   Ouabain structure. *Courtesy of B. Bos*

discovery, ouabain was routinely used to treat atrial fibrillation and congestive heart failure in humans. More recently, ouabain has been replaced by digoxin, a structurally related, but more lipophilic cardiac glucoside.

There are several important observations about the Na^+/K^+ ATPase that had to be factored in before a mechanism of action could be proposed. These include:

1. Na^+/K^+ ATPase is an example of primary active transport and active antiport.
2. Na^+/K^+ ATPase is inhibited by ouabain, a cardiac glycoside.
3. Ouabain binds to the outer surface of the Na^+/K^+ ATPase and blocks K^+ transport into the cell.
4. Na^+ binds better from the inside.
5. K^+ binds better from the outside.
6. ATP phosphorylates an aspartic acid on the enzyme from the inside.
7. Phosphorylation is related to Na^+ transport.
8. Dephosphorylation is related to K^+ transport.
9. Dephosphorylation is inhibited by ouabain.
10. 3 Na^+s are pumped out of the cell as 2 K^+s are pumped in.
11. The Na^+/K^+ ATPase is electrogenic.

The mechanism of the Na^+/K^+ ATPase [19] is based on toggling back and forth between two conformational states of the enzyme, ENZ-1 and ENZ-2. Three Na^+s bind from the inside to the Na^+/K^+ ATPase in one conformation (ENZ-1). This becomes phosphorylated by ATP causing a conformation change to ENZ-2~P. ENZ-2~P does not bind Na^+ but does bind 2 K^+s. Therefore 3 Na^+s are released to the outside and 2 K^+s are bound generating ENZ-2~P (2K^+). Upon hydrolysis of ~P, the Na^+/K^+ ATPase reverts back to the original ENZ-1 conformation that releases 2 K^+s and binds 3 Na^+s from the inside. Ouabain blocks the dephosphorylation step.

Mechanism of the Na^+/K^+ ATPase

$$ENZ\text{-}1\ (3Na^+) + ATP \rightarrow ENZ\text{-}2\text{~}P + 3Na^+\ \text{released outside}$$
$$\text{(inside)} \qquad\qquad\qquad \text{(outside)}$$

$$\downarrow$$

$$ENZ\text{-}2\text{~}P\ (2K^+)\ \leftarrow \qquad ENZ\text{-}2\text{~}P + 2\ K^+$$
$$\text{(outside)} \qquad\qquad\qquad \text{(outside)}$$
$$\downarrow$$
$$ENZ\text{-}1\ (3Na^+) + Pi + 2K^+\text{s released inside}$$
$$\text{(inside)}$$

Secondary Active Transport

Secondary active transport (also known as co-transport) systems are composed of two separate functions. The energy-dependent movement of an ion (e.g. H^+, Na^+ or K^+) generates an electrochemical gradient of the ion across the membrane. This gradient is coupled to the movement of a solute in either the same direction (Symport), or in the opposite direction (Antiport, see Figure 14.13 [18]). Movement of the pumped ion down its electrochemical gradient is by facilitated diffusion. The purpose of both types of co-transport is to use the energy in an electrochemical gradient to drive the movement of another solute against its gradient. An example of Symport is the SGLT1 (sodium glucose transport protein-1) in the intestinal epithelium [20]. SGLT1 uses the energy in a downhill trans-membrane movement of Na^+ to transport glucose across the apical membrane against an uphill glucose gradient so that the sugar can be transported into the blood-stream on the opposite side of the cell.

Bacterial Lactose Transport

The secondary active symport system for lactose uptake in *E. coli* is shown in Figure 14.16 [21]. Lactose uptake is driven through a channel by an H^+ gradient generated by the bacterial electron transport system [22]. The free energy equation for transport described above ($\Delta G = RT \ln [s_o'] / [s_o] + ZF\Delta\Psi$) can be rearranged for cases employing H^+ gradients to:

$$\Delta\mu_{H+} = \Delta\Psi - RT/nF \, \Delta pH$$

Where:
$\Delta\mu H+$ is the proton motive force
$\Delta\Psi$ is the trans-membrane electrical potential

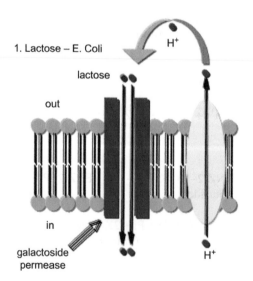

FIGURE 14.16   Lactose transport system in *Escheria coli* [21]. Uptake of lactose is coupled to the movement of H^+ down its electrochemical gradient. This is an example of active transport, co-transport, and active symport.

R is the gas constant

T is the temperature in $^{\circ}K$

n is the solute charge ($+1$ for protons)

F is the Faraday

ΔpH is the trans-membrane pH gradient

It is the *force* on an H^+ (called the proton motive force, $\Delta\mu_{H+}$) that drives lactose uptake. Note that the ability to take up lactose is a combination of the electrical gradient and the pH gradient. Although lactose uptake is directly coupled to H^+ trans-membrane movement, it may be possible to take up lactose even if the pH gradient is zero (i.e. the $\Delta\Psi$ is sufficiently large).

Vectorial Metabolism

Over fifty years ago Peter Mitchell recognized the importance of what he termed 'vectorial metabolism' [23,24]. Water-soluble enzymes convert substrate to product without any directionality. Mitchell proposed that many enzymes are integral membrane proteins that have a unique trans-membrane orientation. When these enzymes convert substrate to product they do so in one direction. This enzymatic conversion is therefore 'vectorial' or unidirectional. Mitchell expanded this basic concept into his famous chemiosmotic mechanism for ATP synthesis in oxidative phosphorylation [25,26]. For this idea Mitchell was awarded the 1997 Nobel Prize in Chemistry. Vectorial metabolism has been used to describe several membrane transport systems. For example, it has been reported in some cases the uptake of glucose into a cell may be faster if the external source of glucose is sucrose rather than free glucose. Through a vectorial trans-membrane reaction, membrane-bound sucrase may convert external sucrose into internal glucose + fructose more rapidly than the direct transport of free glucose through its transport system.

Mitchell defined one type of vectorial transport as Group Translocation, the best example being the PTS (phosphotransferase system) discovered by Saul Roseman in 1964. PTS is a multicomponent active transport system that uses the energy of intracellular phosphoenol pyruvate (PEP) to take up extracellular sugars in bacteria. Transported sugars may include glucose, mannose, fructose, and cellobiose. Components of the system include both plasma membrane and cytosolic enzymes. Energy to drive the system comes from PEP (ΔG of hydrolysis is -61.9 KJ/mol). The high energy phosphoryl group is transferred through an enzyme bucket brigade from PEP to glucose producing glucose-6-phosphate (PEP $\rightarrow$ EI $\rightarrow$ HPr $\rightarrow$ EIIA $\rightarrow$ EIIB $\rightarrow$ EIIC $\rightarrow$ glucose-6-phosphate). The sequence is depicted in more detail in Figure 14.17 [27]. HPr stands for heat-stable protein that carries the high energy ~P from EI (enzyme-I) to EIIA. EIIA is specific for glucose and transfers ~P to EIIB that sits next to the membrane where it takes glucose from the trans-membrane EIIC and phosphorylates it producing glucose-6-phosphate. Although it is glucose that is being transported across the membrane, it never actually appears inside the cell as free glucose, but rather as glucose-6-phosphate. Free glucose could leak back out of the cell, but glucose-6-phosphate is trapped inside, where it can be rapidly metabolized through glycolysis. Group translocation is defined by a transported solute appearing in a different form immediately after crossing the membrane.

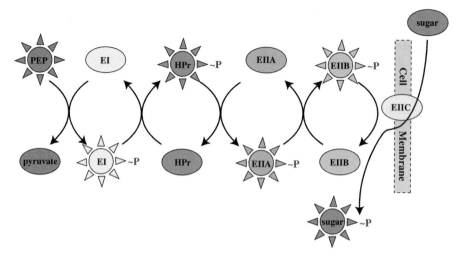

FIGURE 14.17 The bacterial PTS system for glucose transport [27].

E. IONOPHORES

The term ionophore means 'ion bearer'. Ionophores are small, lipid-soluble molecules, usually of microbial origin, whose function is to conduct ions across membranes [28,29]. They are facilitated diffusion carriers that transport ions down their electrochemical gradient. Ionophores can be divided into two basic classes: channel formers and mobile carriers (Figure 14.18) [30]. Channel formers are long lasting, stationary structures that allow many ions to rapidly flow across a membrane. Mobile carriers bind to an ion on one side of a membrane, dissolve in the membrane bilayer, and release the ion on the other side. They can only carry one ion at a time. Four representative ionophores will be discussed, the K^+ ionophore valinomycin, the proton ionophore 2,4-dinitrophenol, synthetic crown ethers, and the channel forming ionophore nystatin (Figure 14.19).

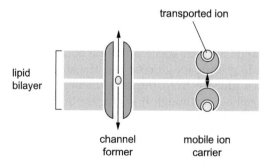

FIGURE 14.18 Two basic types of ionophores, channel formers (left) and mobile carriers (right) [30].

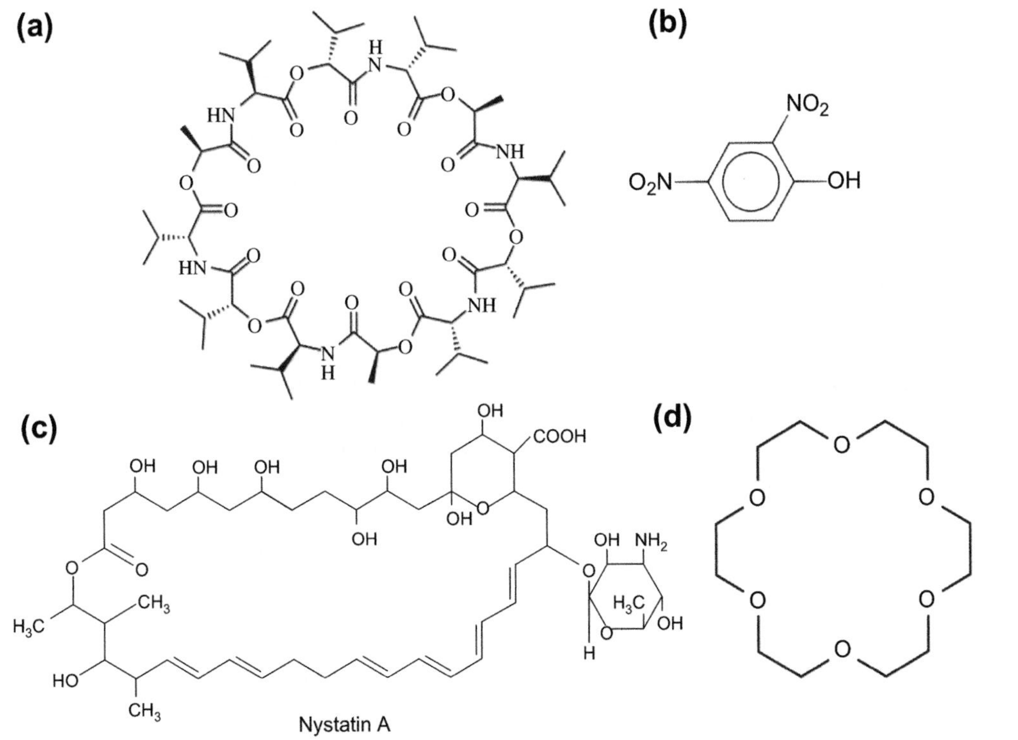

(a)

(b)

(c)

Nystatin A

(d)

FIGURE 14.19 Representative examples of ionophores: (a) Valinomycin (b) 2,4-dinitrophenol (c) Nystatin (d) Crown ether.

Valinomycin

Superficially, valinomycin resembles a cyclic peptide (Figure 14.19). However, upon closer examination the ionophore is actually a 12 unit (dodeca) depsipeptide where amino acid peptide bonds alternate with amino alcohol ester bonds. Therefore the linkages that hold the molecule together alternate between nitrogen esters (peptide bonds) and oxygen esters. The units that comprise valinomycin are D- and L-valine (hence the name 'valinomycin'), hydroxyvaleric acid, and L-lactic acid. The circular structure is a macrocyclic molecule where 12 carbonyl oxygens face the inside of the structure where they chelate a single K^+. The outside surface of valinomycin is coated with 9 hydrophobic side chains of D- and L-valine and L-hydroxyvaleric acid. The polar interior of valinomycin precisely fits one K^+. The binding constant for K^+-valinomycin is 10^6 while Na^+-valinomycin is only 10. This emphasizes the high selectivity valinomycin has for K^+ over Na^+. Valinomycin, therefore, has an oily surface that readily dissolves in a membrane lipid bilayer, carrying K^+ across the membrane down its electrochemical gradient.

Valinomycin was first recognized as a potassium ionophore by Bernard Pressman in the early 1960s [31,32]. He reported that valinomycin, a known antibiotic, stimulated K^+ uptake and H^+ efflux from mitochondria. Many studies showed that valinomycin dissipates essential trans-membrane electrochemical gradients causing tremendous metabolic upheaval in many organisms including microorganisms. It is for this reason that valinomycin was recognized as an antibiotic long before it was identified as an ionophore. Currently several ionophores are added to animal feed as antibiotics and growth enhancing additives [33]. Recently valinomycin has been reported to be the most potent agent against SARS-CoV (severe acute respiratory syndrome-coronavirus), a severe form of pneumonia first identified in 2003 [34].

2,4-Dinitrophenol (DNP)

2,4-Dinitrophenol (DNP) is considered to be the classic uncoupler of oxidative phosphorylation. It is a synthetic lipid-soluble proton ionophore that dissipates proton gradients across bioenergetic membranes (mitochondrial inner, thylakoid, bacterial plasma). An uncoupler is therefore an H^+ facilitated diffusion carrier. Elucidating the role of DNP in uncoupling oxidative phosphorylation was an essential component in support of Peter Mitchell's Chemiosmotic Hypothesis [25]. Electron movement from NADH or $FARH_2$ to O_2 via the mitochondrial electron transport system generates a considerable amount of electrical energy that is partially captured as a trans-membrane pH gradient. The movement of H^+s back across the membrane, driven by the electrochemical gradient, is through a channel in the F_1ATPase (an F-type primary active transport system, see above) that is coupled to ATP synthesis. DNP short-circuits the H^+ gradient before it can pass through the F_1ATPase, thus uncoupling electron transport, the energy source for generating the H^+ gradient, from ATP synthesis. Therefore, in the presence of DNP, electron transport continues, even at an accelerated rate, but ATP production is diminished. The energy that should have been converted to chemical energy as ATP is then released as excess heat.

This combination of properties led to the medical application of DNP to treat obesity from 1933 to 1938 [35]. Upon addition of DNP:

- The patient became weak due to low ATP levels.
- Breathing increased due to increased electron transport to rescue ATP production.
- Metabolic rate increased.
- Body temperature increased due to inability to trap electrical energy as chemical energy in the form of ATP.
- Body weight decreased due to increased respiration burning more stored fat.

DNP was indeed a successful weight loss drug. Two of the early proponents of the use of DNP as a diet drug, Cutting and Tainter at Stanford University, estimated that more than 100,000 people in the United States had tested the weight-loss drug during its first year in use [35]. DNP, however, did have one disturbing side effect — death! Fatality was not caused by a lack of ATP, but rather by a dangerous increase in body temperature (hyperthermia). In humans, 20—50 mg/kg of DNP can be lethal. Although general use of DNP in the United States was discontinued in 1938, it is still employed in other countries and by bodybuilders to eliminate fat before competitions.

Crown Ethers

Crown ethers are a family of synthetic ionophores that are generally similar in function to the natural product valinomycin [36]. The first crown ether was synthesized by Charles Pederson

FIGURE 14.20   Charles Pedersen (1904–1989). *Reprinted with permission.* [63]

(Figure 14.20) while working at DuPont in 1967. For this work Pedersen was co-awarded the 1987 Nobel Prize in Chemistry. Crown ethers are cyclic compounds composed of several ether groups. The most common crown ethers are oligomers of ethylene oxide with repeating units of $(-CH_2CH_2O-)_n$ where $n = 4$ (tetramer), $n = 5$ (pentamer) or $n = 6$ (hexamer). Crown ethers are given names X-crown-Y, where X is the total number of atoms in the ring and Y is the number of these atoms that are oxygen. The term crown refers to the crown-like shape that the molecule takes. Crown ether oxygens form complexes with specific cations that depend on the number of atoms in the ring. For example, 18-crown-6 has high affinity for K^+, 15-crown-5 for Na^+, and 12-crown-4 for Li^+. The crown ether depicted in Figure 14.19d is 18-crown-6. Like valinomycin, the exterior of the ring is hydrophobic, allowing crown ethers to dissolve in the membrane lipid bilayer while carrying the sequestered cation. It is now possible to tailor-make crown ethers of different sizes that can encase a variety of phase transfer catalysts. These crown ethers are used to catalyze reactions inside the membrane hydrophobic interior.

Nystatin

Nystatin is a channel forming ionophore that creates a hydrophobic pore across a membrane [37,38]. Channel-forming ionophores allow for the rapid facilitated diffusion of various ions that depend on the dimensions of the pore. Nystatin, like other channel-forming ionophores amphotericin B and natamycin, is a commonly used anti-fungal agent. Finding medications that can selectively attack fungi in the presence of normal animal cells presents a difficult challenge since both cell types are eukaryotes. Bacteria, being prokaryotes, are sufficiently different from eukaryotes to present a variety of anti-bacterial approaches not amenable to fungi. However, Fungi do have an Achilles Heel. Fungal plasma membranes have as their dominant sterol ergosterol, not the animal sterol cholesterol (see Chapter 5). Nystatin binds preferentially to ergosterol, thus targeting fungi in the presence of animal cells. When present in sufficient levels, nystatin complexes with ergosterol and forms trans-membrane channels that lead to K^+ leakage and death of the fungus. Nystatin is a polyene anti-fungal ionophore that is effective against many molds and yeast including *Candida*. A major use of nystatin is as a prophylaxis for AIDS patients who are at risk for fungal infections.

F. GAP JUNCTIONS

Gap junctions are a structural feature of many animal plasma membranes [39,40]. In plants similar structures are known as plasmodesmata. Gap junctions were introduced earlier in Chapter 11 (see Figure 11.6). Gap junctions represent a primitive type of intercellular communication that allows trans-membrane passage of small solutes like ions, sugars, amino acids, and nucleotides while preventing migration of organelles and large polymers like proteins and nucleic acids. Gap junctions connect the cytoplasms of two adjacent cells through non-selective channels. Connections through adjacent cells are at locations where the gap between cells is only 2–3 nm. This small gap is where the term 'gap junction' originated. Gap junctions are normally clustered from a few to over a thousand in select regions of a cell plasma membrane.

Early experiments involved injecting fluorescent dyes, initially fluorescein (MW 300), into a cell and observing the dye movement into adjacent cells with a fluorescence microscope [41,42]. Currently, Lucifer Yellow has become the fluorescent dye of choice for gap junction studies, replacing fluorescein. The dye at first only appeared in the initially labeled cell. With time, however, the dye was observed to spread to adjacent cells through what appeared to be points on the plasma membrane. These points were later recognized as gap junctions. By varying the size of the fluorescent dye it was shown that there was an upper size limit for dye diffusion. Solutes had to have a molecular weight of less than ~1,200 to cross from one cell to another [41].

Although gap junctions were obviously channels that connected the cytoplasms of adjacent cells, it was years before their structure, shown in Figure 14.21 [43], was determined [44]. Each channel in a gap junction is made up of 12 proteins called connexins. Six hexagonally arranged connexins are associated with each of the adjacent cell plasma membranes that the gap junction spans. Each set of six connexins is called a connexon and forms half of the gap junction channel. Therefore, one gap junction channel is composed of two aligned connexons and 12 connexins. Each connexin has a diameter of ~7nm and the hollow center formed between the six connexins, (the channel), is ~3nm in diameter. Gap junctions allow adjacent cells to be in constant electrical and chemical communication with one another. Of particular importance is the rapid transmission of small second messengers, such as inositol triphosphate (IP_3) and Ca^{2+}.

It appears that all cells in the liver are interconnected through gap junctions. This presents a possible dilemma. If even a single cell is damaged, deleterious effects may be rapidly

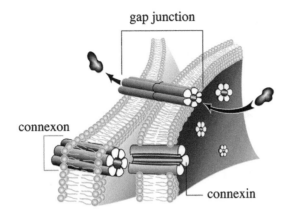

FIGURE 14.21 Gap junction [43]. Six connexins form a connexon and one connexon from each cell unite to form a gap junction.

spread throughout the entire liver. Preventing this is one important function of Ca^{2+}. Extracellular Ca^{2+} is ~10^{-3} M while intracellular levels are maintained at ~10^{-6} M. If a cell is damaged, Ca^{2+} rushes in, dramatically increasing intracellular Ca^{2+}. Gap junction channels close if intracellular Ca^{2+} reaches 10^{-3} M, thus preventing the spread of damage.

Gap junctions are particularly important in cardiac muscle as the electrical signals for contraction are passed efficiently through these channels [45]. As would be expected, malfunctions of gap junctions lead to a number of human disorders including demyelinating neurodegenerative diseases, skin disorders, cataracts, and even some types of deafness.

G. OTHER WAYS TO CROSS THE MEMBRANE

There are several other ways that solutes, including large macromolecules, can cross membranes. These methods, receptor-mediated endocytosis (RME), phagocytosis, pinocytosis, exocytosis, and membrane blebbing involve large sections of a membrane containing many lipids and proteins.

Receptor Mediated Endocytosis

Receptor mediated endocytosis (RME) [46,47] is also known as clathrin-dependent endocytosis because of involvement of the membrane-associated protein clathrin in forming membrane vesicles that become internalized into the cell. Clathrin plays a major role in formation of clathrin coated pits and coated vesicles. Since clathrin was first isolated and named by Barbara Pearse in 1975 [48], it has become clear that clathrin and other coatproteins play essential roles in cell biology. Clathrin is an essential component in building small vesicles for uptake (endocytosis) and export (exocytosis) of many molecules. While the previously discussed methods of membrane transport involved small solutes, RME is the primary mechanism for the specific internalization of most macromolecules by eukaryotic cells.

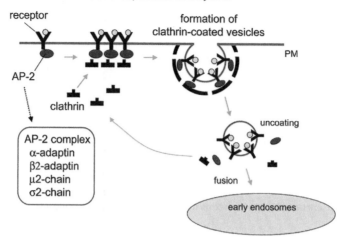

FIGURE 14.22   Receptor mediated endocytosis [49].

RME begins with a ligand binding to a specific receptor that spans the plasma membrane (Figure 14.22, [49]). Examples of these ligands include hormones, growth factors, enzymes, serum proteins, LDL (with attached cholesterol), transferrin (with attached iron), antibodies, some viruses, and even bacterial toxins. After receptor binding, the complex diffuses laterally in the plasma membrane until it encounters a specialized patch of membrane called a coated pit. The receptor-ligand complexes accumulate in these patches as do other proteins including clathrin, adaptor protein, and dynamin. Since coated pits occupy ~20% of the plasma membrane surface area, they are not minor membrane features. The collection of these proteins starts to curve a section of the membrane that eventually pinches off to form an internalized coated vesicle. Clathrin and dynamin then recycle back to the plasma membrane leaving an uncoated vesicle that is free to fuse with an early endosome. After the early endosomes mature into late endosomes, they then go on to the lysosome for digestion. RME is a very fast process. Invagination and vesicle formation takes ~1 min. One single cultured fibloblast cell can produce 2,500 coated pits per minute.

One example of RME has received a great deal of attention because of its essential role in human health, namely maintaining the proper level of cholesterol in the body. Malfunctions in the RME process for uptake of the cholesterol carrying LDL lead to hypercholesterolemia and cardiovascular disease [46,50]. RME and its role in cholesterol metabolism was discovered by Michael Brown and Joseph Goldstein (Figure 14.23) of the University of Texas Health Science Center in Dallas (now the UT Southwestern Medical Center) who received the 1985 Nobel Prize in Physiology and Medicine for their iconic work.

Pinocytosis and Phagocytosis

Two similar transport processes that have been known for a long time are pinocytosis and phagocytosis [51]. Both involve non-specific uptake (endocytosis) of many things from water

FIGURE 14.23   Left: Joseph Goldstein (1940–); Right: Michael Brown (1941–) [64].

and ions through to large macromolecules and, for phagocytosis, even whole cells. Pinocytosis is Greek for 'cell drinking' and involves the plasma membrane invaginating a volume of extra-cellular fluid and anything it contains including water, salts, biochemicals, and even soluble macromolecules. Phagocytosis is Greek for 'cell eating' and involves the plasma membrane invaginating large insoluble solids.

Pinocytosis

Pinocytosis is a form of endocytosis involving fluids containing small solutes. In humans this process occurs in cells lining the small intestine and is used primarily for absorption of fat droplets. In endocytosis the cell plasma membrane extends and folds around desired extra-cellular material forming a pouch that pinches off creating an internalized vesicle (Figure 14.24, [52]). The invaginated pinocytosis vesicles are much smaller than those generated by phagocytosis. The vesicles eventually fuse with the lysosome whereupon the vesicle contents are digested. Pinocytosis involves a considerable investment of cellular energy in the form of ATP and so is many thousand times less efficient than receptor-mediated endocytosis. Also, in sharp contrast to RME, pinocytosis is non-specific for the substances it accumulates. Pinocytosis is not a recent discovery but was first observed decades before the other transport systems discussed above. Its discovery is attributed to Warren Lewis in 1929.

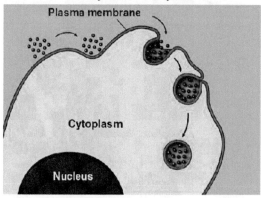

FIGURE 14.24 Pinocytosis, a type of endocytosis. An invagination of the plasma membrane encapsulates many water-soluble solutes ranging in size from salts to macromolecules. *Courtesy of Dr. Gary Kaiser*

Phagocytosis

Phagocytosis is a type of endocytosis that involves uptake of large solid particles, often >0.5 μm [53]. The particles are aggregates of macromolecules, parts of other cells and even whole microorganisms and, in contrast to pinocytosis, phagocytosis (shown in Figure 14.25) has surface proteins that specifically recognize and bind to the solid particles. Figure 14.25 [54] depicts events in phagocytosis. Phagocytosis is a routine process that amoeba and ciliated protozoa use to obtain food. In humans phagocytosis is restricted to specialized cells called phagocytes that include white blood cell neutrophils and macrophages. As with pinocytosis, phagocytosis generates intracellular vesicles called phagosomes that have sequestered solid particles they transport to the lysosome for digestion. Phagocytosis is

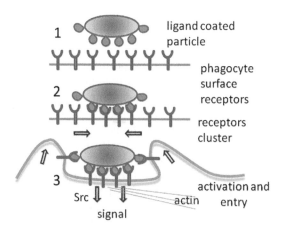

FIGURE 14.25 Phagocytosis, a type of endocytosis that involves uptake of large solid particles [65].

a major mechanism used by the immune system to remove pathogens and cell debris. In fact very early studies of the immune system led Elie Metchnikoff to discover phagocytosis in 1882. For this work Metchnikoff shared the 1908 Nobel Prize in Medicine with Paul Ehrlich.

Exocytosis is the process by which cells excrete waste and other large molecules from the cytoplasm to the cell exterior [55] and therefore is the opposite of endocytosis. Exocytosis generates vesicles referred to as secretory or transport vesicles. In exocytosis intracellular vesicles fuse with the plasma membrane and release their aqueous sequestered contents to the outside at the same time that the vesicular membrane hydrophobic components (mostly lipids and proteins) are added to the plasma membrane (Figure 14.26, [56]). Steady state composition of the plasma membrane results from a balance between endocytosis and exocytosis. The resultant process of plasma membrane recycling is amazingly fast. For example, pancreatic secretory cells recycle an amount of membrane equal to the whole surface of the cell in ~90 min. Even faster are macrophages that can recycle the contents of the plasma membrane in only 30 min.

Before approaching the plasma membrane for fusion, exocytosis vesicles have a prior life that will not be considered here. The vesicles must first dock with the plasma membrane, a process that keeps the two membranes separated at <5–10 nm. During docking, complex molecular rearrangements occur to prepare the membranes for fusion. The process of vesicle fusion and release of aqueous compartment components is driven by SNARE proteins (see Chapter 10) [57]. Therefore by the process of exocytosis:

- The surface of the plasma membrane increases by the size of the fused vesicular membrane. This is particularly important if the cell is growing.

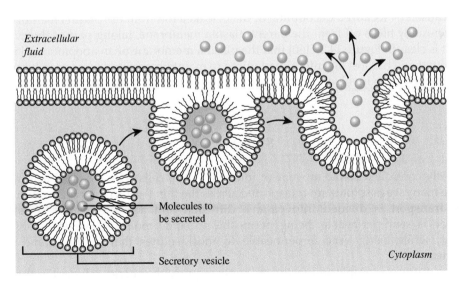

FIGURE 14.26 Exocytosis. Intra-cellular secretory vesicles fuse with the plasma membrane releasing their water-soluble contents to the outside and adding membrane material to the plasma membrane [56].

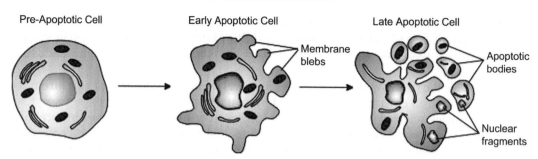

FIGURE 14.27   Membrane blebbing during apoptosis [60].

- The material sequestered inside the vesicle is released to the cell exterior. Included in the vesicle contents may be waste products and intracellular toxins or signaling molecules like hormones or neurotransmitters.
- Proteins embedded in the vesicular membrane become part of the plasma membrane. This makes correct protein orientation in the vesicular membrane absolutely essential. The side of the protein facing the inside of the vesicle before fusion faces the outside of the plasma membrane after fusion.
- A special type of exocytosis called 'kiss-and-run' occurs in synapses [58]. The vesicles only make very brief contact with the plasma membrane whereupon they release their contents (neurotransmitters) to the outside and immediately return to the cytoplasm empty. Since fusion does not occur, the vesicular membrane is not incorporated into the plasma membrane.

Plasma membrane blebbing is a morphological feature of cells undergoing late stage apoptosis (programmed cell death) [59]. A bleb is an irregular bulge in the plasma membrane of a cell caused by localized decoupling of the cytoskeleton from the plasma membrane. The bulge eventually blebs off from the parent plasma membrane, taking part of the cytoplasm with it. It is clear in Figure 14.27 [60] that the plasma membrane of an apoptotic cell is highly disintegrated and has lost the integrity required to maintain essential trans-membrane gradients. Blebbing is also involved in some normal cell processes including cell locomotion and cell division.

SUMMARY

Carefully controlled solute movement into and out of cells is an essential feature of life. There are many ways solutes are transported across the thin (~40 Å) membrane hydrophobic barrier. Transport is divided into passive diffusion and active transport. A biological membrane is semi-permeable, being permeable to some molecules, most notably water (osmosis), while being very impermeable to most solutes that require some form of transporter.

Passive diffusion (simple and facilitated) only requires the energy inherent in the solute's electrochemical gradient and results in its equilibrium across the membrane. In contrast, active transport requires additional energy (i.e. ATP), and results in a non-equilibrium, net accumulation of the solute. Passive transport can involve simple diffusion or facilitated carriers including ionophores and channels. Active transport comes in many, often complex forms. Examples of active transport include primary active transport (uniport), secondary active transport (co-transport, antiport), and group translocation.

Besides the multitude of transport systems, transport can be accomplished by gap junctions, receptor mediated endocytosis, phagocytosis, pinocytosis, exocytosis, and apoptotic membrane blebbing.

The last chapter of this book (Chapter 15) will discuss some biological aspects of membrane structure including how liposomes can be used for drug delivery. Also, the 'bad' dietary fatty acids (*trans* fatty acids) will be contrasted with a 'good' fatty acid (the omega-3 fatty acid docosahexaenoic acid), as they impact human health.

References

[1] Alberts B, Bray D, Lewis J, Raff M, Roberts K, Watson JD. Principles of membrane transport. In: Molecular Biology of the Cell. 3rd ed. New York: Garland; 1994.

[2] Blok MC, van Deenen LLM, de Gier J. Effect of the gel to liquid crystalline phase transition on the osmotic behaviour of phosphatidylcholine liposomes. Biochim Biophys Acta 1976;433:1—12.

[3] Miller FP, Vandome AF, McBrewster J. Ficks Laws of Diffusion. VDM; 2010.

[4] Water Systems: Aqua Technology for the 21st Century. Introductory information on reverse osmosis.

[5] Bangham AD, de Gier J, Greville GD. Osmotic properties and water permeability of phospholipid liquid crystals. Chem Phys Lipids 1967;1:225—46.

[6] Gier De. Osmotic behaviour and permeability properties of liposomes (Review). Chem Phys Lipids 1993;64:187—96.

[7] Baldwin SA, editor. Membrane Transport: A Practical Approach. USA: Oxford University Press; 2000.

[8] Alberts B, Bray D, Johnson A, Lewis J, Raff M, Roberts K, et al. Comparison of passive and active transport. In: Essential Cell Biology. 2nd ed. Taylor Francis: Garland; 2004 [Figure 12.4].

[9] Huang S, Czech MP. The GLUT4 glucose transporter. Cell Metab 2007;5:237—52.

[10] Biologia Medica. 2010. Transporte de Glucosa: GLUT y SGLT Seminarios de Biología Celular y Molecular — USMP Filial Norte. University of San Martin de Porres, Peru.

[11] Lippiat JD. Potassium channels. Methods and protocols. In: Series in Methods in Molecular Biology, vol. 491. New York, NY: Humana; 2009.

[12] Doyle DA, Morais CJ, Pfuetzner RA, Kuo A, Gulbis JM, Cohen SL, et al. The structure of the potassium channel: Molecular basis of K^+ conduction and selectivity. Science 1998;280:69—77.

[13] Goldin AL. Resurgence of sodium channel research. Annu Rev Physiol 2001;63:871—94.

[14] Penzotti JL, Fozzard HA, Lipkind GM, Dudley SC. Differences in Saxitoxin and tetrodotoxin binding revealed by mutagenesis of the Na^+ channel outer vestibule. Biophys J 1998;75(6):2647—57.

[15] Noda Y, Sohara E, Ohta E, Sasaki S. Nephrology. Aquaporins in kidney pathophysiology. Nat Rev 2010;6:168—78.

[16] Hill AE, Shachar-Hill B, Shachar-Hill Y. What are aquaporins for? J Membr Biol 2004;197:1—32.

[17] Vetenskapsakademien Kungl, The Royal Swedish Academy of Sciences. The Nobel Prize in Chemistry. Peter Agre, Roderick MacKinnon, www.nobelprize.org; 2003.

[18] Gerk P. Active Transport of Drugs Across Membranes. Enna SJ and Bylund DB, Editors-in-Chief. xPharm: The Comprehensive Pharmacology Reference. New York: Elsevier; 2007. p. 1—6.

[19] Lopina OD. Na^+, K^+ ATPase: Structure, mechanism, and regulation. Membr Cell Biol 2000;13:721—44.

[20] Wright EM. Renal Na^+-glucose cotransporters. Am J Physiol, Renal Physiol 2001;280:F10—8.

[21] Jakubowski H, editor. Chapter 9. Signal Transduction. A. Energy transduction: Uses of ATP, http://employees.csbsju.edu/hjakubowski/classes/ch331/signaltrans/olsignalenergy.html; 2002.

[22] Martin SA. Nutrient transport by ruminal bacteria: a review. J Anim Sci 1994;72:3019—31.

[23] Mitchell P, Moyle J. Group-translocation: a consequence of enzyme-catalysed group-transfer. Nature 1958;182:372—3.

[24] Mitchell P, Moyle J. Coupling of metabolism and transport by enzymic translocation of substrates through membranes. Proc R Phys Soc Edinburgh 1959;28:19–27.

[25] Mitchell P, Moyle J. Chemiosmotic hypothesis of oxidative phosphorylation. Nature 1967;213:137–9.

[26] Mitchell P. Proton current flow in mitochondrial systems. Nature 1967;214:1327–8.

[27] Herzberg O, Canner D, Harel M, Prilusky J, Hodis E. Enzyme I of the Phosphoenolpyruvate: Sugar Phosphotransferase System. Proteopedia, Weizmann Institute of Science in Israel 2011.

[28] Pressman BC. Biological applications of ionophores. Annu Rev Biochem 1976;45:501–30.

[29] Szabo G. Structural aspects of ionophore function. Fed Proc 1981;40:2196–201.

[30] Alberts B, Johnson A, Lewis J, Raff M, Roberts K, Walter P. Principles of membrane transport [Figure 11.5]. In: Molecular Biology of the Cell. 4th ed. New York: Garland; 2002.

[31] Moore C, Pressman BC. Mechanism of action of valinomycin on mitochondria. Biochem Biophys Res Commun 1964;15:562–7.

[32] Pressman BC. Induced active transport of ions in mitochondria. Proc Natl Acad Sci USA 1965;53: 1076–83.

[33] Page SW. The role of enteric antibiotics in livestock production. Canberra, Australia: Avcare; 2003.

[34] Cheng YQ. Deciphering the biosynthetic codes for the potent anti-SARS-CoV cyclodepsipeptide valinomycin in Streptomyces tsusimaensis ATCC 15141. Chembiochem 2006;7:471–7.

[35] Tainter ML, Stockton AB, Cutting WC. Use of dinitrophenol in obesity and related conditions: a progress report. J Am Med Assoc 1933;101:1472–5.

[36] Huszthy P, Toth T. Synthesis and molecular recognition studies of crown ethers. Periodica Polytechnica 2007;51:45–51.

[37] Borgos SEF, Tsan P, Sletta H, Ellingsen TE, Lancelin J-M, Sergey B, et al. Probing the Structure–function relationship of polyene macrolides: Engineered biosynthesis of soluble nystatin analogues. J Med Chem 2006;49:2431–9.

[38] Lopes S, Castanho MARBJ. Revealing the orientation of nystatin and amphotericin B in lipidic multilayers by UV-Vis linear dichroism. Phys Chem B 2002;106:7278–82.

[39] Revel JP, Karnovsky MJ. Hexagonal array of subunits in intracellular junctions of the mouse heart and liver. J Cell Biol 1967;33:C7–C12.

[40] Peracchia, C. (Ed.). Gap junctions: Molecular basis of cell communication in health and disease. In: Current Topics in Membranes and Transport. Series Editor: D. Benos. Elsevier Publishing, New York; 1999

[41] Simpson I, Rose B, Loewenstein WR. Size limit of molecules permeating the junctional membrane channels. Science 1977;195:294–6.

[42] Imanaga I, Kameyama M, Irisawa H. Cell-to-cell diffusion of fluorescent dyes in paired ventricular cells. Am J Physiol Heart Circ Physiol 1987;252:H223–32.

[43] Echevarria W, Nathanson MH. Gap junctions in the liver. In: Chapter: Channels and Transporters. Molecular Pathogenesis of Cholestasis. Madame Curie Bioscience Database; 2003.

[44] Cao F, Eckert R, Elfgang C, Nitsche JM, Snyder SA, Hulsen DF, et al. A quantitative analysis of connexin-specific permeability differences of gap junctions expressed in HeLa transfectants and Xenopus oocytes. J Cell Sci 1998;111:31–43.

[45] Jongsma HJ, Wilders R. Gap junctions in cardiovascular disease. Circ Res 2000;86:1193–7.

[46] Goldstein JL, Brown MS, Anderson RGW, Russell DW, Schneider W. Receptor-mediated endocytosis: Concepts emerging from the LDL receptor system. Annu Rev Cell Biol 1985;1:1–39.

[47] Wileman T, Harding C, Stahl P. Receptor-mediated endocytosis. Biochem J 1985;232:1–14.

[48] Pearse BM. Clathrin: a unique protein associated with intracellular transfer of membrane by coated vesicles. Proc Natl Acad Sci USA 1976;73:1255–9.

[49] Grant BD, Sato M, 2006, http://www.wormbook.org/chapters/www_intracellulartrafficking/intracellulartrafficking.html.

[50] Brown MS, Goldstein JL. Receptor-mediated control of cholesterol metabolism. Science 1976;191:150–4.

[51] Aderem A, Underhill DM. Mechanisms of phagocytosis in macrophages. Annu Rev Immunol 1999;17:593–623.

[52] http://faculty.southwest.tn.edu/rburkett/Menbranes%20&%20cell%20function.htm.

[53] Underhill DM, Ozinsky A. Phagocytosis of microbes: Complexity in action. Annu Rev Immunol 2002;20:825–52.

[54] Ernst JD, Stendahl O, editors. Phagocytosis of Bacteria and Bacterial Pathogenicity. Cambridge University Press; 2006. p. 6.

[55] Li L, Chin L-S. The molecular machinery of synaptic vesicle exocytosis. Cell Mol Life Sci 2003;60:942—60.

[56] Cell and Cell Structure. IV. Membrane Transport Processes. 2001. Benjamin Cummings. An imprint of Addison Wesley Longman. Cell and Cell Structure. IV. Membrane Transport Processes.

[57] Sudhof TC, Rothman JE. Membrane fusion: Grappling with SNARE and SM proteins. Science 2009;323:474—7.

[58] Wightman RM, Haynes CL. Synaptic vesicles really do kiss and run. Nat Neurosci 2004;7:321—2.

[59] Coleman ML, Sahai EA, Yeo M, Bosch M, Dewar A, Olson MF. Membrane blebbing during apoptosis results from caspase-mediated activation of ROCK1. Nat Cell Biol 2001;3:339—45.

[60] O'Day D. 2011. Human Development, Bio 380F. Modified from: Walker NI, Harmon BV, Gobe GC, Kerr JF. 1988. Patterns of cell death. Methods Achiev Exp Pathol 13, 18—54.

[61] Paroxysm: http://en.wikipedia.org/wiki/Adolf_Gaston_Eugen_Fick; public domain.

[62] Hannes Röst: http://en.wikipedia.org/wiki/Rod_MacKinnon; public domain.

[63] Goodman C. Jean-Marie Lehn. Nature Chemical Biology 2007;3:685.

[64] The Scientist: http://classic.the-scientist.com/fragments/election/scientist_voting.jsp; reprinted with permission

[65] GrahamColm: http://en.wikipedia.org/wiki/Phagocytosis; CC-BY-SA-3.0

Membranes and Human Health

OUTLINE

A. Liposomes as Drug-Delivery Agents 339
   1. Gout 339
   2. Liposomes 340
   3. Leishmaniasis 342
   4. Development of Liposomes that Avoid the RES 343
   5. Stealth Liposomes 344
   6. Thermo-Liposomes 344
   7. pH-Sensitive Liposomes 346

B. Effect of Dietary Lipids on Membrane Structure/Function 348
   1. Trans Fatty Acids 348
   2. Docosahexaenoic Acid (DHA) 350

Summary 352

References 353

A. LIPOSOMES AS DRUG-DELIVERY AGENTS

The complexity of membrane structure and functions outlined in this book supports a wealth of potential targets to alleviate human afflictions. Indeed, drug companies have taken notice and made membranes a major area for future drug development. A thorough survey of these approaches is far too large to address here. Instead a few representative examples are discussed below.

1. Gout

For decades it has been appreciated that most life processes are in some way related to membranes. Many examples of clever experiments designed to test possible disease mechanisms of action adorn the literature. An early example by Gerald Weissmann (Figure 15.1) in 1972 [1] typifies this approach. (Gerald Weissmann is also known for coining the term 'liposome'.) His experiment involved understanding the molecular mechanism of gout.

FIGURE 15.1　Gerald Weissmann (1930–). *Courtesy of John Simon Guggenheim Memorial Foundation.*

For centuries many famous historical figures suffered from gout. Included in this list are Henry VIII, Issac Newton, Thomas Jefferson, Benjamin Franklin, Charles Darwin, and even 'Sue', the famous Tyrannosaurus Rex skeleton. So, what is responsible for gout? It is known that gout is related to a rich diet, perhaps one that is based on red meat dripping with cholesterol. Men 'take the gout', but young women do not. However, post-menopausal women do get gout. Also, it is a historical fact that eunuchs (castrated males) do not 'take the gout'. Weissmann therefore proposed that gout was related to cholesterol and testosterone, but not estrogen. Gout is also associated with accumulation of uric acid crystals. Weissmann proposed that uric acid crystals attach to the lysosomal membrane containing cholesterol and testosterone, making the organelle very leaky to the sequestered hydrolytic enzymes. Once released from the lysosome, the enzymes destroy the cell, inducing gout. To test this hypothesis, Weissmann made 'boy' (containing testosterone) and 'girl' (containing estrogen) cholesterol-enriched liposomes. Upon the addition of uric acid the 'boy' liposomes, but not the 'girl' liposomes, became leaky to a sequestered solute. This clever experiment accounts for the basic characteristics of gout. Years later, Weissmann developed two liposome-encapsulated drugs (Abelcet and Myocet, see Table 15.1).

2. Liposomes

Soon after their discovery by Alec Bangham in 1961, it became evident that the basic properties of liposomes (discussed in Chapter 13) make them ideally suited for a plethora of medical applications [2,3]. Liposomes are tiny, sealed lipid vesicles that have a sequestered aqueous space for water-soluble (hydrophilic) drugs and a surrounding, largely impermeable lipid bilayer membrane that can simultaneously house lipid-soluble (hydrophobic) drugs and an external facing surface that can be modified with specific ligands (Figure 15.2, [4]). Valuable liposome properties include:

1. They can be made from thousands of natural and artificial lipids.
2. They exist in lipid bilayers and so resemble biological membranes.
3. They are generally not antigenic.
4. They can be made in a wide range of sizes.
5. They can be made with a range of different sequestered volumes.

TABLE 15.1 Examples of Liposome-Sequestered Drugs Currently On the Market.

Name	Drug carried	Disease targeted
Myocet	Doxorubicin	Breast cancer
Doxil, Caelyx	Doxorubicin	Several cancers
LipoDox	Doxorubicin	Several cancers
Thermodox	Doxorubicin	Liver cancer
DaunoXome	Daunorubicin	Kaposi's sarcoma
Abelcet	Amphoteracin B	Fungal infections
Ambisome	Amphoteracin B	Fungal infections
Estrasorb	Estradiol	Menopausal therapy
Marqibo	Vincristine	Metastatic malignant melanoma
Visudyne	Verteporfin	Eye diseases
DepoCyt	Cytarabine	Meningitis
DepoDur	Morphine sulfate	Postoperative pain
Arikace	Amikacin	Lung infections
Lipoplatin	Cisplatin	Epithelial malignancies
LEP-ETU	Paclitaxel	Ovarian, breast & lung cancer
Epaxal	Hepatitis A vaccine	Hepatitis A
Inflexal V	Influenza vaccine	Influenza

6. They can be unilamellar or multilamellar.
7. They can be positive, negative or neutral.
8. They can be made with a wide range of permeability to sequestered solutes.
9. They can simultaneously accommodate water-soluble and lipid-soluble drugs.
10. They are non-toxic and biodegradable.
11. Various ligands can be attached to their surfaces.
12. Lectins, antigens, and receptors can be attached to their surfaces.
13. There are many ways to make liposomes (discussed in Chapter 13).

From this impressive list, it would appear that liposomes can be constructed for just about any drug delivery purpose. But are liposomes really a panacea? Unfortunately, nothing in life is as simple as one would hope. Upon testing liposomes as drug delivery agents, it was soon discovered that when injected intravenously, most liposomes are very rapidly taken up by cells of the reticulo-endothelial system (RES) and are sent to the lysosome for destruction [5]. Through the years, many attempts have been made to avoid the RES, and thus enhance liposome circulation time and the possibility to selectively target many parts of the human body. There are now a growing number of liposome-sequestered drugs that are medically used to combat a variety of human afflictions and many more are in clinical trials. A representative

FIGURE 15.2 Drug delivery using liposomes. Liposomes are tiny, sealed lipid vesicles that have a sequestered aqueous space for water-soluble (hydrophilic) drugs and a surrounding, largely impermeable lipid bilayer membrane that can simultaneously house lipid-soluble (hydrophobic) drugs and an external facing surface that can be modified with specific ligands. *Reproduced with permission* [4].

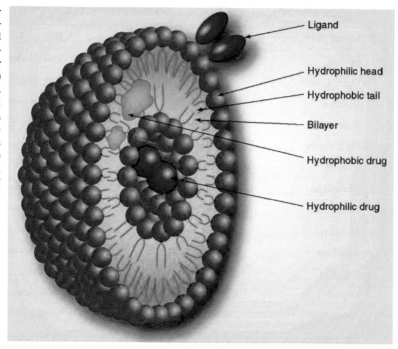

Ligand

Hydrophilic head

Hydrophobic tail

Bilayer

Hydrophobic drug

Hydrophilic drug

list of liposome-sequestered drugs is shown in Table 15.1. A few applications of liposome-sequestered drugs follow.

3. Leishmaniasis

Whereas the RES presents a substantial hurdle for most intravenous applications of targeting with drug-encapsulated liposomes, it is actually advantageous for certain diseases. The best known of these is leishmaniasis, a serious, mostly tropical disease [6]. While rare in the United States, leishmaniasis has been reported in immigrant populations, returning tourists, and military personnel from the Persian Gulf. Leishmaniasis (also known as kala-azar, black fever, and Dumdum fever) is the second greatest parasitic killer in the world (after malaria) and affects about 12 million people worldwide, with 1.5–2 million new cases reported each year. Sadly, leishmaniasis is often fatal. Historically, the most effective anti-leishmaniasis drugs are antimonial compounds that unfortunately have about the same inherent toxicity as arsenates. The cure can be as bad as the disease!

In the late 1970s Carl Alving (Figure 15.3) proposed an unusual solution that was based on the life cycle of the leishmaniasis protozoan parasite [7]. The disease is spread by the bite of a female sandfly, whereupon the protozoan is rapidly taken up by the victim's macrophage (part of the RES) where they multiply. Alving noticed that when liposomes were injected intravenously, they were rapidly taken up by the same RES macrophages. He reasoned that sequestering the antimonial drugs into liposomes would target the drugs directly to the parasite living

FIGURE 15.3   Carl Alving.

in the macrophages, thus greatly decreasing the drug's systemic toxicity. Alving encapsulated the antimonial drugs meglumine antimoniate and sodium stibogluconate into liposomes composed of phospholipids with disaturated chains (DPPC), cholesterol, and an
anionic lipid. When encapsulated into liposomes, both drugs were more than 700 times
more active than either of the free (unencapsulated) drugs when tested in hamsters [7].

4. Development of Liposomes that Avoid the RES

The purpose of developing liposome-sequestered drugs is to increase the therapeutic index
of the drug while minimizing its side effects. For most potential applications of drug-seques-
tered liposomes, being largely removed by the RES in a first pass through the circulatory
system is a fatal flaw. Therefore countless attempts have been made to avoid the RES and
thus enhance liposome circulation time [8,9]. The simplest method is to first inject empty liposomes (free of drug) to satiate the RES before the drug-encapsulated liposomes are injected.
Most attempts to make long-circulating liposomes have involved changing liposomal properties by altering their lipid composition, size, and charge. As a general rule, small liposomes
have longer circulation lifetimes than do larger liposomes [10]. Unfortunately, small unilamellar vesicles (SUVs) have much smaller sequestered volumes than large unilamellar vesicles
(LUVs), (discussed in Chapter 13), and so have diminished drug-carrying capacities. Even
worse are the multi-lamellar vesicles (MLVs) that are both large and have a very limited
sequestered aqueous space. Most of an MLV's sequestered space is occupied by lipid bilayers.
In 1982 Senior and Gregoriadis [10] reported that saturated chain PCs and SM (both lipids
have high T_ms, see Chapter 5) have longer circulating half-lives than unsaturated PCs. Cholesterol, a phospholipid-'condensing' sterol, also increased liposome circulation time. On the
other hand, positively charged liposomes are useless as drug carrying agents since they are
toxic. A major advance in producing long-circulating liposomes came from studies on the
surface of erythrocytes that manage to avoid the RES for their entire life of ~127 days. These
liposomes had their surfaces modified with gangliosides (particularly GM1, discussed in
Chapters 5 and 7) and sialic acid derivatives [11]. Therefore, the modified liposomes had
surface properties similar to erythrocytes. GM-liposomes have additional advantages in being
useful to deliver drugs by oral administration and being able to cross the blood–brain barrier.
The explanation of GM1 in RES avoidance relates to its flexible chain that occupies the space

immediately adjacent to the liposome surface [9]. This space ('periliposomal layer') makes it difficult for a macrophage to bind to a liposome.

5. Stealth Liposomes

The importance of the GMI 'periliposomal layer' led to the idea of replacing the ganglioside with a non-toxic synthetic polymer such as polyethylene glycol (PEG, Figure 15.4). PEG-liposomes are usually referred to as 'Stealth Liposomes' [12] due to their ability to avoid the RES and are heavily employed in many types of drug delivery. PEG is a linear polyether diol:

An important feature of PEG is the ability to be readily synthesized from short to very long polymers and polymer length affects circulation longevity. PEG is normally anchored to the liposome by attachment to the primary amine head group of PE (Figure 15.5):

In one example, Allen et al. [13] made liposomes from SM/PC/CHOL/DSPE-PEG where the PEG length varied from short to long. They reported increased circulation longevity for liposomes made from long length PEGs (PEG-1900 and PEG-5000) compared to short PEGs (PEG-750 and PEG-120). While PEG increases circulation longevity, by itself it cannot specifically target a particular cell. In fact, PEG actually hinders binding of the liposome to the delivery site. Most PEG-liposomes are further modified by attachment of biological species including monoclonal antibodies or fragments (immunoliposome), vitamins, lectins, specific antigens, peptides, growth factors, glycoproteins, carbohydrates or other appropriate ligands [14]. The ideal situation would be to create a drug-carrying liposome that would be invisible to the RES but would specifically bind to the diseased tissue or cell. The 'biological species' is attached to the free end of the PEG in DSPE-PEG by a maleimide group. Transferrin is often the ligand of choice for specific delivery of anticancer drugs. Modified liposomes are also commonly used in diagnostic imaging and in vaccines (virosomes). Other synthetic polymers, including poly(vinyl pyrrolidone) (PVP) and poly(acryl amide) (PAA), have been tested as replacements for PEG, with some limited success.

6. Thermo-Liposomes

It is now well accepted that by using systemic chemotherapy for solid tumors, it is almost impossible to achieve therapeutic drug levels without damaging healthy organs and tissues

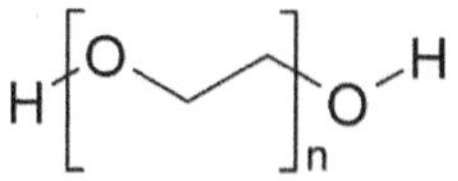

FIGURE 15.4   Polyethylene glycol [57].

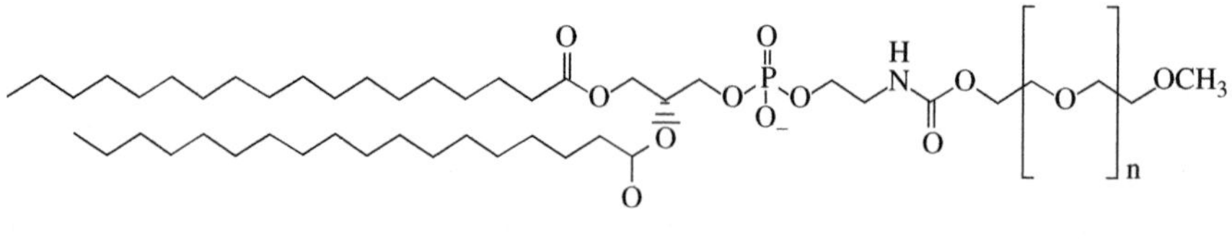

FIGURE15.5   DSPE-PEG. *Reprinted with permission* [58].

[15]. Hence, there is a very keen interest in liposomes as vehicles for drug delivery. It is believed that liposomes can meet the four basic requirements of a successful drug delivery system: 'Retain (the sequestered drug), Evade (the RES), Target (the diseased cell), and Release (the drug at the appropriate location)' [16]. We have already discussed the role of liposomes in 'Retain', 'Evade' and 'Target'. Finally, we will consider the issue of 'Release', the process whereby the drug is rapidly deposited at the target site.

In the late 1970s, Yatvin et al. [17] pioneered the use of mild hyperthermia to release liposome-sequestered drugs at the site of solid tumors. They based their approach on the large increase in leakiness that liposomes exhibit when heated through their main lipid phase transition temperature (T_m, see Chapter 14) [18]. Permeability of DPPC liposomes was discussed in Chapter 14. Liposomes in the gel state are poorly permeable while those in the liquid crystalline state exhibit considerable permeability. However, maximum permeability is achieved at the T_m, where equal amounts of gel and liquid crystalline state domains co-exist. Domain interfaces are locations of extremely high permeability. Temperature-sensitive, or thermo-liposomes, are in the impermeable gel state at physiological temperature (37°C), but undergo a sharp phase transition a few degrees above this. At T_m they rapidly become leaky, dropping most of their sequestered drug load. The liposome transition temperature cannot be too high as mammalian cells start to show damage at ~42°C [19]. So how does one make liposomes with T_ms between ~40−44°C? Many thermo-liposomes have as their major bilayer component DPPC since the T_m of this phospholipid is 41.3°C (see Chapter 5).

In their initial report, Yatvin et al. [17] used, as a model system, neomycin-sequestered liposomes to affect *E. coli* protein synthesis and cell survival in culture. Their initial experiments employed sonicated SUVs composed of DPPC/DSPC (3:1). This formulation may seem strange since DSPC has a higher T_m (56°C) than that of DPPC (41.3°C) and the liposomes would have a T_m of around 44°C. However, it is known that tightly curved SUVs exhibit permeability maxima ~4°C below their actual T_m. Therefore the effective (permeability maxima) temperature was a very acceptable ~42°C. It was demonstrated that *E. coli* protein synthesis was inhibited and cell killing enhanced by heating the *E. coli* with neomycin-containing liposomes to this reduced temperature. Maximum drug release and cell killing with the thermo-liposomes was between 42−46°C. This experiment established the potential of thermo-liposomes as useful drug carriers. A large number of drug-bearing thermo-liposomes were soon developed and tested in animal models [20]. The first generation of thermo-liposomes concentrated on avoiding the RES by lipid composition and size manipulations. In a typical example, Lindner et al. [21] made 175 nm thermo-liposomes from DPPC, DSPC, and a novel lipid DPPGOG (1.2-dipalmitoyl-*sn*-glycero-3-phosphoglyceroglycerol). These thermo-liposomes had a long circulation time and trapped drugs were released under mildly hyperthermic conditions (41−42°C). Thermo-liposomes were taken a step further by making them 'Stealth' with attached PEG. For example, Needham et al. [15] developed a series of 'Stealth' thermo-liposome drug carriers to target solid tumors. The liposomes were made from various mixtures of DPPC, MPPC (1-palmitoyl-2-hydroxy-*sn*-glycero-3-phosphocholine (a lyso-PC)), HSPC (hydrogenated soy *sn*-glycero-3-phosphocholine), cholesterol, and DSPE-PEG-2000. The sequestered drug was doxorubicin and the tumor was a human squamous cell carcinoma xenograft line (FaDu). DPPC and cholesterol provided the gel state, MPPC was responsible for the narrow phase transition (39−40°C), and DSPE-PEG-2000 made the thermo-liposomes invisible to the RES. The liposomes proved to be very susceptible to mild hyperthermia where they rapidly (10s of seconds) released their drug load. Finally, the 'Stealth' thermo-liposomes are being further modified by adding components (biological species including monoclonal antibodies or fragments, vitamins, lectins, specific antigens, peptides, growth factors, glycoproteins, carbohydrates or other appropriate ligands [14]) that will specifically target and bind the liposomes to the diseased cells.

A final aspect of delivering drugs to target cells or tissues involves the methods of generating localized, but limited, heat. This is referred to as 'mild hyperthermia.' The target tumor is locally heated by focusing electromagnetic or ultrasound energy on the tumor. In some cases radiofrequency electrodes are directly implanted into the tumor. More frequently, localized heating is achieved non-invasively by microwave antennas or ultrasound transducers — focused ultrasound (FUS) [22]. Upon heating, the liposomes rapidly release their sequestered anti-cancer drug directly inside the tumor.

7. pH-Sensitive Liposomes

In the previous sections we have discussed some of the many challenges associated with the use of liposomes as drug delivery agents. Another problem involves the ultimate fate of the liposomes after they have been internalized into a cell through endocytosis. Once internalized into an endosome, the liposomes and associated drugs are targeted to the lysosome where they are destroyed. In most cases it would be tremendously advantageous if the drug (or plasmid DNA or RNA) could be released from the liposome, and escape from the endosome into the cell's cytoplasm before encountering the lysosome. Early endosome studies showed that their interior pH is mildly acidic (pH is ~5), offering a possible target for liposome drug release [23]. This observation led to the development of a wide variety of 'pH-sensitive liposomes' [24,25]. pH-sensitive liposomes are stable (non-leaky) at physiological pH (pH ~7.4), but become unstable (leaky) at low pH (pH ~5.0). The pH-sensitive liposomes must also avoid the RES, be stable in the presence of blood plasma, and have the ability to fuse with the endosome membrane in order to eventually release their load into the cytoplasm.

Many pH-sensitive liposome compositions have been tested, but most have been mixtures of a lipid containing a pH titratable group and, as the bulk lipid, an unsaturated chain PE [26]. The titratable group is responsible for pH sensitivity and is often an acylated amino acid, a phospholipid derivative (usually a PE), a free fatty acid, a cholesterol derivative or a double chain amphiphile. The earliest pH-sensitive liposomes [27] were composed of PE as the major lipid component and single-chain amphiphiles such as fatty acids or N-acyl amino acids as the pH-sensitive group. Unfortunately, these early liposomes were destroyed in the plasma. Although liposome stability was shown to be increased by incorporation of cholesterol, the sterol unfortunately decreased liposome fusion to the endosome membrane, creating yet another problem. Therefore the initial objective of a functional pH-sensitive liposome was to produce a cholesterol-free liposome that remained stable in the plasma yet still retained fusogenicity and pH sensitivity.

One popular paradigm for the design of pH-sensitive liposomes is based on lipid-anchored compounds that exist in two different conformations, one existing in mildly acidic conditions (the compound is protonated) and the other in neutral or slightly basic conditions (the compound is dissociated). Most of these compounds have been lipid-linked to homocysteine or succinate. The original 1980 report of a pH-sensitive liposome by Yatvin et al. [27]

employed homocysteine linked to the membrane through a palmitic acid. Homocysteine exists in a cyclic structure that disrupts the liposomal membrane under acidic conditions but does not affect membrane permeability near neutral pH (Figure 15.6, Panel (a)). These liposomes were only useful between pH 7.4 and pH 6.0. It was predicted that these liposomes

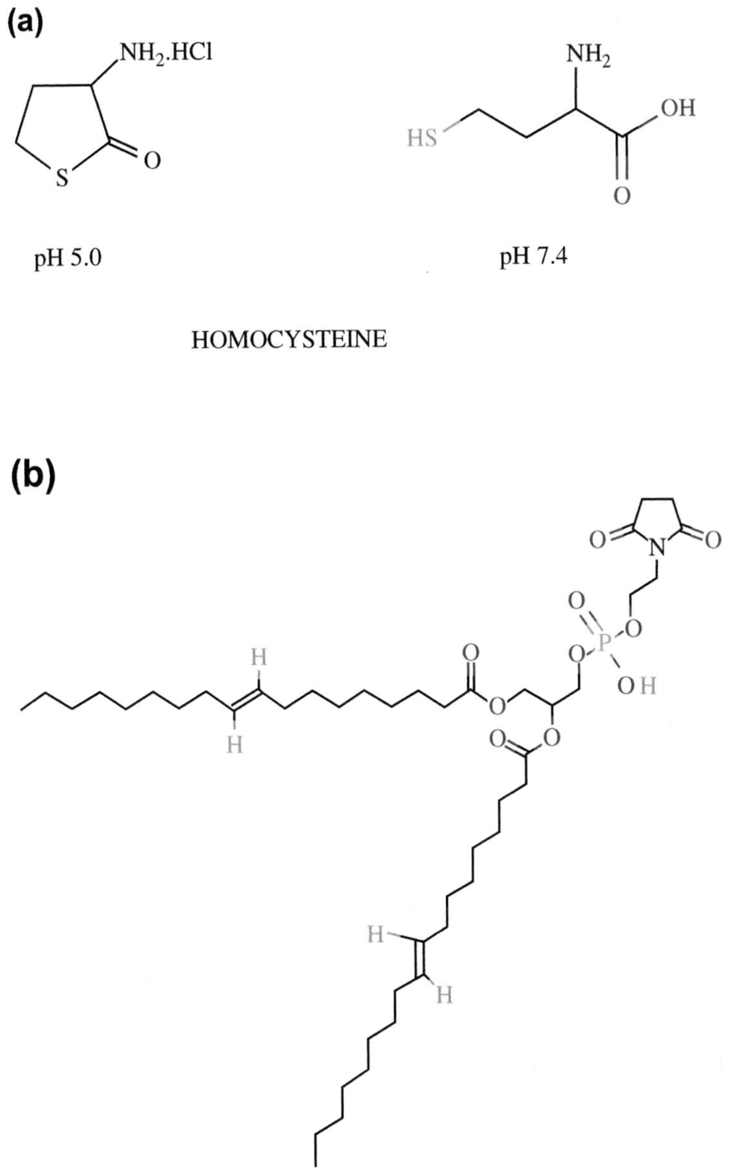

(a)

pH 5.0   pH 7.4

HOMOCYSTEINE

(b)

FIGURE 15.6 **Panel (a)**. Structure of homocysteine at pH 5.0 and pH 7.4. *Reprinted with permission* [59]. The low pH cyclic conformation disrupts the liposome membrane, increasing leakiness. The neutral pH conformation does not affect liposome membrane permeability [60]. **Panel (b)**. N-succinyl 1,2-dioleoyl-3-phosphoglycerol. *Reprinted with permission* [61].

could be useful in targeting drugs to acidic tissues including tumors and sites of inflammation or infection.

For the last 20+ years the most successful pH-sensitive liposomes have incorporated succinate. The first use of this type of pH-sensitive liposome was reported by Leventis et al. (1987) [28]. This group used a series of double chain amphiphiles, including 1, 2-dioleoyl-3-succinylglycerol (DOSG) that, when combined with POPE (16:0, 18:1 PE) at pH 7.4 formed non-leaky liposomes. Importantly, these liposomes were fusogenic and became leaky under mildly acidic conditions, but were stable in serum. A few years later, Collins et al. (1990) [29] made a series of pH-sensitive liposomes from three similar diacyl-succinylglycerols (including DOSG) and PE. These workers advanced the earlier report [28] by attaching an anti-H2k^k antibody isolated from the murine hybridoma cell line 11-4.1 to the N-hydroxysuccinimide ester of palmitic acid, thus converting the pH-sensitive liposome into a pH-sensitive 'immunoliposome'. Later additions attached a phosphate to the glycerol *sn*-3 position (Figure 15.6, Panel B). Finally, the phosphate was replaced by a PE (N-succinyl-1,2-dioleoylphosphatidylethanolamine) and PEG was attached to the PE. Therefore the final product was a pH-sensitive, stealth, immunoliposome with tremendous cytotoxic potential. This example emphasizes the versatility of liposomes for drug targeting.

B. EFFECT OF DIETARY LIPIDS ON MEMBRANE STRUCTURE/FUNCTION

The old adage 'you are what you eat' certainly applies to the effect of dietary lipids on membrane structure and function. At the vanguard of dietary fatty acids are both 'bad guys,' *trans* fatty acids (TFAs) and 'good guys,' docosahexaenoic acid (DHA). Although many other lipids also affect membranes, this chapter will only contrast TFA and DHA, the alpha and omega of the fatty acid business, because of their topical interest in human health.

1. *Trans* Fatty Acids

Dietary *trans* fatty acids (TFAs) terrify the health-conscious public, and for good reason. Since the 1950s, TFAs have been closely linked to coronary heart disease and arteriosclerosis [30]. A 2004 report of the Harvard School of Public Health estimated that partially hydrogenated fat, the major dietary source of TFAs, may be responsible for between 30,000 and 100,000 premature coronary deaths per year in the United States alone [31]. To date, there have been numerous large epidemiology studies that have strongly supported this conclusion [32]. Much less certain are reports that TFAs may also be weakly carcinogenic [33] and may even affect normal brain function [34]. TFAs are now universally recognized to be 'bad guys', but why? TFAs are very similar to *cis* fatty acids that are far more common and are essential for life. So why have TFAs turned to the dark side?

For millennia, TFAs have been a normal component of human membranes. For example, the sphingosine backbone of sphingolipids contains a *trans* double bond in its fixed hydrophobic chain. The simplest TFA, elaidic acid (18:1$^{\Delta 9t}$), is naturally present in ruminant fat, meat, and dairy products, and so is commonly found in low levels in human membranes [35]. The fact that elaidic acid has been in the human diet for so long cannot account for the large increase in heart disease observed over the past century. It is doubtful that elaidic acid causes these health problems. Instead, the culprit is likely to be partially hydrogenated plant oils. The first successful hydrogenation of plant oils was reported in 1897 [35]. An industrial process was developed to harden fluid plant oils and to decrease their susceptibility to oxidation, opening their application to food processing. Since then there has been a steady increase in the amount of TFAs appearing in the human diet, and with it, a concomitant increase in heart disease. Partial hydrogenation of plant lipids produces a bewildering array of TFAs, the result of *cis* double bond reduction and migration up and down the chain and partial conversion of normal *cis* to deleterious *trans* double bonds. The process creates a wide range of geometric and positional fatty acid isomers. In recent years, the increase in heart disease has resulted in banning *trans* fats from many parts of the world.

The question, however, remains as to why *cis* fatty acids are essential for human health, while *trans* fatty acids are so harmful. One possible answer, related to the theme of this book, involves incorporation of unnatural *trans* fatty acids into membrane phospholipids, where they replace the natural *cis* lipids, thus altering the structure and function of the membranes. All fatty acids, including TFAs, can be incorporated into phospholipids and thereby affect the hydrophobic interior of membranes [36,37].

A first approach to investigating how *cis* and *trans* fatty acids differ in their effect on membrane physical properties can be extracted from the main phase transition temperatures (T_ms) reported in Chapter 4, Tables 4.3 and 4.7. For the 18-carbon series:

Fatty acid	Designation	T_m	ΔT_m
Stearic	Saturated (18:0)	69.6°C	0°C
Oleic	*cis* ($18:1^{\Delta 9c}$)	16.2°C	−53.4°C
Elaidic	*trans* ($18:1^{\Delta 9t}$)	43.7°C	−25.9°C

Both unsaturated fatty acids (oleic and elaidic) exhibit T_ms that are lower than that of saturated stearic acid. However, the change in T_ms (ΔT_m) is much larger for the *cis* than for the *trans* fatty acid. Therefore, at least by T_m analysis, although both elaidic and oleic acid have 18-carbons and a single $\Delta 9$ double bond, elaidic acid is more similar to the saturated stearic acid than it is to oleic acid. From the T_ms it can be concluded that *cis* double bonds have a larger effect on lipid packing than do *trans* double bonds that exhibit considerable saturated fatty acid-like properties.

Roach et al. [38] extended the T_m measurements on oleic versus elaidic acids to several PC model membranes (monolayers and bilayers). They compared the effect of oleic versus elaidic and linoleic versus linelaidic on homo-chain and hetero-chain PCs using molecular dynamics, lateral lipid packing, thermotropic phase behavior, 'fluidity', lateral mobility, and permeability. In all cases the *cis* unsaturated chains induced much larger membrane perturbations than did the *trans* unsaturated analogs. Once again, the *trans* double bond acyl chains behaved more like a saturated chain than a *cis* unsaturated chain. Corroborating this conclusion is the molecular dynamics simulations of Pasenkiewicz-Gierula and co-workers who could detect no significant difference between 16:0-18:1$^{\Delta 9c}$ PC and 16:0-18:1$^{\Delta 9t}$ PC at the aqueous interface [39], but did observe a difference between the two PCs within the bilayer hydrophobic interior [40]. They concluded that the *trans*-PC was more similar to the disaturated DMPC control than to the *cis*- PC. A similar conclusion was obtained from a diet study of serum lipoprotein levels [41]. The effect of dietary *trans* elaidic acid was more similar to saturated stearic acid than to *cis* oleic acid. In addition there has been a plethora of membrane enzymes and receptors whose activity has been shown to decrease with TFAs [42].

Although it is now clear that TFAs are incorporated into membrane phospholipids where they affect many membrane properties, questions still abound. Are TFAs mistakenly identified as a saturated fatty acid and placed in the *sn*-1 chain of a phospholipid, or are they recognized as an unsaturated fatty acid and placed in the *sn*-2 position? Either option seems possible. Emken et al. [43] reported that in human erythrocytes and platelets, three times more elaidic acid than oleic acid accumulates in the *sn*-1 position of PCs. Since elaidic acid resembles a saturated fatty acid, its accumulation into the *sn*-1 position is not surprising and would likely have only a minimal effect on normal membrane structure and function [44]. Supporting this, Larque et al. [36] reported that as dietary TFA levels rise in liver microsomes and mitochondria, saturated fatty acid levels drop. Being a natural food product, elaidic acid in low amounts is probably not responsible for human heart problems. Instead, it is the enormous number of positional and geometric isomers found in partially hydrogenated oils that are likely responsible for causing heart disease. When incorporated into membrane phospholipids, TFAs must replace existing saturated or natural *cis* unsaturated acyl chains.

It is more likely that addition of polyunsaturated TFAs to the *sn*-2 position of phospholipids alters membrane structure and function far more than incorporation of elaidic acid in the *sn*-1 position [45]. For example, if dietary TFAs replace natural DHA in the *sn*-2 position of brain membrane phospholipids, it is likely that electrical activity of neurons and hence brain function will be affected. However, other completely different possibilities also exist. For example, it is possible that only one of the multitude of TFA isomers, existing at miniscule levels, is doing 99+ percent of the harm. Identifying such a unique *trans* isomer will be a difficult endeavor.

2. Docosahexaenoic Acid (DHA)

Docosahexaenoic acid (DHA) (22:6(n-3), see Chapters 4 and 10) is the longest (22 carbons) and most unsaturated (6 *cis* double bonds) fatty acid commonly found in mammalian membranes [46]. In recent years DHA (and other omega-3 fatty acids) has received a great deal of attention due to its reputed involvement in alleviating a wide variety of human afflictions [47] (listed in Table 15.2, [48]). This list, which spans the entire gambit of human disorders, can be roughly divided into six non-exclusive categories; heart disease, cancer, immune problems, neuronal functions, aging and 'other' hard to categorize problems such as migraine headaches, malaria, and sperm fertility. Historically the primary source of DHA in the human diet has been oily, coldwater fish. In fact, the initial link of fish oils to a human health problem (ischemic heart disease in Greenland Eskimos) originated in the pioneering work of Bang and Dyerberg from the 1970s [49]. DHA supplementation is currently in vogue.

TABLE 15.2 A Partial List of Human Afflictions that have been Linked to DHA.

ADHD	Depression	Multiple sclerosis
Aggression	Dermatitis	Neurovisual development
Alcoholism	Diabetes	Nephropathy
Alzheimer's disease	Dyslexia	Periodontitis
Arthritis	Eczema	Phenylketonuria
Asthma	Fertility	Placental function
Atrial fibrillation	Gingivitis	Psoriasis
Autism	Heart disease	Respiratory diseases
Bipolar disorder	Hypersensitivity	Schizophrenia
Blindness	Inflammatory response	Sperm fertility
Blood clotting	Kidney disease	Suicide
Bone mineral density	Lupus	Ulcerative colitis
Brain development	Malaria	Visual acuity
Cancer	Methylmelonic acidemia	Zellweger's syndrome
Crohn's disease	Migraine headaches	Elasticity
Cystic fibrosis	Mood and behavior	

TABLE 15.3 Lipid Physical Properties Affecting Membrane Structure and Function that Occur When a Single Double Bond is Added to a Saturated Acyl Chain.

Phase transition temperature	Membrane permeability
Phase preference	Lateral diffusion rate
'Fluidity'	Flip-flop rate
Area/molecule	Lipid–lipid affinity
Lipid packing	Lipid protein affinity
Packing free volume	Interaction with cholesterol
Compressibility	Lipid microdomain formation and stability
'Squeeze out'	Susceptibility to phospholipases
Bilayer thickness	Susceptibility to peroxidation

DHA, omega-3 fatty acid, and fish oil capsules are commonly found in grocery and drug stores. DHA is also included in many infant formulas, since it is known to accumulate in the brain and eyes of the fetus, where it affects early visual acuity and enhances neural development. DHA is also used as a component of parenteral (intravenous) and enteral (feeding tube) nutrition.

An obvious question is how such a simple molecule can affect so many seemingly unrelated processes. To accomplish this, DHA must be exerting its influence at some fundamental level that is common to many types of cells and tissues. Possible non-exclusive modes of action for DHA include: (1) eicosanoid biosynthesis; (2) protein activity through direct interaction; (3) protein activity through indirect interactions involving transcription factors; (4) lipid peroxidation products; (5) membrane structure and function; and (6) lipid raft structure and function [50,51]. Previous chapters in this book have discussed the enormous change in lipid physical properties that affect membrane structure and function, occurring when a single double bond is added to a saturated acyl chain. Table 15.3 shows a list of these properties. Since double bonds have such an enormous effect on membranes it seems logical to predict that the DHA with 6-double bonds might be the most influential membrane fatty acid of all. Indeed, many biophysical studies on lipid monolayer and bilayer properties have been reported [50,52,53]. The problem is that none of the listed properties are unique to DHA, but instead are characteristic to some extent of all polyunsaturated fatty acids. In most, but not all, examples DHA does exert a larger effect than other less unsaturated fatty acids, but this difference is probably not sufficient to account for DHA's unusual health benefits. One possibility for DHA's molecular mode of action is in affecting important cell signaling processes by altering the composition, size, structure, and stability of lipid rafts. Initial investigations have demonstrated that DHA does indeed affect lipid raft structure and cell signaling [54–56].

It is likely that no single membrane property is sufficient to account for DHA's health benefits, but rather a combination of as yet unidentified DHA-affected properties is required. But what exactly are these properties? The answer to this conundrum will require further research into the molecular aspects of membrane structure and function. It can be predicted that in the near future the next major advance in membranes will involve understanding exactly what a lipid raft is at the molecular level. How many types of lipid rafts and lipid non-rafts are there and how do they control cellular events? Once the very fundamental questions about membrane structure are better understood, it should be possible to design new paradigms to benefit the human condition.

SUMMARY

The complexity of membrane structure and function outlined in this book suggests a wealth of potential targets to alleviate human afflictions. This chapter investigates how liposomes can be modified for targeted drug delivery. The liposomes must be non-leaky and able to avoid the reticulo-endothelial system (RES). Several types of drug-carrying liposomes are discussed including 'Stealth' liposomes (coated with polyethyleneglycol), Thermo-liposomes (made from lipid mixtures with a phase transition temperature ~2–4°C above physiological), pH-sensitive liposomes (containing a titratable group, often derivatives of homocysteine or succinate) and several targeted liposomes (with specific antibodies, lectins, receptors, and vitamins attached to the liposome surface). Also discussed are the effects of dietary fatty acids on membrane structure and function as it influences human health. In this regard, harmful *trans* fatty acids (TFAs) are contrasted with the beneficial omega-3 fatty acid, docosahexaenoic acid (DHA).

References

[1] Weissmann G, Rita GA. Molecular basis of gouty inflammation: interaction of monosodium urate crystals with lysosomes and liposomes. Nature New Biol 1972;240(101):167–72.

[2] Duzgunes N. Liposomes. Elsevier Academic Press; 2009.

[3] Chrai SS, Murari R, Ahmad I. Liposomes (A Review): Part Two: Drug Delivery Systems. BioPharm 2002:40–9.

[4] Medscape: http://img.medscape.com/article/734/055/734055-fig3.jpg.

[5] Gregoriadis G, Ryman BE. Lysosomal localization of β-fructofuranosidase-containing liposomes injected into rats. Some implications in the treatment of genetic disorders. Biochem J 1972;129:123–33.

[6] Myler PJ, Fasel N, editors. Leishmania: After the Genome. Norfolk, UK: Caister Academic Press; 2008.

[7] Alving CR, Steck EA, Chapman Jr WL, Waits VB, Hendricks LD, Swartz Jr GM, et al. Therapy of leishmaniasis: Superior efficacies of liposome-encapsulated drugs. Proc Natl Acad Sci USA 75(6) 1978;75:2959–63.

[8] Immordino ML, Dosio F, Cattel L. Stealth liposomes: review of the basic science, rationale, and clinical applications, existing and potential. Int J Nanomedicine 2006;1(3):297–315.

[9] Drummond DC, Meyer O, Hong K, Kirpotin DB, Papahadjopoulos D. Optimizing liposomes for delivery of chemotherapeutic agents to solid tumors. Pharm Rev 1999;51(4):691–744.

[10] Senior J, Gregoriadis G. Stability of small unilamellar liposomes in serum and clearance from the circulation: The effect of the phospholipid and cholesterol components. Life Sci 1982;30:2123–36.

[11] Gabizon A, Papahadjopoulos D. Liposome formulations with prolonged circulation time in blood and enhanced uptake by tumors. Proc Natl Acad Sci USA 1988;85:6949–53.

[12] Lasic DD, Martin EJ, editors. Stealth Liposomes. CRC Press, Taylor & Francis; 1995.

[13] Allen TM, Hansen C, Martin F, Redemann C, Yau-Young A. Liposomes containing synthetic lipid derivatives of poly(ethylene glycol) show prolonged circulation half-lives in vivo. Biochim Biophys Acta 1991;1066:29–36.

[14] Sapra P, Allen TM. Ligand-targeted liposomal anticancer drugs (Review). Prog Lipid Res 2003;42:439–62.

[15] Needham D, Anyarambhatla G, Kong G, Dewhirst MW. A new temperature-sensitive liposome for use with mild hyperthermia: characterization and testing in a human tumor xenograft model. Cancer Res 2000;60:1197–201.

[16] Needham D. Materials engineering of lipid bilayers for drug carrier function. MRS Bull 1999;24:32–40.

[17] Yatvin MB, Weinstein JN, Dennis WH, Blumenthal R. Design of liposomes for enhanced local release of drugs by hyperthermia. Science 1978;202:1290–3.

[18] Blok MC, van Deenen LLM, de Gier J. Effect of the gel to liquid crystalline phase transition on the osmotic behaviour of phosphatidylcholine liposomes. Biochim Biophys Acta 1976;433:1–12.

[19] Crile G. Selective destruction of cancers after exposure to heat. Annu Surg 1962;156:404–7.

[20] Kong G, Dewhirst MW. Hyperthermia and liposomes: a review. Int J Hyperthermia 1999;15:345–70.

[21] Lindner LH, Eichhorn ME, Eibl H, Teichert N, Schmitt-Sody M, Issels RD, et al. Novel temperature-sensitive liposomes with prolonged circulation time. Clin Cancer Res 2004;10:2168–78.

[22] Koning GA, Eggermont AMM, Lindner LH, ten Hagen TLM. Hyperthermia and Thermosensitive Liposomes for Improved Delivery of Chemotherapeutic Drugs to Solid Tumors. Pharm Res 2010;27:1750–4.

[23] White J, Matlin K, Helenius A. Cell fusion by Semliki Forest, influenza, and vesicular stomatitis viruses. J Cell Biol 1981;89:674–9.

[24] Chu C-J, Szoka FC. pH-Sensitive liposomes. J Liposome Res 1994;4:361–95.

[25] Drummond DC, Zignani M, Leroux I. Current status of pH-sensitive liposomes in drug delivery. Prog Lipid Res 2000;39(5):409–60.

[26] Connor J, Yatvin MB, Huang L. Proc Natl Acad Sci USA 1984;81:1715–8.

[27] Yatvin MB, Kreuz W, Horowitz BA, Shinitzky M. pH-Sensitive liposomes: Possible clinical implications. Science 1980;210:1253–5.

[28] Leventis R, Diacovo T, Silvius JR. pH-Dependent Stability and Fusion of Liposomes Combining Protonatable Double-Chain Amphiphiles with Phosphatidylethanolamine. Biochemistry 1987;26:3267–76.

[29] Collins D, Litzinger DC, Huang L. Structural and functional comparisons of pH-sensitive liposomes composed of phosphatidylethanolamine and three different diacylsuccinylglycerols. Biochim Biophys Acta 1990;1025:234–42.

[30] Denke MA. Serum lipid concentrations in humans. In: Trans fatty acids and coronary heart disease risk. Am J Clin Nutr 1995;62:693S–700S.

[31] Ascherio A, Stampfer MJ, Willett WC. Trans fatty acids and coronary heart disease. Harvard School Public Health 2004:1–8.

[32] Zaloga GP, Harvey KA, Stillwell W, Siddiqui R. Trans fatty acids and coronary heart disease. Nutr Clin Pract 2006;21:505–12.

[33] Slattery ML, Benson J, Ma KN, Schaffer D, Potter JD. Trans fatty acids and colon cancer. Nutr Cancer 2001;39:170–5.

[34] Phivilay A, Julien C, Tremblay C, Berthiaume L, Julien P, Giguere Y, et al. High dietary consumption of trans fatty acids decreases brain docosahexaenoic acid but does not alter amyloid-beta and tau pathologies in the 3xTg-AD model of Alzheimers disease. Neuroscience 2009;159:296–307.

[35] Emken EA. Nutrition and biochemistry of trans and positional fatty acid isomers in hydrogenated oils. Annu Rev Nutr 1994;4:339–76.

[36] Larque E, Garcia-Ruiz PA, Perez-Llamas F, Zamora S, Gil A. Dietary trans fatty acids alter the composition of microsomes and mitochondria and the activities of microsome Δ6- fatty acid desaturase and glucose-6-phosphatase in livers of pregnant rats. J Nutr 2003;133:2526–31.

[37] Morgado N, Galleguillos A, Sanhueza J, Garrido A, Nieto S, Valenzuela A. Effect of the degree of hydrogenation of dietary fish oil on the trans fatty acid content and enzymatic activity of rat hepatic microsomes. Lipids 1998;33:669–73.

[38] Roach C, Feller SE, Ward JA, Shaikh SR, Zerouga M, Stillwell W. Comparison of cis and trans fatty acid containing phosphatidylcholines on membrane properties. Biochemistry 2004;43:6344–51.

[39] Murzyn K, Rog T, Jezierski G, Takaoka Y, Pasenkiewicz-Gierula M. Effects of phospholipid unsaturation on the membrane/water interface: a molecular simulation study. Biophys J 2001;81:170–83.

[40] Rog T, Murzyn K, Gurbiel R, Takaoka Y, Kusumi A, Pasenkiewicz-Gierula M. Effects of phospholipid unsaturation on the bilayer nonpolar region: A molecular simulation study. J Lipid Res 2004;45: 326–36.

[41] Mensink RP, Katan MB. Effect of dietary trans fatty acids on high density and low density lipoprotein cholesterol levels in healthy subjects. N Engl J Med 1990;323:439–45.

[42] Alam SQ, Ren YF, Alam BS. Effect of dietary trans fatty acids on some membrane-associated enzymes and receptors in rat heart. Lipids 1989;24:39–44.

[43] Emken EA, Rohwedder WK, Dutton HJ, Dejarlais WJ, Adlof RO. Incorporation of deuterium-labeled cis- and trans-9-octadecenoic acids in humans: plasma, erythrocyte, and platelet phospholipids. Lipids 1979;14:547–54.

[44] Wolff RL, Entressangles B. Steady-state fluorescence polarization study of structurally defined phospholipids from liver mitochondria of rats fed elaidic acid. Biochim Biophys Acta 1994;1211:198–206.

[45] Lichtenstein AH. Dietary trans fatty acid. J Cardiopulm Rehabil 2000;20:143–6.

[46] Salem NJ, Kim H-Y, Yergey JA. Docosahexaenoic acid: membrane function and metabolism. In: Simopolous AP, Kifer RR, Martin RE, editors. Health Effects of Polyunsaturated Fatty Acids in Seafoods. New York: Academic Press; 1986. p. 319–51.

[47] Stillwell W, Shaikh SR, Lo Cascio D, Siddiqui RA, Seo J, Chapkin RS, et al. Docosahexaenoic acid: an important membrane-altering omega-3 fatty acid. In: Huamg JD, editor. Frontiers in Nutrition Research. New York, NY: NOVA Science Publishers; 2006. p. 249–71, [Chapter 8].

[48] Stillwell W. Docosahexaenoic acid: a most unusual fatty acid. Chem Phys Lipids 2008;153:1–2.

[49] Dyerberg J, Bang HO, Hjorne N. Fatty acid composition of the plasma lipids in Greenland Eskimos. Am J Clin Nutr 1975;28:958–66.

[50] Stillwell W, Wassall SR. Docosahexaenoic acid: membrane properties of a unique fatty acid. Chem Phys Lipids 2003;126:1–27.

[51] Stillwell W. The role of polyunsaturated lipids in membrane raft function. Scand J Food Nutr 2006;50 (Suppl. 2):107–13.

[52] Salem Jr N, Litman B, Kim HY, Gawrisch K. Mechanisms of action of docosahexaenoic acid in the nervous system. Lipids 2001;36:945–59.

[53] Feller SE, Gawrisch K. Properties of docosahexaenoic acid-containing lipids and their influence on the function of rhodopsin. Curr Opin Struct Biol 2005;15:416–22.

[54] Stillwell W, Shaikh SR, Zerouga M, Siddiqui R, Wassall SR. Docosahexaenoic acid affects cell signaling by altering lipid rafts. Reprod Nutr Dev 2005;45:559–79.

[55] Kim W, Fan Y, Barhoumi R, Smith R, McMurray DN, Chapkin RS. n-3 Polyunsaturated fatty acids suppress
      the localization and activation of signaling proteins at the immunological synapse in murine CD4+ T cells by
      affecting lipid raft formation. J Immunol 2008;181:6236–43.

[56] Shaikh SR. Diet-induced docosahexaenoic acid non-raft domains and lymphocyte function. 2010. Prosta-
      glandins Leukotrienes Essent. Fatty Acids 2010;28:159–64.

[57] Hoffmeier, Klaus: http://en.wikipedia.org/wiki/File:Polyethylene_glycol.png; public domain.

[58] Biotechnolog.pl: http://www.biotechnolog.pl/artykul-208.htm

[59] Courtesy of Santa Cruz Biotechnology, Inc; http://www.scbt.com/datasheet-205294-dl-homocysteine-
      thiolactone-hydrochloride.html

[60] Edgar181:http://en.wikipedia.org/wiki/File:Homocysteine_racemic.png; public domain.

[61] Lookchem: http://www.lookchem.com/cas-977/97782-02-0.html

A

Chronology of Membrane Studies

The study of membranes has a very long and rich history. Below is a compendium of many (~100) of the most important observations concerning membranes. Such a list must terminate several years before its compilation date since it takes some considerable time before it is clear which newer studies will have a lasting impact and which studies will reach a dead end.

Year	Discoverer	Discovery
~540 BC	Thales of Miletus	First to emphasize the fundamental importance of water.
450 BC	Hippo of Samos	Perceived life as water.
77 AD	Pliny the Elder	Described the commonly used fishing trick of floating oil on water.
1665	Robert Hooke	Early microscopist. Coined the term 'cell'.
1769	F.P. de la Salle	Discovered cholesterol in gallstones.
1773	William Hewson	Discovered osmotic swelling and shrinking of erythrocytes. Proposed existence of a plasma membrane.
1774	Benjamin Franklin	First scientific lipid monolayer study.
1781	Henry Cavendish	Discovered the chemical composition of water.
1806	Vauquelin & Robiquet	Discovered the first amino acid, asparagine, from asparagus.
1823	Michel Chevreul	Discovered stearic acid and oleic acid in pork fat.
1828	Friedrich Wohler	Synthesized an organic molecule (urea) from an inorganic molecule (ammonium cyanate). Proved organic molecules are not 'special'.
1836	C. H. Schultz	Visualized erythrocyte plasma membrane using iodine as a stain. First to use the term 'membrane'.
1839	T. Schwann	Established the 'Cell Theory'.
1852	George Stokes	Discovered fluorescence.
1855	Karl von Nageli	Membrane is barrier to osmosis.
1855	Adolf Fick	Described what is now known as Fick's Laws of Diffusion.
1871	Hugo de Vries	Measured membrane permeability of ammonia and glycerol.
1877	Wilhelm Pfeffer	Proposed the first membrane theory. A membrane is thin and semi-permeable.
1882	Elie Metchnikoff	Discovered phagocytosis.

(Continued)

—cont'd

Year	Discoverer	Discovery
1888	F. Reinitzer	Discovered the liquid crystalline phase.
	Herrmann Stillmark	Discovered lectins.
	Walter Nernst	Developed theory of electrical potentials (basis of electrophysiology) that explains ion flux across membranes.
1889	Agnes Pockels	Developed methodology for lipid monolayer studies while working in her kitchen.
1890	Lord Raleigh	Determined the size of triolein using lipid monolayer methodology.
1898	Carmillo Golgi	Discovered the Golgi apparatus.
1899	Charles Ernest Overton	First true 'membranologist'. Membranes have a lipid-like barrier. Proposed passive and active transport.
1913	E. McCollum & M. Davis	Discovered vitamin A in butterfat and cod liver oil.
1917	Irving Langmuir	Credited with making lipid monolayers a 'precise' science. His work led to the seminal 1925 Gorter/Grendel experiment.
1920	Latimer & Rodebush	Suggested water H-bonding is major driving force for membrane stabilization.
1922	Evans & Bishop	Discovered vitamin E.
1925	Gorter & Grendel	The most important paper ever published on membranes. First experimental evidence for a lipid bilayer.
	Leathers & Raper	In their book *The Fats* suggested phospholipids were essential components of membranes.
	H. Fricke	Used electrical impedance measurements to determine the thickness of an erythrocyte membrane.
1926	James Sumner	Isolated the first enzyme, urease.
	J.B. Perrin	Developed fluorescence polarization (FP).
1929	Warren Lewis	Discovered pinocytosis.
1930	J.D. van der Waal / Fritz London	London extended the van der Waals force to include induced dipole–dipole interactions.
1931	Ernst Ruska	Invented the electron microscope.
1935	Danielli & Davson	Proposed the 'Pauci-Molecular' model for membrane structure. Based on lipid bilayer.
1938	Izmailov & Shraiber	Discovered TLC (thin layer chromatography), a major technique for lipid separation.
1942	E. Klenk	Identified gangliosides from brain.
	M. Pangborn	Discovered cardiolipin.
1945	Cole & Marmount	Developed voltage clamp technique.
	Porter, Claud & Pullam	First to observe the endoplasmic reticulum.
1948	Linus Pauling	Proposed the α-helix.
1949	W.C. Griffin	Proposed HLB (hydrophile/lipophile balance) parameter for detergents.
1950	A.J.P. Martin	Developed gas-liquid chromatography, the major technique used for fatty acid analysis.
1951	Oliver Lowery	Discovered the first sensitive method to quantify proteins.

—cont'd

Year	Discoverer	Discovery
1952	Gunnar Blix	After 15 years of work on this sugar, Blix coined the term 'sialic acid'.
	Frederick Sanger	Sequenced the first protein, insulin.
1953	George Palade	Observed Caveolae using electron microscopy.
1955	Lathe and Ruthven	Discovered size exclusion (gel filtration) chromatography.
	Christian de Duve	Discovered lysosomes.
1957	J.D. Robertson	Proposed the 'Unit Membrane' model for membrane structure.
	Jens Skou	Discovered the plasma membrane Na^+/K^+ ATPase.
	Folch, Lees, Stanley	Developed method to extract lipids from membranes that is still in use today.
	Frederick Crane	Discovered Coenzyme Q in beef heart.
1960	Watson & O'Neill	Developed the technique of differential scanning calorimetry (DSC).
1961	Peter Mitchell	Proposed the Chemiosmotic Hypothesis for oxidative phosphorylation.
1962	Mueller et al.	Made the first stable large lipid bilayer called a planar bimolecular lipid membrane (BLM).
	Kauzman & Tanford	Developed the hydrophobic effect theory that explains membrane stability.
1963	Palade & Farquahr	First description of Gap Junctions.
1964	R.T. Holman	Proposed omega nomenclature for fatty acids.
	H. Fernandez-Moran	First definitive evidence that a membrane protein (mitochondrial F_1 ATPase) is 100% asymmetrically distributed across a membrane.
	A.D. Bangham	Reported the first production of liposomes.
	Saul Roseman	Discovered the PTS sugar transport system.
	Bernard Pressman	Valinomycin recognized as a potassium ionophore.
1967	Charles Pederson	Discovered crown ethers.
	Christian de Duve	Discovered peroxisomes.
1968	D. Zilversmit	Discovered phospholipid exchange proteins (PLEPs).
1969	Braun & Rodin	Discovered palmitoylation of a membrane protein.
	Huang	Reconstituted membrane protein into LUV. Important for transport studies.
1970	L. Frye & M. Edidin	First measurement of lateral diffusion in membranes.
	Hladky & Haydon	Discovered Gramicidin A trans-membrane channel.
1972	Singer & Nicolson	Proposed the 'Fluid Mosaic' model for membrane structure.
	Mark Bretscher	First report of partial lipid asymmetry in membranes.
1974	M. Sinensky	Proposed homeoviscous adaptation (HVA).
1975	Henderson & Unwin	First electron microscopy-derived structure of a membrane protein, bacteriorhodopsin at 7Å resolution (has seven transmembrane alpha helices).
	Tomita & Marchesi	Sequenced glycophorin, first integral membrane protein sequenced.
	Barbara Pearse	Discovered clathrin.
1976	J. Axelrod	Membrane lateral diffusion rates determined by FRAP.
	M. Low	Described GPI-anchored proteins.
	H. Ikezawa	
	Brown & Goldstein	Formulated Receptor-Mediated Endocytosis hypothesis for cholesterol metabolism.
	Neher & Sakmann	Developed patch clamp technique, able to measure single channel conductance.

(Continued)

—cont'd

Year	Discoverer	Discovery
1977	Demel et al.	Used DSC to determine the affinity of cholesterol for various phospholipids: SM > PS, PG > PC > PE.
	Yasutomi Nishizuka	Discovered protein kinase C.
1978	Kamiya et al.	Discovered first prenylated protein.
1982	Kyte & Doolittle	Propose hydropathy scale to predict orientation of an integral membrane protein.
	Karnovsky & Klausner	Proposed lipid microdomain concept, foreshadowing lipid rafts.
	Binnig & Rohrer	Invented AFM.
	Aitken et al. Carr et al.	Discovered first myristoylated protein.
1984	R.A. Schmidt	Reported that many proteins are anchored to membranes via long chain isoprenoids.
	Seigneuret & Devaux	Discovered the first ATP-dependent flippase.
1985	Diesenhofer et al.	First high resolution x-ray structure of a membrane protein, a bacterial reaction center.
1986	Akiharo Kusumi	Developed Single Particle Tracking to measure lateral movement of a single protein in a membrane.
1992	Peter Agre	Discovered the water channel, aquaporin.
1997	Kai Simons	Proposed the 'Lipid Raft' model for membrane structure.
1998	Rod MacKinnon	Determined x-ray crystallography structure of the K^+ channel.
2005	E. Fahy et al.	Published a comprehensive classification system for lipids based on lipidomics.

Index

Note: Page numbers with "f" denote figures; "t" tables.

A

Active transport (ATP)
  bacterial lactose transport, 322–323, 322f
  cell concentrations, 317
  direct active transport/uniport, 319
  glucose transport, 323, 324f
  Na$^+$/K$^+$ ATPase, 319–321, 320f
  secondary active transport, 321–322
  types, 317, 318f, 319t
  vectorial/unidirectional metabolism, 323
Anti-freeze proteins (AFPs), 233
ATP-dependent flipase, 143

B

Bacterial proline transporter, 294–295
Bacteriorhodopsin, 96–97, 98f
Biological membrane
  biochemical/physiological process, 4
  composition of, 10–11, 10t
  definition, 1
  domain size
    esoteric instrumentation, 8
    light microscope, 9–10, 9f
    membrane bilayer, 9
  eukaryote cell structure
    animal cell components, 4, 5f
    compartmentation, 4
    endomembrane system, 4, 6f
    endoplasmic reticulum, 7
    Golgi apparatus, 7
    lysosome, 7
    mitochondria, 7–8
    nuclear envelope, 6
    peroxisome, 7, 8f
    plasma membrane, 4–6
  'lipid raft', 3
  'thread of life', 2
  trans-membrane ion gradients, 'typical'
    mammalian cell, 3, 3t
  typical liver cell, 2, 2f
Branched chain (isoprenoid) fatty acids,
    54–55, 54f

C

Carbohydrate asymmetry
  lectins, 139
  lipid asymmetry
    chemical modification, 140–143, 140f, 141f
    enzymatic modification, 141–143, 142f
    function, 139–140
    methods, 140
  membrane-bound carbohydrates, 139
Coenzyme Q (CoQ), 71f, 80–81
Critical micelle concentration (CMC), 266–269, 268f
Cytochrome c, 92

D

Detergent resistant membranes (DRMs), 127
DHA, *see* Docosahexaenoic acid (DHA)
Differential scanning calorimetry (DSC)
  cholesterol, 161–162, 162f
  cooperativity unit, 161–163
  DPPC *vs.* DPPE, 160, 161f
  lipid 'contaminants', 160–161
  membrane suspension, 159–160
  van't Hoff enthalpy, 162
Diffusion rate, 10
2,4-Dinitrophenol (DNP), 326
1,6-Diphenylhexatriene (DPH), 168f, 169–170
Docosahexaenoic acid (DHA), 350–352, 351t

F

Fatty acids
  18-carbon series, T_ms, 49, 49t
  chain length, 47
  complex lipids, 279–280
  definition, 46
  double bonds, 48–49
  fatty acyl methyl esters, 281
  gas chromatography, 280–281, 280f
  isoprene, 54, 54f
  for liver phospholipids, 281, 282f
  MUFA, 49–50, 50t
  omega nomenclature, 51–52, 52f
  omega-6 series, 52–53

Fatty acids (*Continued*)
prenylated proteins, 54—55
PUFAs, 50—51, 50t, 51f
saturated fatty acids, 47—48, 47t, 48t
trans double bonds, 53—54
'typical' mammalian fatty acid composition, 53, 53t
unsaturated fatty acid, 47
Fluid mosaic model
flip-flop rates, 157, *see also* Lipid trans-membrane diffusion
fluidity, 170—172
lateral diffusion
FRAP, *see* Fluorescence recovery after photobleaching (FRAP)
Frye-Edidin experiment, 145, 146f, 147f, 150—151
lateral mobility, 144—145
phospholipid diffusion, 151
SPT, 148—150, 150f
lipid melting behavior
DSC, *see* Differential scanning calorimetry (DSC)
FP, *see* Fluorescence polarization (FP)
FT-IR spectroscopy, 163—164
'gauche kinks', 159, 159f
lamellar phase, 157
'melting' temperature, 157—158, 158f
membrane asymmetry
carbohydrate asymmetry, *see* Carbohydrate asymmetry
erythrocyte lipid asymmetry, 143—144, 143f
protein asymmetry, 137—138, 137f, 138f
membrane thickness
planar bimolecular lipid membranes, 134—136, 135f
trans-membrane electrical measurements, 134
X-ray diffraction, 136—137, 136f
size and time domains, 132—133
biophysical techniques, 133—134, 133f
electromagnetic spectrum, 131, 132f
Fluorescence polarization (FP)
anisotropy, 166—168
^{13}C-NMR-derived T_1S, 170f
DPH, 168f, 169—170
fluorophore, 165, 166f
FP measurement, 166—168, 167f
gel-to-liquid crystal transition, 168f
NMR order parameters, 166—168, 169f
invisible radiation, 164—165
time domains, 165—166
Fluorescence recovery after photobleaching (FRAP)
data plot, 147f, 148
diffusion rate, 148
fluorescent lectin, 146—148
laser technology, 145—146

Fourier transform infrared (FT-IR) spectroscopy, 163—164, 164f
FP, *see* Fluorescence polarization (FP)

G
Gangliosides, 111
Glucose transporter (GLUT), 311—312, 313f
G-protein-coupled receptors (GPCR), 96—97

H
Homeoviscous adaptation (HVA)
acyl chain double bonds, 233
acyl chain length, 232
anti-freeze proteins, 233
divalent metals, 233—234, 234t
Don Juan Pond, 234—235, 234f
history of, 232
isoprene lipids, 233
phospholipid head groups, 233
sterols, 233
Human health
dietary lipids effect
DHA, 350—352, 351t
TFAs, *see* Trans fatty acids (TFAs)
drug-delivery agents, *see* Liposomes

I
Isolation methods
density gradient centrifugation
continuous gradient, 250—251, 251f
densest solution, 249—250, 250f
discontinuous gradient, 249—250, 250f
DNA density, 251—252
Ficoll™, 253
iodixanol, 252f, 253
OptiPrep, 253
particle density, 252, 252f
Percoll™, 253
sedimentation coefficient, 249
Svedberg Unit, 249
differential centrifugation, 248—249
homogenization
bath sonicater, 245f, 246
bead beaters, 246, 246f
Brinkman Polytron blender, 242—244, 244f
cup horn sonicater, 245—246, 245f
definition, 241
enzyme usage, 242
Freeze/Thaw, 248
French Press, 247, 247f
intracellular organelles, 245
osmometer, 244—245
Parr Pressure Bomb, 246—247, 247f
shear force, 242, 243f

tip sonicater, 245, 245f
ultrasonication, 245
Virtis Tissue Homogenizer, 242−244, 244f
Waring blender, 242−244, 244f
membrane marker enzymes, 260, 261t
membrane structural markers, 260, 260t
microsomes
  characteristics, 240
  Ficoll, 240−241
  sequence of, 241
non-centrifugation methods
  affinity chromatography, 253−254
  anion exchange chromatography, 256−257
  antibody-affinity chromatography, 256
  lectin-affinity chromatography, 254−255, 255t
  ligand-receptor affinity chromatography, 256
  magnetic beads, 257
  membrane purification, 259−260
  micro- and nanospheres, 258
  Millipore Filters, 259
  silica particles, 259
  'tricks', 253
  two phase partitioning, 258−259

L
Langmuir film balance, 217−218, 217f
  compressibilities, II-A, 218−219, 220f
  monolayer surface pressure, 217−218
  pressure-area (II-A) isotherms, 218, 219f
  surface tension measurements, 218
Large unilamellar vesicle (LUVs), 292t
  Amicon concentration filter device types,
    291, 292f
  disadvantage, 291
  ethanol, 290
  ether vaporization, 290, 291f
Lipid-anchored proteins
  glycosylphosphatidylinositols, 99−101, 101f, 103
  myristic acid, 99−101, 100f
  palmitic acid, 99−102, 100f
  prenylated hydrophobic acyl chains,
    99−102, 100f
Lipid membrane properties
  acyl chains, 177−178
  Chevreul, Eugene, 176−177, 177f
  α-cyclodextrin, 176, 176f
  DSC study, 178−179, 178f
  lipid interdigitation
    biological relevance, 210
    5-DOXYL stearic acid, 209, 209f
    interdigitated-lipid shape, 206−208, 208f
    perturbations, 209−210
    symmetrical chain phospholipids, 206, 207f,
      208−210

lipid−protein interactions, *see* Lipid−protein
    interactions
lipid raft detergent extractions, 181−182, 182f
membrane 'fluidity' and permeability, 177
non-lamellar phases
  amphipathic lipids, 183
  cubic phase, 186−187
  dimensionless parameter, 184
  inverted hexagonal phase, 185−186, 185f, 186f, 187f
  lipid polymorphism, 182−183
  lipid shapes, 183−184, 183f
  ^{31}P NMR, *see* 31P NMR membrane phase
  spontaneous radius of curvature, 184−185, 184t
phase diagrams, 190
  single-component lipid phase diagrams, 190−191, 191f
  three-component lipid phase diagram, 192−194, 193f
  two-component lipid phase diagrams, 191−192,
    192f, 193f
X-ray diffraction study
  cholesterol's solubility, 179, 180t
  monohydrate crystals, I-q plots, 179−180, 180f
  normalized scattering peaks, 179−180, 181f
Lipidomics, 46
Lipid phosphate phosphohydrolases (LPPs), 64
Lipid−protein interactions
  annular lipids
    bilayer lipids, 194, 194f
    binding sites, 196
    Ca^{2+} ATPase, 197, 198f
    electron spin resonance, 195−196, 195f
    'typical' plasma membrane, 196, 197f
  co-factor (non-annular) lipids
    Ca^{2+} ATPase, 198−199
    membrane proteins, 197−198
    Na^+/K^+ ATPase, 199, 200f
    PKC, 199
    potassium channel KcsA, 199−200
  hydrophobic match
    biological function, 205−206
    cholesterol-rich domains, 203
    lipid diversity, 203
    Mouritsen and Bloom 'Mattress' model,
      203−205, 204f
    single span α-helices, 205
  non-anionic co-factor lipids
    anionic head group lipids, 200
    cytochrome c oxidase, 202−203
    β-hydroxybutyrate dehydrogenase, 202
    rhodopsin, 200−202, 201f
Lipid trans-membrane diffusion
  chemical method, 152−154, 153f
  PLEP, *see* Phospholipid exchange protein method
    (PLEP)
  process of, 151−152, 152f

Liposomes
  bathing solution, 287
  characteristics, 340–341, 342f
  DSPE-PEG, 344, 344f
  gel filtration chromatography, 285–286, 285f
  gout, 339–340, 340f, 341t
  leakage rate measurement, 286
  leishmaniasis, 342–343, 343f
  liposome-sequestered drugs, 343–344
  multilamellar vesicles, 284–285, 284f
  PEG, 344, 344f
  pH-sensitive liposomes, 344, 347f
  properties, 340–341
  reticulo-endothelial system, 341–342
  smectic mesophases, 284–285
  thermo-liposomes, 344–346
Long-range membrane properties
  bilayer lipid packing
    area/molecule, 219–220
    condensation, 220–221
    lipid monolayers, see Langmuir film balance
    lipid packing free volume, 215–217,
      216t, 217t
    squeeze out, 221–222, 222f
    surface elasticity moduli, 221
  freeze fracture electron microscopy,
    222–223, 223f
  HVA
    acyl chain double bonds, 233
    acyl chain length, 232
    anti-freeze proteins, 233
    divalent metals, 233–234, 234t
    Don Juan Pond, 234–235, 234f
    history of, 232
    isoprene lipids, 233
    phospholipid head groups, 233
    sterols, 233
  macrodomains and microdomains, see Membrane
    domains
LUVs, see Large unilamellar vesicle (LUVs)

M

Magnificent Seven protein, 96–97
Membrane domains
  macrodomains
    caveolae, 226–227
    clathrin-coated pits, 226, 226f
    domain, definition, 223–224
    gap junctions, 224–225, 225f
    Halobacterium halobium, 224
  microdomains
    atomic force microscope, 230–232, 231f
    Ca^{2+} ATPase, 227–228
    Ca^{2+}-induced anionic phopholipid, 229, 229f

DOPC membrane, 229–230, 230f
    liquid crystalline/gel phase separations,
      228–229, 228f
Membrane history
  Fluid mosaic model, 13, 20
  lipid bilayer membrane
    Gorter, Evert, 25–27, 25f, 26f
    Hewson, William, 20, 21f
    Overton, Charles Ernest, 23–24, 24f
    Pfeffer, Wilhelm, 23, 23f
    plasma membrane, 20
    Schultz, C.H., 20–21
    von Nageli, Karl, 21–23, 22f
  oil on water
    Franklin, Benjamin, 15–16, 15f
    Langmuir, Irving, 18–19, 19f
    Lord Rayleigh, 16, 17f
    Pliny the Elder, 13–15, 14f
    Pockels, Agnes, 16–18, 18f
Membrane lipids
  'amphipathic' lipids, 43
  classification systems, 45–46, 45t, 46t
  definition, 43
  fatty acids
    18-carbon series, T$_m$s, 49, 49t
    chain length, 47
    definition, 46
    double bonds, 48–49
    isoprene, 54, 54f
    MUFA, 49–50, 50t
    omega nomenclature, 51–52, 52f
    omega-6 series, 52–53
    prenylated proteins, 54–55
    PUFAs, 50–51, 50t, 51f
    saturated fatty acids, 47–48, 47t, 48t
    trans double bonds, 53–54
    'typical' mammalian fatty acid composition,
      53, 53t
    unsaturated fatty acid, 47
  functions, 44
  species types, 44
Membrane models
  Benson And Green's Lipoprotein Subunit models,
    120–121, 120f, 122f
  cellular junk, 117–118
  Danielli-Davson Pauci-molecular model,
    118–119, 119f
  erythrocyte membrane, 117–118
  Robertson Unit Membrane model, 119–120, 119f
  Simons' Lipid Raft model, 126f
    acyl chains, 126–127
    caveolae, 127f, 128
    lipid clusters, 125
    protein–protein interactions, 127

Singer-Nicolson Fluid Mosaic model, 123f
  cytoskeleton, 125
  dark bands, 122–124
  features of, 124t, 125
  membrane, definition, 121
  for membrane structure, 122–124, 123f
Membrane polar lipids, *see also* Membrane lipids
  acyl chain bonds
    phosphatidylcholine (PC), 58, 59f
    phosphatidylethanolamine (PE), 57–58, 58f
    phosphatidylserine (PS), 59–60, 60f
    saponification, 58–59
    triacylglycerol, 58
  detergent effect, 57
  galactosyl diacylglycerols, 77f
  hydrolysis of, 78f
  lipid distribution, 76
  in low abundance
    CoQ, 71f, 80–81, 81f
    free fatty acids and lysolipids, 70f, 78–79
    lipid-soluble vitamins, 70f, 79–80, 80f
  phosphatidic acid, 57, 58f
  phospholipids
    alkyl ether and alkenyl ether linkages, 65f, 67
    average lipid composition, mammalian liver cell, 66, 72t
    bond types, 60, 64f
    cardiolipin (CL), 61, 63f, 65f, 70–71, 70f
    fatty acids, 61
    LPPs, 64
    PA, 63–64
    PCs and PEs, melting point, 66t
    phosphatidylglycerol (PG), 60–61, 62f, 68–69
    phosphatidylinositol (PI), 60, 61f
    phosphoinositides, 68
    phosphoric acid, 62, 63f
    PS, 66
    sphingosine, C-1 position, 67t
    triacylglycerol hydrolysis, 64f, 67
    zwitterions, 65
  plant lipids, 69f, 76–77, 78t
  sphingolipids, 67f, 68f, 71–73
  sphingomyelin, 73f
  sterols
    cholesterol, 74f, 75–76
    functions, 69f, 74–75
    pKas, 68f, 74
Membrane proteins
  amino acids
    amphipathic nature, 88, 88f
    hydropathy plot, 90, 90f
    Kyte-Doolittle hydropathy index values, 88, 89t
    liquid ammonia production, 86
    'lumpers' and 'splitters', 87, 87t

non-polar and polar amino acids, 87, 87t
    Vauquelin, L.N., 86, 86f
    water-soluble globular proteins, 90
    X-ray crystallography, 88
  amphitropic proteins, 93
  brute force method, 93–94
  chaotropic agents function, 93–94
  endo and ecto proteins, 94, 94f
  myelin lipids, 85–86
  peripheral proteins, 91–92, 92f
  protein classification system, 91, 91t
  trans-membrane protein
    bacteriorhodopsin, 96–97, 98f
    β-barrel, 97–99, 99f
    glycophorin, 95–96, 96f
    lipid-anchored proteins, *see* Lipid-anchored proteins
    membrane lipid bilayer sections, 94–95, 95f
    tyrosines and tryptophans, 95
Membrane reconstitution
  Ca^{2+} ATPase, 293–294, 297f
  detergents
    Brij-35, 269
    cholate, 269–270, 270f
    HLB and CMC values, 266, 266t, 269
    functional membrane proteins, 269
    functions, 266
    SDS, 271
    structures of, 266–268, 267f
    surface tension measurements, CMC, 268, 268f
  energetic, 294–295
  lateral and trans-membrane asymmetry, 265
  lipid bilayer membranes
    detergent dilution, 289–290, 290t
    high pressure extruder, 288, 289f
    liposomes, *see* Liposomes
    LUVs, *see* Large unilamellar vesicle (LUVs)
    planar bimolecular lipid membrane, 282–284, 283f
    SUV, sonication, 288
    ULV, 287–288
  lipid isolation, 274f
    acetone extraction, 275
    fatty acid analysis, *see* Fatty acids
    Folch method, 276
    HPLC, 278–279, 279t
    TLC, 276–278, 277f, 277t, 278f
  lipid rafts
    characteristics, 299–300
    definition, 295
    disadvantage, 295–296
    GUVs, 297, 298f
    micro-domains, 295
    phase separations, 297

Membrane reconstitution (*Continued*)
    polyunsaturated fatty acyl chains, 299
    reputed raft proteins, 298–299
    Smart Prep, 296–297
    types, 296
  orientations of, 293
  protein/detergent micelles, 292–293
  protein isolation
    acetone powders, 271
    Bradford Protein Assay kit, 275
    centrifugation and affinity chromatography,
      272
    chaotropic agents, 271–272
    detergent level effect, 272, 272t
    His-tag technique, 272–273
    multi-step protein purification process,
      273, 273f
    revolutionary colorimetric procedure, 274f, 275
    specific activity, 275
    water-soluble protein, 271
  proteoliposome production, 294
Membrane sugars
  carbohydrate code, 109–110
  carbohydrates, 108, 109f
  di-saccharide maltose, 109, 110f
  glycocalyx, 108, 108f
  glycolipids, 111–112, 112f, 113t
  glycoproteins, 112–114, 113f
  GPI-anchored proteins, 114–115
  hydrated carbon, 107, 108f
  sialic acid, 110–111
Membrane transport, 305
  ATP
    bacterial lactose transport, 322–323, 322f
    cell concentrations, 317
    direct active transport/uniport, 319
    glucose transport, 323, 324f
    Na$^+$/K$^+$ ATPase, 319–321, 320f
    secondary active transport, 321–322
    types, 317, 318f, 319t
    vectorial/unidirectional metabolism, 323
  facilitated diffusion
    GLUT, 311–312, 313f
    Michaelis-Menton saturation kinetics,
      311, 312f
    potassium channels, 312–313, 314f, 315f
    sodium channel, *see* Sodium channel
  exocytosis, 333, 333f
  gap junctions, 328–329, 329f
  Fick's first law, 307, 308f
  ionophores
    channel formers and mobile carriers, 324, 324f
    crown ethers, 326–327, 327f
    DNP, 326

    nystatin, 327–328
    types, 324, 325f
    valinomycin, 325–326
  LUVs, 307
  membrane blebbing, 334, 334f
  osmosis and osmotic pressure, 307–309,
    309f
  passive diffusion, 309–311, 310f, 310t
  permeability, 306, 306f
  phagocytosis, 330–333, 332f
  pinocytosis, 330–331, 332f
  RME, 329–330, 330f, 331f
  solute's molecular weight and diffusion coefficient,
    307, 308t
Monoenoic fatty acids (MUFA), 49–50, 50t
Myelin basic protein (MBP), 92

P
PEG, *see* Polyethylene glycol (PEG)
Phospholipid exchange protein method (PLEP)
  flippase, 156
  flopases, 156–157
  for NBD, 155–156, 156t
  protein-free lipid vesicle, 154, 155f
  scramblases, 156–157
  single lipid species, 155–156
Plasmalogens, 59–60, 60f
^{31}P NMR membrane phase
  lamellar phase spectra, 187–188, 188f
  membrane lipids and phase, 188–190,
    189t
  molecular species, 188–190
  X-ray diffraction, 187–188
Polyenoic fatty acids (PUFAs), 50–51, 50t, 51f
Polyethylene glycol (PEG), 344, 344f
Protein kinase C (PKC), 199

R
Receptor mediated endocytosis (RME), 329–330,
  330f, 331f

S
Saxitoxin (STX), 314–315, 316f
Simons' Lipid Raft model, 126f
  acyl chains, 126–127
  caveolae, 127f, 128
  lipid clusters, 125
  protein–protein interactions, 127
Singer-Nicolson Fluid Mosaic model, 123f
  cytoskeleton, 125
  dark bands, 122–124
  features of, 124t, 125
  membrane, definition, 121
  for membrane structure, 122–124, 123f

Single particle tracking (SPT), 148–150, 150f
Single trans-membrane α-helix, 95–96, 96f
Sodium channel
  action potential phase, 313–314
  aquaporins, 316–317, 317f, 318f
  free energy change, 315
  STX, 314–315, 316f
  TTX, 314, 315f
Sodium dodecylsulfate (SDS), 271
SPT, *see* Single particle tracking (SPT)
Stealth liposomes, 344, 344f
STX, *see* Saxitoxin (STX)

T
Tay-Sachs disease, 111–112
Tetrodotoxin (TTX), 314, 315f
Trans fatty acids (TFAs)
  membrane phospholipids, 350
  oleic *vs.* elaidic acids, 349–350
  sphingosine, 348–349
  unnatural trans fatty acids, 349

U
Ubiquinone, 71f, 80–81
Unilamellar vesicles (ULVs), 287–288, 287t

W
Water
  Cassini spacecraft, 30, 31f
  extra-terrestrial life, 30
  frozen ice surface, 30
  hydrophobic effect
    bond energies, 40, 40t
    definition, 37–38
    hydrocarbon hexane, 38
    membrane bilayer stability, 37f, 38
    Tanford, Charles, 37–38, 39f
    van der Waal, J.D., 38, 40f
  properties, 29, 32–34
  structure
    charge distribution, 31, 31f
    chemical composition, 31
    cluster size, 32, 33t
    hydrogen bonding, 32, 32f
  surface tension
    amphipathic molecules, 36, 37f
    Basilisk/'Jesus lizard', 34, 34f
    Du Nouy Ring method, 35–36, 36f
    ethanol, 34
    hexane dispersion, 34–35, 35f
    'model membrane', 35–36
    Wilhelmy Plate method, 35–36, 36f

Printed and bound by CPI Group (UK) Ltd, Croydon, CR0 4YY

03/10/2024

01040328-0002